高等学校工程管理专业规划教材

工程合同管理（双语）
Contract Administration for Construction

卢有杰　编著

中国建筑工业出版社

图书在版编目(CIP)数据

工程合同管理(双语)/卢有杰编著.—北京：中国建筑工业出版社，2007
高等学校工程管理专业规划教材
ISBN 978-7-112-08911-6

Ⅰ.工… Ⅱ.卢… Ⅲ.建筑工程－经济合同－管理－高等学校－教材 Ⅳ.TU723.1

中国版本图书馆 CIP 数据核字（2007）第 068799 号

高等学校工程管理专业规划教材
工程合同管理（双语）
Contract Administration for Construction
卢有杰 编著

*

中国建筑工业出版社出版、发行（北京西郊百万庄）
各地新华书店、建筑书店经销
北京密云红光制版公司制版
廊坊市海涛印刷有限公司印刷

*

开本：787×1092 毫米 1/16 印张：23½ 字数：570千字
2007 年 7 月第一版 2014 年 6 月第二次印刷
定价：**32.00** 元
ISBN 978-7-112-08911-6
（15575）

版权所有 翻印必究
如有印装质量问题，可寄本社退换
（邮政编码100037）

本书介绍了国际公认的以市场经济为背景的工程合同管理的主要内容，具体包含有建筑业及其市场结构；建设项目的参与者及其环境；建筑及其他工程的程序；各种工程采购方式及选择原则；招标程序及文件；合同与合同管理的概念，合同双方在合同管理中各自的任务与责任；风险管理；承包商现场组织及建筑师/工程师施工阶段的合同管理；工程量与变更计量及计价；索赔，以及合同结尾。

合同格式，以及合同双方的权利、义务与责任取决于建筑市场买方在卖方竞争基础上选定的工程或咨询与设计服务采购方式，而采购方式取决于建筑市场（竞争）结构以及有关的风险，本书的选材与编排即遵循这一思路。激烈的市场竞争催生了丰富多彩的工程采购方式与合同关系，20世纪八九十年代编写的施工合同管理教材已不能满足现代工程合同管理的新需要。本书在全面介绍各种不同类型合同的基础上，以 FIDIC（国际咨询工程师联合会）1999 年出版的红皮书条文为例，详细解释了在有建筑师/工程师参与的情况下合同双方的权利、义务与责任，以及具体的合同管理工作。

风险及其管理是合同管理的重要内容，本书单设一章，详细阐述分担风险的原则、做法，以及相应的合同条文。合同结尾是合同管理的重要环节，为此也单设一章。

我国建筑市场已成为国际建筑市场最重要的组成部分，以英文编写的合同文件用得越来越多。为提高我国管理国际公认的工程合同的能力，本书以英文编写。同时，为了更好地帮助读者透彻地理解本书内容，书中对重要的概念、术语、句子与段落进行了详细解释，并有译文，有利于读者自学。

* * *

责任编辑：张　晶　刘平平
责任设计：赵明霞
责任校对：安　东　王雪竹

FOREWORD

Contract administration in China has become an increasingly important component which the students in construction management, civil engineering, and even architecture shall add to their body of knowledge. More and more of them are expected to be given by their future employers a consulting or managerial position in procurement and contracting of building and engineering works, as opposed to a technical one in the previous 1980s and 1990s.

The author has been delivering the courses entitled Contract Documents for Civil Engineering Construction and Contract Administration for Construction to the students in construction management, structural engineering and hydro-engineering at Tsinghua University since 1992.

The author prepares his lecture materials based on a number of the textbooks which were used in Hong Kong and the United Kingdom in 1990s and other available English literature which is related to construction contract, procurement of works, consultant services, project management, etc..

The number of students who choose to attend his lectures has been increasing in recent years, on one hand. On the other hand, various systems have emerged in the world for procurement of works and related engineering services since 1990s, which can be, to a great extent, attributable to the advances in technology, especially the hi-tech, accelerated globalization and the employers' diversified needs arising out of procurement of works and related engineering services. Much of the developments in procurement systems are characterized by 'one single point of responsibility' and 'fast-track', which has created new contractual relationships among the parties involved in increasingly complex and sizable construction activities.

In response to the new contractual relationships the world-wide influential professional institutions, such as the American Institute of Architects (AIA) in the United States, the Institution of Civil Engineers (ICE) in United Kingdom, FIDIC in Lausanne, Switzerland, etc. embarked upon updating standard forms of contract in middle 1980s throughout late 1990s.

ICE issued the New Engineering Contract (NEC) in 1993 which was the first attempt at reform in over 100 years. The NEC is based on best modern project management principles, designed to cater for the various relationships created by the new procurement systems, and uses simple procedures and language, with terminology that can be used not only in civil engineering, but also more general engineering.

FIDIC has taken actions, too, by updating their previous rainbow books. Their commitments brought a new FIDIC suite of conditions of contract to publication in 1999. The new documents departed from the traditional style and structure inherited from the ICE Conditions of Contract and were based on the allocation of the design function: the new Red Book is for building and engineering works designed by the employer (referred to as the Red Book in this book); the new Yellow Book for electrical and mechanical plant and building and engineering work designed by the contractor; and the Silver Book is for turnkey projects.

It is these dramatic changes that have pressed the author to update not only the materials to be delivered to the students but also the way of delivery and it is the increasing interest of the students in procurement management and contract administration that encourages him to make the updated materials available to more readers in the form of a book, not only to the students at Tsinghua.

While compiling the book, with the changes in mind, the author has made attempts to put contract administration under the framework of project management by both the employer's part and the contractor. As a matter of fact the FIDIC rainbow books can serve as a project management handbook for both parties to a contract because they cover all the knowledge areas and processes needed for construction project management.

In view of the fact that FIDIC's four new books match each other, with each topic being covered in similarly-worded provisions in each book except where otherwise necessary, the author believes that there is no need to introduce them all to the students and other Chinese readers who do not have specific interest in any one of the four books. It is of the author's opinion that the knowledge and understanding of any one of the four books will suffice and therefore he presents the concepts, terms, procedures, documents, obligations, rights, etc. by reference to the FIDIC Red Book which is used more than other three books.

Examples, sample formats of notice, sample contract documents and case studies are given in this book to help the students and the general readers understand the subjects.

In addition certain words, phrases, expressions and sentences are explained in detail with Chinese translations to help the students and other Chinese readers to capture the concepts, ideas and practice which are internationally accepted but strange to them.

It is hoped that this book can contribute to enhance the capacity of Chinese contractors to undertake the building and engineering works both abroad and at home for which standard tender and/or contract documents in English are used.

The author's sincere gratitude is due to the persons or organizations whose books, articles, reports or lecture notes are used in this book. The author will apologize for his inadvertent use of materials without permission or acknowledgement.

The author welcomes comments and instances of the problems in this book for inclusion in its future editions.

Lu Youjie
Heqingyuan, Tsinghua University

Contents

Chapter 1　Introduction 1
  1.1 Construction industry 1
  1.2 Construction project 4
  1.3 Construction markets 8
  1.4 Participants in construction industry 14
  1.5 Building and engineering procedure 23
  1.6 Context of construction projects 34

Chapter 2　Procurement of Works 41
  2.1 Procurement of works 41
  2.2 Procurement management 45
  2.3 Procurement planning 48
  2.4 Procurement methods 51
  2.5 Contracting planning 63

Chapter 3　Tendering Procedure 69
  3.1 Request for seller responses 69
  3.2 Tender documents 75
  3.3 Inviting for tenders 104
  3.4 Contractor's planning 105
  3.5 Preparation of tender 113
  3.6 Selecting a contractor 127

Chapter 4　Contract Administration 131
  4.1 Contract formation 131
  4.2 Pre-contract arrangements 137
  4.3 Contract administration 141
  4.4 Contract documents and their priority 148
  4.5 Contract administration during construction 151
  4.6 Division of responsibility 157
  4.7 Inputs to contract administration 166
  4.8 Tools and techniques for contract administration 167

Chapter 5 Risk Management ········· 182
   5.1 Risk with construction projects ········· 182
   5.2 Risk identification ········· 184
   5.3 Risk analysis ········· 188
   5.4 Allocation of risks ········· 189
   5.5 Risk response planning ········· 190
   5.6 Contract strategy ········· 194
   5.7 Allocation of risks under the *Red Book* ········· 196

Chapter 6 Contractor's Site Organization and A/E's Site Supervision ········· 211
   6.1 Arrangements prior to commencement ········· 211
   6.2 Contractor's organizational structure ········· 214
   6.3 Site layout ········· 224
   6.4 Planning and monitoring ········· 225
   6.5 Resource scheduling ········· 227
   6.6 Costing and accounting arrangements ········· 228
   6.7 Usage and costing of contractor's equipment ········· 229
   6.8 Safety aspects ········· 231
   6.9 A/E's site supervision ········· 232

Chapter 7 Measurement, Valuation and Payment ········· 238
   7.1 Method of measurement ········· 238
   7.2 Measurement of work executed and cost records ········· 242
   7.3 Daywork ········· 249
   7.4 Certificates and payments ········· 250
   7.5 Adjustments of contract price ········· 262
   7.6 Delay damages ········· 267

Chapter 8 Contractors' Claims and Their Settlement ········· 276
   8.1 Introduction ········· 276
   8.2 Definition and types of claims ········· 277
   8.3 Procedure for claims ········· 285
   8.4 Justifying claims by contract ········· 292
   8.5 Additional loss or expense ········· 309
   8.6 Disruption due to variations ········· 310
   8.7 Preparation of claims ········· 319

  8.8 Reducing of claims ·· 327

Chapter 9 Contract Closure ·· 328
  9.1 Discharge of a contract ··· 328
  9.2 Discharge of contract under FIDIC *Red Book* ·· 335
  9.3 Contract closure ·· 342

Appendix Clause Headings of the *Red Book* ·· 355
Index ··· 359
References ·· 366

Chapter 1　Introduction

1.1　Construction industry

Construction industry is a sector of the economy that transforms various resources into constructed economic and social infrastructure and facilities. It embraces all phases of the transformation process, namely, planning, designing, financing, procuring, constructing, maintaining and operating[1].

Nowadays, construction includes building and engineering work. Engineering means not only civil engineering, but also structural engineering, mechanical engineering, electrical engineering, etc.. The sector's physical output ranges widely, including both buildings and engineering works. The firms in construction industry erect, repair and demolish all types of building and engineering structures.

Construction is one of the oldest of all industries, retaining its role as a core economic activity from the early days of human civilization to this day. With its close link to public works and hence to the implementation of fiscal policy, it has always been considered as a strategically important industry for creating employment and sustaining growth. For the developing economies, the construction sector carries particular importance because of its link to the development of basic infrastructure, training of local personnel, transfer of technologies, and improved access to information channels.

Words, Phrases and Expressions 1.1

[1]　economy　n. (a)国民经济(the structure of economic life in a country, area, or period);(b)经济制度(an economic system)

[2]　sector　n. 部门 (a sociological, economic, or political subdivision of society)

[3]　industry　n. 行业 (a department, branch or sector of a craft, art, business, or manufacture; especially the one that employs a large personnel and capital, especially in manufacturing);产业 (a group of productive or profit making enterprises)

[4]　construction　n. 建筑，建造，构筑，营造，施工 (the process, art or manner of constructing something);建筑物，构筑物 (a thing constructed)

[5]　construction industry (building industry, construction) 建筑业，过去曾有"营造业"、"建筑工业"之称。

[6]　resources　n. 资源 (a source of supply or support; a natural source of wealth or revenue)

[7]　facility/facilities　n. 设施 (something such as a hospital, an educational institution, or

a post office that is designed, built, installed, or established to serve a particular function)

[8] infrastructure n. 基础设施 (the system of public works or the basic military installations, communication and transport facilities, etc. of a country, state or region)

[9] embrace vt. 包含,含有,有 (to include or contain)

[10] phase n. 阶段 (a distinguishable part in a course, development, or cycle)

[11] stage n. 阶段 (a period or step in a progress, activity, or development) stage 和 phase 都可译成"阶段",但 phase 强调事物外表逐渐变化,而 stage 强调前后接续的时间段。

[12] plan n. 计划 (a method for achieving an end)

[13] plan vt. (planned, planning) 规划,制定计划 (to arrange a plan or scheme for; to make plans)

[14] financing n. 筹集资金 (the act or process of an instance of raising or providing funds)

[15] instance n. 要求,建议 (urgent or earnest solicitation; instigation, request)

[16] procure vt. (procured, procuring, procurement) 采购 (to get possession of, especially obtain by particular care and effort) procurement of works 采购工程,工程采购

[17] maintain vt. (maintained, maintaining) 维护,保养 (keep in good repair, 此处 repair 的意思是"使用中的相对状况" (relative condition for usage or being used). *The building is out of* ~ (*in good* ~) 这栋楼已损坏(状况良好)。

[18] operate vt. (operated, operating) 使用,经营,管理 (to put or keep in operation)

[19] physical adj. 物质的,确实的,实在的,有形的,实际的 (of or pertaining to matter or things existing in natural; having material existence; perceptible especially through the senses and subject to the laws of nature)

[20] subject to 依照

[21] output n. (总)产量,产品与服务总量,(总)产出,(总)成果 (production, especially the quantity or amount produced in a given time)

[22] physical output 实际成果

[23] building n. (a) 建筑物 (a relatively permanent roof and walls, used for living, manufacturing, social gathering, etc.); (b) 建筑,建造,修筑,营造 (the act or business of constructing houses, etc.)

[24] engineering n. 设计与制造 (design and manufacture of complex products)

[25] civil engineering n. 土木工程 (the process, art, or manner of building roads, railways, dams, canals, docks, etc.)

[26] work n. (a) 工作 (exertion or effort directed to produce or accomplish something); (b) 工程 (a intermediate or final product of exertion or effort directed to construct a

building or a engineering structure, can be a part thereof.)

[27] works *n*. 工程 (a result of building or engineering; structure, such as a bridge, a road, etc; the operation on site required to produce a building or a engineering or other structure. Works includes, not only the building or the structure itself at various stages of construction, but also all ancillary works necessary such as scaffolding, site huts, temporary roads etc. even though they may not form part of the finished building or structure. Most standard forms of contract draw a distinction between the "the Works" and "work". Under the FIDIC Conditions of Contract for Construction First Edition 1999, "the Works" means the permanent works and the temporary works, or either of them as appropriate (sub-paragraph 1.1.5.8). In contrast, "work" means "work carried out under the contract".)

[28] civil engineering works *n*. 土木工程(构筑物)(a result of civil engineering)

[29] cover *vt*. (covered, covering) 包含，包括，包罗 (include; comprise; extend over) *His researches covered a wide field*. 他的研究范围很广。

Notes 1.1

1. 疑难词语解释

[1] infrastructure 基础设施 This term focuses on economic infrastructure and includes services from: (1) public utilities—power, telecommunication piped water supply, sanitation and sewerage, solid waste collection and disposal, and piped gas. (2) public works—roads and major dam and canal works for irrigation and drainage. (3) other transport sectors—urban and interurban railways, urban transport, ports and waterways, and airports. Infrastructure is an umbrella term for many activities referred to as "social overhead capital" by some development economists. Neither term is precisely defined, but both encompass activities that share technical features (such as economies of scale) and economic features (such as spillovers from users to nonusers).

【译文】 这里"基础设施"主要指经济基础设施，包括以下方面的服务：(1) 公用事业——电力、电信、自来水、卫生设施和排污，固体废弃物的收集与处理及管道煤气等。(2) 公共工程——公路、大坝和灌溉及排水用的渠道工程等。(3) 其他交通部门——城市及城市间铁路，城市交通，港口和水路以及机场等。基础设施是意义广泛的术语，包括很多活动，有些发展经济学家将这些活动称为"社会管理资本"。"基础设施"和"社会管理资本"这两个词都没有精确的定义，但其包括的活动具有共同的技术特征（如规模经济）和经济特征（如从使用者向非使用者的溢出）。

[2] For the developing economies, the construction sector carries particular importance because of its link to the development of basic infrastructure, training of local personnel, transfer of technologies, and improved access to information channels.

【译文】 建筑业同基础设施建设，本国人员的培训，引进技术，以及扩大利用信息渠道的机会有紧密的联系，所以在发展中国家特别重要。

【解释】 access 的意思是"获得，取得，接近（或进入）的方法（或权利、机会等）"。例如，The project team must have access to an appropriate set of technical skills and knowledge. 项目班子必须能够取得一套适当的技能和知识；access to data 使用资料（的权利），查阅资料（的权利）；access to market 进入市场（的机会）。

1.2 Construction project

1.2.1 Construction project and its product

A project is the temporary endeavor made to create unique product, service or result and is refered to as a constuction project when the product is a building or an engineering works. A construction project is obviously different from its product, usually a building or an engineering works, on one hand.

On the other hand, a construction project is often a part of another larger project, such as a private capital project or a public project. A capital project usually not only needs buildings or engineering works, but also needs inputs other than buildings or engineering works, such as land, machinery, production plant, tools and fixtures, staff and workers for running the completed product of the project, Figure 1.1 depicts the relations between a construction project and a capital, infrastructure or other public project. The chart is referred to as Work Breakdown Structure (WBS) of the project. The construction project is actually one of the project's subprojects.

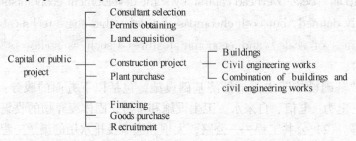

Figure 1.1 WBS of a capital or public project

A construction project can comprise a number of sections, while a section in turn consists of parts, and so forth. Figure 1.2 is WBS of a construction project.

1.2.2 Phases and life cycle of a construction project

Projects can be divided into phases to provide better management by appropriate

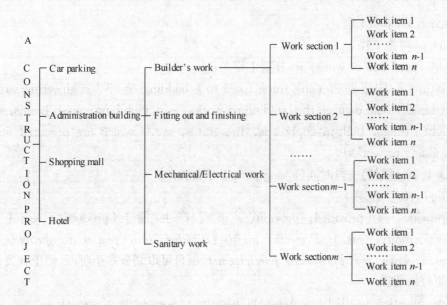

Figure 1.2 WBS of a construction project

participants. The phases collectively are referred to as the project life cycle.
For example, a construction project can be divided into eight phases: identification or formulation, feasibility study, technical and commercial investigation, design, construction, cnmmissioning, operation and decommissioning. When feasibility study shows that a new office building is viable, the employer engages an architectural firm to define it and then do the design. On completion of the design, the employer may need a construction firm to erect the office building. The architect's design project and the contractor's construction project, however, will have their own series of phases. The design project starts from conceptual development, through definition and implementation, to closure. The architect can even treat designing the facility and supporting the construction as separate projects, each with its own set of phases.

Words phrares and Expressinns 1.2

[1] project *n*. 项目,我国台湾将其译成"专案"
[2] construction project 施工项目，建设项目，工程项目
[3] product *n*. 产物,成果（something produced by nature or by man）
[4] private sector 民营部门，民间部门（不宜译成"私营经济"、"私人部门"）
[5] capital project 资本项目，投资项目
[6] Work Breakdown Structure (WBS) 工作分解结构 (A product-oriented "family tree" that organizes and defines the total scope of the project. Each descending level represents an increasingly detailed description of the project elements. 围绕成果展开的"家谱"，该谱组织并确定了项目整个范围。自上而下，每下降一层，对于项目各组成部分的说明就深入一步。)

[7] section *n.* 单项工程

[8] part *n.* 单位工程

[9] work item(item of work) *n.* 分项工程

[10] fixture *n.* 固定器物(any thing fixed to a building or civil engineering works; its removal would damage the building or works, e.g. plumbing, wash basins, electric light fittings, built-in cupboards, fire-grates, etc., which are procured with the building or works)

[11] cycle *n.* (很长)一段时间(a long period of time)

[12] life cycle 生命期

[13] provide *vt.* (provided, providing) 形成,产生,便于(produce a useful result, opportunity, etc.) *Projects can be divided into phases to provide better management by appropriate participants.* 项目可以划分多个阶段,便于有关参与者更好地管理。

[14] identification *n.* 识别(establish the identity of somebody of something)

[15] feasibility study 可行性研究(A study of a proposed project to indicate whether the proposal is attractive enough to justify more detailed preparation. Of limited detail, but the amount of detail included in feasibility studies varies widely. 对于提出的项目做的一种研究,指出该项建议是否具有足够的吸引力值得更详细地准备。详细程度有限,但各不同可行性研究包括的细节数量却彼此相去甚远。)

[16] commission *vt.* 试运行(Start up the product completed by a project after construction is complete and inspected and tested as corrected to the design. 在完成施工,经过检查、测试并按设计纠正之后启用竣工的项目成果。)

[17] decommission *vt.* 停止运行并拆除、处置(discharge or release from operation or service and/or dispose of)

[18] employer *n.* 雇主,业主(This is the term used in building and engineering contracts to describe the contracting party for whose benefit the works is carried out. This usage must be distinguished from the more general one, meaning employer of labour. Where the works is to an existing building the term 'building owner' is sometimes used as an alternative.)

[19] engage *vt.* 聘用,雇用(obtain the right to employ)

[20] architectural firm 建筑师事务所,香港称之为"则师行"

[21] architect *n.* 建筑师(One who designs and supervises the construction of buildings. He prepares drawings and specifications, inspects sites, obtains tenders, and handles legal negotiations needed before work can start. His functions now extend into town planning and the study of the social and work activities that need buildings. In some countries, laws have been made to protect the title 'architect' so that only those who are registered with a legally constituted institution may practice as architects. To qualify for registration a person must pass an architectural examination of university

degree level as well as one in professional practice.)

[22] tender *n*. (投)标书 (The terms "bid" and "tender" and their derivatives ("bidder/tenderer", bid/tendered", "bidding/tendering", etc.) are synonymous in this book.)

[23] contractor *n*. 承包商,承建商;建筑公司,我国近代曾称之为"营造厂" (Contractor is one who contracts to carry out and complete construction works in return for a financial consideration. The contractor is responsible for the planning of the work and for the acquisition and deployment of resources on site to achieve the completion of the project. Subject to any term to the contrary he is liable to the employer for the defaults of his subcontractors, including those nominated by the employer through his architect or engineer. An alternative term used for lesser works is 'builder', which has been judicially interpreted as 'a person who builds upon his own land or that of another for profit' and could be applied to an individual self-employed tradesman. Building contractors and civil engineering contractors, as the terms imply, are respectively concerned with works of building or civil engineering construction; these functions are not, however, exclusive.)

[24] default (a) *vt*. 不还债,不负责任(fail to pay a debt or perform a duty when required to do so) (b) *n*. 不还债,不负责任(act of defaulting)

[25] to the contrary 有相反的意思;相反的(地),一般做后置定语,修饰其前面的名词。*Subject to any term to the contrary he is liable to the employer for the defaults of his subcontractors, including those nominated by the employer through his architect or engineer.* 按照任何与此相反的条款,他都要为其分包商不履行合同义务的行为向雇主负责,包括雇主通过建筑师或工程师指定的分包商。

[26] subcontract *n*. 分包合同(a contract between a party to an original contract and third party; especially one to provide all or a specified part of the work or materials required in the original contract/a part of a contract, often specialist work such as asphalting, which is done by a separate firm from the (main) contractor. The (main) contractor is responsible for the work, pays the subcontractor, and is paid by the A/E or the employer for it.)

[27] subcontractor *n*. 分包商 (an individual or business firm contracting to perform part or all of another's contract 签约以履行另一合同的部分或全部的个人或商行/a specialist employed by a (main) contractor to perform a subcontract. Subcontractors must usually be approved by the architect or engineer before the contractor engages them.)

[28] contract to do *sth*. (缔约)承揽(某事) (undertake by contract) *The builder contracts to build a storage for wastes for a hospital*. 这家建筑公司承包了某医院废物堆放间。

[29] conceptual development 深化概念设计

[30] closure *n*. & close out *vt*. 结束,收尾

Notes 1.2

[1] project life cycle 项目生命期。生命繁衍不息,但其个体则是短暂的,或者说"(短期性)一次性"。正因如此,项目如同生命。不少书籍上有"项目周期"、"项目寿命周期"、"项目生命周期"或其他类似词语。诚然,自从有了人类,已经有了一个又一个的项目,将来的项目还将继续。然而,具体的项目一旦取得了预定的成果就结束了,不会再来一遍。否则,就同项目的"(短期性)一次性"相矛盾。为什么不断有人使用"周期"二字呢?或许这同理解 project life cycle 的含义有关。life 译成"生命"实在好,因为项目同生命一样,有始有终。但是,cycle 应如何理解呢?《新英汉词典》这样写:"①周期;循环;一转;②(一段)长时期,(一个)时代;③……"。cycle 的确有"周期"之意,但在这里不能取此意,而应取"(一段)长时期",简称"期"。因为人生的 cycle 一般是几十年,足够长了。

作者建议今后使用"生命期"表达"项目所有阶段的全体"这一概念。

[2] The employer engages an architectural firm to define it and then do the design.

【译文】 业主聘请一家建筑师事务所确定办公楼的形体与功能,然后设计。

【解释】 define 的意思是 make clear the outline or form of *sth*. 即清楚地表示某物轮廓与形式。

1.3 Construction markets

1.3.1 Definition of construction markets

Market is a term that is used in economics to signify an institution through which buyers and sellers interact and engage in exchange. The interactions between buyers and sellers may be very simple, or they may be quite complex. For example, in rural areas farmers simply bring their produce and/or goods to a central place and trade them.

Construction markets refer to the institution through which the demanders of building and engineering works interact with the suppliers of the services needed to complete and deliver the needed works and the buyers pay the sellers for the sellers' services when delivered. The construction markets are much more complex than many other markets, including the services to exchange, the buyers and sellers and the regulatory systems.[3]

A seller can't make a profit unless some buyers want to buy the goods or services that the seller is selling. This logic leads to notion of consumer sovereignty. The construction markets now in China are dictated ultimately by the tastes and preferences of the buyers who "vote" by buying or not buying.

The services supplied on the global construction market can be discussed under two major headings, that is, *construction and related engineering services* and *architectural and engineering services*, according to the World Trade Organization.[4][5]

1.3.2 Construction and related engineering services

It is well understood that *construction and related engineering services* and *architectural and engineering services* are distinct but closely interrelated service sectors. It is recognized that the construction-related engineering services involves services provided by professional engineers. The two categories of services overlap each other. Construction firms very often provide both service categories.

Construction and related engineering services covers the activities as follow.[4]
(1) General construction work for buildings
(2) General construction work for civil engineering
(3) Installation and assembly work
(4) Building completion and finishing work
(5) Other

This item includes pre-erection work at construction sites, as well as special trade construction work, such as foundation work, water well drilling, roofing, concrete work, steel bending and erection, and masonry work. It also covers renting services related to equipment for construction or demolition of buildings or civil engineering works, with operator.

Construction services may be carried out by general contractors who complete the entire work for the proprietor of the project, or by specialized subcontractors who undertake parts of the work.

Public-sector financing and public procurement play an important role in the consumption of construction services, generating as much as half of the total demand for the services.

1.3.3 Architectural and engineering services

The sub-sectors of architectural and engineering services are briefly discussed below under four headings.[5]
(1) Architectural services
(2) Engineering services
(3) Integrated engineering services
(4) Urban planning and landscape architectural services

The above does not explicitly cover services to be provided by surveyors or topographical engineers.

Architectural firms provide blueprints and designs for buildings and other structures, while engineering firms provide planning, design, construction and management services for building structures, installations, civil engineering works and industrial processes etc.. Consulting engineers are involved in all stages of a project, and thus their services overlap substantially with those of other professionals. Thus, architectural and engineering services are strongly integrated or inter-related with physical construction activity and/or other business services.

Architects and engineers occupy a position upstream of the building and construction process. Therefore, demand for architectural and engineering services is closely related to that of construction and overall industrial investment, both of which are in turn linked to the economic cycle. Demand for those services may be medium-term leading indicators for the construction industry.

1.3.4 Organization of construction industry

The sector consists of different types of specialist firm working together on temporary sites to produce buildings and engineering works. Once they have fulfilled their contractual undertakings, each firm moves on to other work with different firms on new sites.

One way of looking at the structure of the construction sector is to look at the number of firms involved in different aspects of the industry. In general, the contributions of specialist trades, general builders, building and civil engineering contractors and civil engineers differ, decreasing in that order.

Another way of viewing the sector is to look at its market structures. Standard industrial economics identifies four basic market structures: perfect competition, monopoly, oligopoly and monopolistic competition.[6]

1.3.4.1 Perfect competition

Perfect competition represents a totally competitive structure. Although perfect competition can not be observed in reality, it can be viewed as a benchmark against which all other market structures can be compared.

The main characteristics of a perfectly competitive industry are as follow:

☐ there is an extremely large number of buyers and sellers.

☐ each buyer and seller is so small, relative to the overall amount that is traded, that the market price is completely unaffected by a change in the behaviour of any one of them. Buyers and sellers are therefore referred to as price takers.

☐ the products that are traded are completely identical (or homogeneous), regardless of who produces it. If a firm chooses to raise its price above the market price, it will lose its all customers since they will simply buy an identical product from a (cheaper) source elsewhere.

☐ there is no uncertainty or industrial secrecy—it is assumed that all economic agents, whether buyers or sellers, are endowed with perfect information about the market.

☐ there are no impediments to prevent a firm from entering or exiting the industry. For example, no product is afforded patent protection.

1.3.4.2 Pure monopoly

Strictly speaking, "monopoly" refers to a situation where just one firm is operating in a particular industry. The goods or services produced by the pure monopolist are assumed to have no close substitutes and are demanded by a large number of buyers who have no market power. Despite being the only producer of a given commodity, the pure monopolist does not have total power over the market.

1.3.4.3 Oligopoly and duopoly

Oligopolistic industries are characterized by a small number of firms accounting for a large proportion (or all) of total output. In the case of duopoly, this is limited to two firms. Oligopolistic market structures are common throughout the developed world in industries that provide firms with the opportunity to exploit size economies.

1.3.4.4 Monopolistic competition

Monopolistic competition exists when a large number of firms are operating in a particular market but, unlike perfect competition, each producer offers the customer a slightly differentiated product. This differentiation may be explicit in the good or service; lodged in the mind of the consumer by persuasive advertising; or because the firms competing against each other are located in different geographical areas.

This means that, unlike firms in a perfectly competitive industry, it is possible for a firm to increase the price of a product without losing all its customers. There are numerous examples of this type of market structure in the construction industry. For example, although all qualified plumbers can offer the same basic service, it is possible for a plumber based in Beijing to charge a higher price than a plumber in Tianjin without losing customers in Beijing.

Of course few industries can be expected to fit precisely into any one of the above categories. However, there is strong evidence to suggest that there are examples of both oligopolistic and monopolistic competition.

Words, Phrases and Expressions 1.3

[1] signify *vt.* (signified, signifying) 表示（某人观点、意图、目的等）(make known one's views, intentions purpose, etc.)

[2] institution *n.* 制度 (long established law, custom, or practice)

[3] interact *vt.* (interacted, interacting) 互相作用，互相影响，进而得到某种结果；讨价还价；回合 (act on each other) (*Market is a term that is used in economics to signify an institution through which buyers and sellers interact and engage in exchange.* 市场是在经济学中表示买卖双方讨价还价，进行交换这种制度的一个术语。)

[4] produce *n.* 农产品 (the thing produced by farming)

[5] regulatory systems （政府对经济和社会活动）调节的制度

[6] notion *n.* 观念，概念 (idea, opinion) *This logic leads to notion of consumer*

sovereignty. 根据对这种关系的认识，就提出了"买方市场"这一概念。）

[7] consumer sovereignty 买方市场（The idea that consumers ultimately dictate what will be produced (or not produced) by choosing what to purchase (and what not to purchase). 消费者通过选择买什么（和不买什么）最终决定要生产什么（或不要生产什么）的思想。）

[8] construction and related engineering services 施工及设计服务

[9] involve *vt.*（*a*）免不了，需要，不可缺少（require as a necessary accompaniment; entail）*It is recognized that the construction-related engineering services involves services provided by professional engineers*. 人们已经认识到，同施工有关的工程设计服务少不了专业工程师提供的服务。/（specialization of the service involved 必不可少服务的专业化）*Although development within other sectors of the economy has often involved expanding the size of operations, and hence increasing the level of industrial concentration, the construction industry continues to be geographically fragmented with a large number of small firms*. 虽然国民经济其他部门内部发展经常免不了扩大经营规模，进而提高了产业集中度，但是，建筑业仍然是大量的小公司分散在各地。）请读者注意，*involved* 常以过去分词形式当作名词后置定语。（*b*）使……参与（engage as a participant）*He has involved twelve workers in building his house*. 他雇了12个工人盖自己的房子。/*The twelve workers involved in building his house are all from the same place*. 给他盖房子的都是一个地方的。/*Consulting engineers are involved in all stages of a project*. 咨询工程师参与了项目的所有阶段。

[10] architectural and engineering services 建筑及工程设计服务

[11] general construction work for building 一般建筑物施工

[12] general construction work for civil engineering 一般土木工程施工

[13] install *vt.*（installed, installing）安装（place, fix (apparatus) in position for use. ~ heating or lighting system. 安装供暖或照明设备。）

[14] assemble *vt.*（assembled, assembling）装配（put or fit together the parts of sth.）

[15] installation and assembly work 安装与装配工程，安装工程

[16] installation(s) 装置

[17] construction *n*. 建筑物（整体或部分），构筑物（整体或部分）；构造（a structure, something constructed）

[18] building completion and finishing work 建筑装修工程（completion 是完成装修工程开始时还未完成的工作或工程，例如为了便于装修而在墙上预留的洞口。）

[19] trade *n*. 行业，例如木工、石筑、砌砖等。（an occupation, especially one of skilled manual or mechanical work, such as carpentry, masonry, bricklaying, etc.）

[20] special trade construction work 专业施工工程

[21] roofing *n*. 屋面材料（material for a roof）；屋面防水工程（make a roof water proof）

[22] general contractors 一般承包商

[23] specialized subcontractors 专业承包商

[24] public *adj.* (opposite of *private*) *private* 的反义词，公共的（of, for, connected with, relating to, owned by, done for or done by, known to people as a whole）
[25] public sector 公共部门
[26] public-sector financing 公共部门提供的资金；公共部门为……提供资金
[27] public procurement 公共采购
[28] architectural services 建筑师提供的服务，简称"建筑服务"
[29] engineering services 工程设计服务
[30] surveyors or topographical engineers 测量员或地形图测量员
[31] blueprints and designs 蓝图和式样
[32] industrial processes 生产过程
[33] consulting engineers 咨询工程师
[34] professional 专业技术人员
[35] economic cycle 经济周期
[36] contractual undertakings（书于）合同（之中的）承诺
[37] specialist trades 专业工种
[38] general builders 建造一般建筑物者，一般建造师
[39] industrial economics 产业经济学
[40] market structures 市场结构
[41] perfect competition 完全竞争
[42] monopoly *n.* 垄断
[43] oligopoly *n.* 寡头垄断
[44] monopolistic competition 垄断竞争
[45] represent *vt.* 表示，表现，象征，是（be, give, make, a picture, sign, symbol）*perfect competition represents a totally competitive structure.* 完全竞争是一种全面展开竞争的（市场）结构。
[46] benchmark *n.* 基准（sth. that serves as a standard by which others may be measured or judged）
[47] price takers 价格接受者，指无讨价还价能力
[48] industrial secrecy 行业秘密
[49] economic agents 经济主体，参与经济活动者
[50] oligopolistic industries 寡头垄断行业
[51] size economies（economies of scale）规模经济（A characteristic of a production technology whereby unit costs decline with increasing output over a large range. Economies of scale are a major source of natural monopoly. 生产技术的一个特点是，单位成本能够在很大范围内随着产出的增加而下降。规模经济是自然垄断的主要根源。）
[52] suggest *vt.* 表明，使想到（bring an idea, possibility, etc. into the mind; call to mind by thought or association. *However, there is strong evidence to suggest that there are*

examples of both oligopolistic and monopolistic competition. 然而，大量迹象表明，寡头和垄断竞争都有先例。）

[53]　major construction projects 大型建设项目（不宜译成"主要的建设项目"）

[54]　favourable financial packages 有利的财务安排

Notes 1.3

[1]　administer(administration)与manage(management)区别

这两个词译成汉语时，都可以写成"管理"。但英文原义不同。administer 的原义是 manage or supervise the execution; use or conduct of, 而 manage 的原义是 succeed in accomplishing; exercise executive administrative, and supervisory direction of. 两者的大致区别如下：

administer（administration）指根据法律、程序或方针指导、监督和矫正资源的使用，以及人的行为；而 manage（management）指利用各种资源，通过努力，以各种方式完成某种成果，实现某种目的或达到某些目标。前者按照既定规使对象不逾矩；照看已有之物。后者则努力设法，以获得成功。可以将前者比做"守业"，后者"创业"。正是因为有这种区别，与"合同管理"对应的应当是 contract administration，而不是 contract management。当然，contract management 也常出现在某些文献中，但那是误用。

[2]　疑难词语解释

1) In general, the contributions of specialist trades, general builders, building and civil engineering contractors and civil engineers differ, decreasing in that order.

【译文】　一般说来，专业承包商、一般建筑承包商、建筑与土木工程承包商和土木工程师的贡献是不同的，大小顺序也就是上面列举的那样。

2) Internationally active companies constitute still only a handful of the total, but recently, international activity is reportedly rising, along with a certain concentration of the industry.

【译文】　活跃在国际之间的公司仅占总数的很小一部分，但是，据报道，近些年来，国际业务正在增加，而且建筑业越来越集中（在少数大公司手里）。

1.4　Participants in construction industry

The participants in the construction industry business include the planners, architectural and engineering designers, employers, contractors, material and equipment suppliers, construction workers, supervisors, financiers, accountants, lawyers, insurers and operators (see Figure 1.3).

The government is involved in the industry as purchaser, financier, regulator and adjudicator.

The production responsibility is divided among the employers, designers, contractors, subcontractors, material suppliers, equipment dealers, funding institutions and services such as transport, electricity and water.

The principal parties to a building or engineering contract are the employer and the contractor. Either the architect or the engineer is not a party to the contract.

The employer initiates the project and is responsible for providing funds for its execution. He is defined as the "Employer" in the *FIDIC Conditions of Contract*, and in other forms of contract he is also referred to as the "Purchaser".

The architect or engineer is appointed by the employer to have overall architectural and engineering responsibility for the investigation and design of a building or engineering project, and to make sure that the contractor's performance is satisfactory.

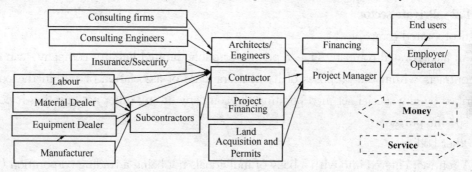

Figure 1.3 Supply-chain of construction activities

The contractor is the organization or individual entrusted with the construction of the works. The contractor is primarily a resource manager of men, materials, equipment, money and time and a coordinator of the activities of many participants who are not directly responsible for the final product and over whom he has little control.

1.4.1 The employer

The buyer of the services of consultancy, design, construction, construction management, project management and so forth is called an employer, if he wants to have a building built, or a promoter, if he wants to have a civil engineering work erected. For the sake of simplicity, the term "employer" is used throughout this book to refer to the buyer.

The architect or engineer usually refers to the employer as the client; the contractor may refer to him as the employer as he is called under some forms of contract. The employer may have virtually any legal status: he may be a sole trader, a partnership, a limited liability company or part of local or central government, or any other incorporated or unincorporated body.

Employers can be broadly identified as between public and private categories.[7]

1.4.1.1 Public sector

(1) Central government

Contracts are entered into with government through its departments. Under such contracts payment is made out of public funds.

(2) Local government

Local authorities can enter into contracts and raise funds for payments due. A contract validly entered into carries an implied undertaking that the authority possesses or will collect the requisite funds; lack of funds is no defense to an action for payment.

(3) National or public industries, corporations, or other bodies

Statutory boards, corporations, basic industries and utilities (e. g. railway, postal services, coal mining, electric power, water supplies and some port authorities) are generally publicly owned.

1.4.1.2 Private sector

(1) Incorporated companies

An incorporated company (i. e. a private or public limited liability company) can enter into contracts within the purposes of its memorandum of association or within the powers prescribed by any special act incorporating the company or any other act granting it powers for a specific purpose.

(2) Other bodies

A contract entered into with a body of individuals not being a trading corporation (e. g. a partnership or a club committee) must be considered on its merit in respect of the authority of one or more of the individuals to bind the rest in personal liability for the purpose of the contract and for making payments due under it.

1.4.1.3 Project manager

The employer may decide to appoint a project manager. He may be a consultant experienced in developing and managing major projects from feasibility study through to commissioning. If the project is large enough or complex the project manager should appoint a team to cover the various engineering disciplines involved, or should arrange for their appointment by the employer.

1.4.2 The A/E

The term "the Architect/Engineer (A/E)" refers to the person or firm that provides professional advice on the investigation for, and the design and construction of, buildings or civil engineering works.

The employer may decide, based on the project's type and extent, to utilize the services of an A/E in his employment or may engage consulting engineers to provide engineering advice to enable him to assess the feasibility and/or relative merits of various alternative schemes to meet his requirements.

It is usual for the A/E's functions to include the development, design and technical direction of the building or civil engineering works, preparation of specifications, bills of quantities and other contract documents. In particular, during the execution of the works, his duties, as the employer's contract administrator, include the inspection of materials and workmanship and a variety of administrative duties such as the measuring and valuing of the work and the pricing of varied work and so forth.

1.4.3 International organizations

A number of the international organizations have been involved in the construction activities in the developing countries, such as China. They have played an important role in promoting development of the construction industry and the economy as a whole in the countries.

1.4.3.1 FIDIC[8]

FIDIC, the International Federation of Consulting Engineers (the acronym stands for the French version of the name, i. e. *Fédération Internationale des Ingénieurs-Conseils*), represents globally the consulting engineering industry. As such, the FIDIC promotes the business interest of firms supplying technology-based intellectual services for the built and natural environment.

FIDIC activities are carried out by committees, task forces and forums appointed by the executive committee and mainly composed of volunteers. The executive committee is elected by the general assembly to carry out the FIDIC's work. Members propose to the general assembly for election each year a president, a president-elect or vice-president and a treasurer. The executive committee is responsible for the secretariat that carries out the federation's work.

Companies and organizations belonging to FIDIC national member associations may announce themselves as FIDIC members, and use the FIDIC logo.

Founded in 1913, FIDIC membership today numbers 74 member associations representing some 1 million professionals.

Members are generally national member associations. Companies and organizations may join FIDIC as affiliate members if their country either has no national member association or they do not qualify to belong to the national association and yet have interests related to FIDIC's.

1.4.3.2 World Bank[9]

The World Bank or World Bank Group is a vital source of financial and technical assistance to developing countries around the world. It is not a bank in the common sense. It is made up of two unique development institutions owned by 184 member countries—the International Bank for Reconstruction and Development (IBRD) and the International Development Association (IDA).

Each institution plays a different but supportive role in the World Bank's mission of global poverty reduction and the improvement of living standards. The IBRD focuses on middle income and creditworthy poor countries, while IDA focuses on the poorest countries in the world. The two institutions together provide low-interest loans, interest-free credit and grants to developing countries for education, health, infrastructure, communications and many other purposes.

1.4.3.3 Asian Development Bank[10]

The Asian Development Bank (ADB) is a multilateral development financial institution owned by 66 members, 47 from the Asia and 19 from other parts of the world.

ADB is aimed at improving the welfare of the people in Asia and the Pacific, particularly the 1.9 billion who live on less than $ 2 a day. Despite many success stories, Asia and the Pacific remain home to two thirds of the world's poor population.

ADB's vision is a region free of poverty. Its mission is to help its developing member countries reduce poverty and improve the quality of life of their citizens.

ADB's main instruments for providing help to its developing member countries are policy dialogue, loans, technical assistance, grants, guarantees and equity investments.

ADB's annual lending volume is typically about $ 6 billion, with technical assistance usually totaling about $ 180 million a year.

ADB is headquartered in Manila. It has 26 other offices around the world: 19 resident missions in Asia, 3 sub-regional offices in the Pacific, representative offices in Frankfurt for Europe, Tokyo for Japan, and Washington D.C. for North America, and a special liaison office in Timor-Leste.

ADB has more than 2,000 employees from over 50 countries.

1.4.4 The contractor

Under the *FIDIC Conditions of Contract*[11][12], there is usually only one contractor, sometimes referred to as the main contractor; on some works he may sublet or subcontract parts of the works to specialist or other contractors who become known as subcontractors. Building and civil engineering contractors may be broadly classified under two headings: general contractors and specialist contractors.

1.4.4.1 General contractors

General contractors are those who, on account of their knowledge and experience, are able to undertake responsibility for the construction of the whole of a project.

1.4.4.2 Specialist contractors

Those who confine their activities to selected classes of work are referred to as specialist contractors. In some cases their operations are protected or covered by patents, but generally they rely on special resources and experience.

A specialist contractor may perform his work by subcontract to a general contractor who will act as coordinator and is called subcontractor. Alternatively the specialist contractor may perform his work by direct contract with the employer, and is referred to as nominated subcontractor.

1.4.4.3 Contractor's primary obligations[12]

As stated in sub-clause 4.1 of the FIDIC *Conditions of Contract*[12], the contractor shall design (to the extent specified in the contract), execute and complete the works in accordance with the contract and with the Engineer's instructions, and shall remedy any defects in the works.

Attention shall be drawn to the word "*complete*". Nowadays the contractor's obligation to complete the works is accepted in most construction contracts. However, the obligation to complete was omitted in many standard general conditions of contract, which was criticized[13]. Although it may be argued that it is implicit in construction contracts drafted based on common law (see 1.6.4.2 for more details) that the contractor must complete the works, it would have been much clearer to set out this requirement explicitly in the conditions of contract. This is because the remaining clauses in the general conditions of contract assume such an obligation and, perhaps more importantly, because such an obligation is not necessarily implied in all legal jurisdictions.

Under common law, it is the contractor's duty to complete the works whatever happens. This duty results from a case which occurred in the United Kingdom in 1867 where it was held that the contractor was not entitled to recover anything from the employer in respect of any portion of the machinery which he had erected and then was destroyed in a fire prior to the completion of the work.

It has become accepted in common law group that by virtue of the express undertaking to complete the works the contractor would be liable to carry out his work again free of charge in the event of some accidental damage occurring before completion even in the absence of any express provisions for protection of the work.

Words, Phrases and Expression 1.4

[1] contractor's performance 承包商的表现,承包商履行合同的结果
[2] client *n*. 委托人(the architect or engineer usually refers to the employer as the client)
[3] legal status 法律身份
[4] sole trader 个体经营者
[5] partnership *n*. 合伙企业
[6] board n. 局、署、处(an official group of persons who direct or supervise some activity)
[7] statutory boards 各法定局、署、处,例如"盐业管理局"、"航道管理局"等。
[8] charters *n*. 章程 (a government document outlining the conditions under which a corporation is organized.)
[9] act n. 法案 (law made by a legislative body)
[10] authority *n*. 当局,权力机关(a person or body of persons in whom such power is vested.)

[11] liability n. 债务,(归还债务的)责任

[12] incorporated company 立了章程已在工商管理部门登记的公司

[13] action n. 诉讼(legal process, bring an action against *sb*. for *sth*., seek judgement against him in a law court for *sth*.);

[14] sanction (*a*) n. 权利或许可;批准 (right or permission given by authority to do something) (*b*) vt. 批准,认可 (give a sanction to; agree to)

[15] contract vt. 发包(hire by contract; procure goods, services or works on a contract basis—often used with out)

[16] be subject to 应服从（的）,受制于……（的）,受……制约（的）(owing obedience to) *Local authorities are subject to the ordinary legal liabilities as to their powers to contract and their liability to be sued.* 地方当局在其发包权力方面以及在他人要求其还债时承担普通的法律责任。

[17] statutory adj. 法定的(fixed, done, required by statute) *statutory boards* 法定委员会

[18] cover vt. 包罗,够用 (include, comprise, be adequate for) *If the project is large enough or complex the project manager should appoint a team to cover the various engineering disciplines involved, or should arrange for their appointment by the employer.* 如果项目大而复杂,项目经理应建立包罗各必要工程门类的团队,或设法让雇主聘请之。

[19] specification n. 技术规格（用于货物采购）,技术要求,技术要求说明书,技术条款,设计说明书（用于工程和技术咨询服务采购）(The document that prescribes the requirements with which the product or service has to conform. Note: A specifications should refer to or include drawings, patterns or other relevant documents and should also indicate the means and the criteria whereby conformity can be checked. 规定产品或服务必须遵循哪些要求的文件。说明:技术要求可指或包括图纸、样品、模型或其他有关的文件,还可指明如何检查是否符合要求的手段和标准)。

[20] bill(s) of quantities 工程量清单,工程量表

[21] contract documents 合同文件

[22] execute the works/execution of the works 将（设计文件上的工程）付诸实施,施工

[23] contract administrator 合同管理者,合约管理者

[24] price vt. 确定或计算……的价格或价钱 (set a price on; find out the price of)

[25] varied work 变更了的工作 *price a varied work/pricing of a varied work* 确定变更了的工作的价钱

[26] *Fédération Internationale des Ingéieurs-Conseils* 国际咨询工程师联合会（原文为法文）,缩写为 FIDIC

[27] memorandum 契约、合同或协议的要点

[28] memorandum of association 合营协议

[29]　memorandum of understanding 理解备忘录

[30]　subcontract *vt*. 将……分包出去(engage a third party to perform under a subcontract all or part of work included in original contract-sometimes used with out) He may sublet or subcontract parts of the works to specialist or other contractors who become known as subcontractors.

[31]　sublet *vt*. 与 subcontract 基本同义 *He may sublet or subcontract (out) parts of the works to specialist or other contractors who become known as subcontractors.* 他可以将部分工程分包给专业公司或叫做"分包商"的其他承包商。

[32]　specialist contractors 专业承包商

[33]　contractor's primary obligations 承包商的基本义务

[34]　common law 普通法 (The common law system originated in England after the Norman conquest of 1066 but it was mainly created by the judges appointed by the Crown in the twelfth and thirteenth centuries. It is therefore judge-made, sometimes referred to as case law.)

[35]　common law group 采用普通法体系的国家

Notes 1.4

[1]　两点说明

1) 关于 Specification 的译法

Specification 是重要的招标文件（签约后则成合同文件）之一，由招标文件编制者根据拟招标工程的具体情况编写。因此，宜将 Specification 译成"技术要求"，"技术要求说明书"或"技术条款"，译成"规范"则不妥。在现代汉语中，"规范"指政府部门制订并颁发的有关设计、制造、加工、施工、试验、验收等等的规定、要求和标准。在发达国家，某些权威的行业和专业协会也制订颁发"规范"。但是一般雇主、工程设计或施工企业都不能对"规范"进行补充和修改。因此，若将 Specification 译为"规范"，则 FIDIC 于 1999 年出版的施工合同条件 "1.1.1.5 'Specification' means the document entitled specification, as included in the Contract, *and any additions and modifications to the specification* in accordance with the Contract. Such document specifies the Works." 中的 *any additions and modifications to the specification* 应如何理解呢？另外，Specification 除了指明工程中使用的材料、设备、工法、质量标准、应遵守的规范、标准之外，还常常以大量篇幅说明承包商合同条件规定以外的一切特殊责任；设计依据的资料；雇主对工程所用材料、设备的产地、生产厂家的要求；对生产、加工、施工、质量检查等等的具体要求。而这些内容无论如何也不能理解成"规范"。

2) 关于 loan(贷款)和 credit(信贷)的区别

The original World Bank consisted solely of the International Bank of Reconstruction and Development (IBRD). The IBRD supported productive projects for which other financing was not available on reasonable terms. While this arrangement addressed the problems of stronger developing countries that could service principal and interest payments, it left out poorer nations, which could not. Realizing that a special funding mechanism was needed for the poorer nations joining the World Bank in large numbers, the Bank's member countries created International Development Association (IDA) in 1960. IDA supports the same types of projects as those financed by IBRD. They are managed by the same staff and have the same review procedures. The difference is that the loans are interest-free and require only a small annual service. Repayment of principal stretches over 35-40 years with a ten-year grace period. Only the poorest countries are eligible for assistance from IDA. These funds are known as credits to distinguish them from IBRD loans, mostly going to countries with annual per capita incomes below $690. Some developing countries receive a blend of IDA and IBRD loans. Today more than 50 low-income countries, with more than 2.5 billion people, are eligible for IDA lending. Credits from IDA also called "soft loans".

【译文】 世界银行原来只有国际复兴开发银行。国际复兴开发银行支持无其他融资办法的生产性项目。这种做法虽然也解决了能够偿还贷款本金和利息的较强发展中国家的问题，但漏掉了穷国，穷国还不起贷款本金或利息。在认识到有必要为大量加入世界银行的穷国建立特别的筹资渠道之后，国际复兴开发银行成员国于1960年创建了国际开发协会。国际开发协会支持的项目类型与国际复兴开发银行一样。这些项目由国际复兴开发银行的同一批人员按照同样的审批程序进行管理。不同之处在于贷款不收利息，每年只收少量的手续费。贷款本金的偿还长达35～40年，宽限期10年。只有最穷的国家才能获得国际开发协会的贷款支持。这种资金叫做信贷，以示同国际复兴开发银行贷款的区别，大多数都贷给了人均年收入低于690美元的国家。有些发达国家从国际复兴开发银行和国际开发协会得到混合贷款。现在，有50多个低收入国家，25亿人口符合从国际开发协会申请信贷的条件。国际开发协会信贷也叫做"软贷款"。

[2] 疑难句解释

1) As stated in sub-clause 4.1 of the FIDIC Conditions of Contract, the contractor shall design (to the extent specified in the contract), execute and complete the works in accordance with the contract and with the Engineer's instructions, and shall remedy any defects in the works.

【译文】 FIDIC 第 4.1 款写明，承包商应按照合同规定与工程师的指示对合同规定的工程部分或范围进行设计，实施与完成工程并弥补其中的缺陷。

【解释】 请注意，承包商单单按照合同要求施工是不够的，必须将其"完成"。如果是办公楼，其他都有了，即使仅门前台阶面层还未做，也不能叫完成。因为人们进不了楼，也就是说，这栋楼还未具备合同文件要求的功能。所以，合同条件中必须要有"完成"。当然，"弥补工程中的缺陷"的要求也不能遗漏。

2) Although it may be argued that it is implicit in construction contracts drafted based on common law (see 1.6.4.2 for more details) that the contractor must complete the works, it would have been much clearer to set out this requirement explicitly in the conditions of

contract. This is because the remaining clauses in the general conditions of contract assume such an obligation and, perhaps more importantly, because such an obligation is not necessarily implied in all legal jurisdictions.

【译文】 人们虽然会说，根据普通法编制的施工合同隐含规定承包商必须完成工程，但是，在合同条件中明文列入这一要求要清楚得多。这是因为一般合同条件中的其余条文要以此义务为前提，还因为并非所有其他法律管辖区也一定将这一义务隐含在合同之中，也许这后一点更重要。

3) It has become accepted in common law group that by virtue of the express undertaking to complete the works the contractor would be liable to carry out his work again free of charge in the event of some accidental damage occurring before completion even in the absence of any express provisions for protection of the work.

【译文】 采用普通法体系的各国已经普遍认为，承包商若明确承诺完成工程，即使合同无任何保护工程的明文，也要在工程于完成之前因某种事故而受损时自费返工。

4) A contract entered into with a body of individuals not being a trading corporation (e.g. a partnership or a club committee) must be considered on its merit in respect of the authority of one or more of the individuals to bind the rest in personal liability for the purpose of the contract and for making payments due under it.

【译文】 必须从某群体中一人或多人在某合同的目的与该合同规定的款项支付方面是否有权制约该群体中其他人个人责任的角度考虑与不是营业公司的该群体(例如合伙或俱乐部委员会)签订合同是否有利。

1.5 Building and engineering procedure

The procedure from initiation to completion of a building or engineering works is conveniently illustrated by the flow chart shown in Figure 1.4.[7] It shows the three parties contractually involved in such a work and the principal activities in their logical sequence.

In practice many of the activities overlap. The contractor is generally paid in stages, for work done at agreed payment times. Sometimes design and construction develop more or less in parallel, and it is not unknown for construction to start before a contract has been signed or even before the employer has acquired all his land and raised all his funds. Procedures and contractual methods vary with time and circumstances. Figure 1.4 shows a procedure which the design-tender-construction (employer/engineer/contractor) arrangements will follow.

No matter what the contractual relationship between the parties to a contract may be Figure1.4 embodies a principle of working that is well established in many parts of the world and the principle seems to remain valid for most purpose-built engineering works. For example, the A/E, whether an employee of the employer or a consulting engineer must retain his independence.

The contractor's engineer performs any of the functions shown in the A/E column in Figure1.4 and he has a similar problem. A prudently run direct labour organization will operate

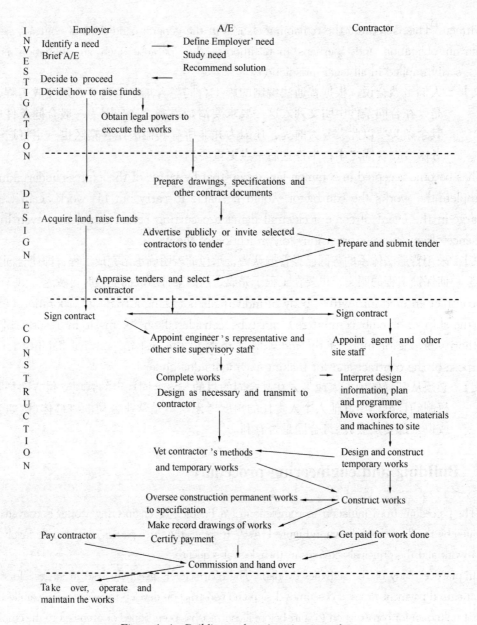

Figure 1.4 Building and engineering procedure

some kind of internal tendering procedure to keep a check on prices and provide a basis for cost control. The organization will appoint someone in the role of engineer's representative to ensure that his colleague who functions as an agent executes the design correctly and to the required quality standards.

1.5.1　Initiation/promotion

1.5.1.1　Legal authority[7]

Normally various legal authorizations are to be obtained prior to commencement of a building or engineering works on site. The employer whose project has to be approved under statutory regulations should comply with specific procedures as laid down in the regulations. These requirements call for special experience on the A/E's part, and place a duty on the employer to initiate investigations in good time to allow all economic and other considerations to be given the attention they deserve. Lack of appreciation of the time required for proper investigation has been the cause of rejection of some projects, or of much additional expense that could have been avoided.

1.5.1.2 Employer's responsibilities and duties

The employer's responsibilities and duties are normally written into a client/consultant agreement with the A/E and the contract for construction and are generally:

☐ to define the functions that the building or engineering works is to perform,
☐ to provide information and data in his possession and required by the A/E,
☐ to obtain necessary permits or licenses to make construction legal,
☐ to finance his project, and
☐ to acquire the land to seat the building and/or engineering works.

1.5.2 Selection of the A/E

The selection of an A/E should start with assessing the qualifications and experience of possible contenders. Final selection may be through inviting proposals or competitive fee offers for comparison. Whenever the employer engages an architect or consulting firm for a specific project a formal agreement should be completed between the two, which sets out the duties and responsibilities of each party and the fees and expenses to be paid.

It is a matter of importance that the person appointed as the A/E should have qualifications and experience appropriate to the project. The A/E may either have official status in government or public service, public company or private enterprise, or be a senior member of a firm of consulting engineers and who, from his established reputation, is considered well qualified for advising on the development of a particular project.

It is internationally accepted that the A/E shall be responsible to the society, seek solutions that are compatible with the principles of sustainable development, and at all times uphold the dignity, standing and reputation of the consulting industry. [8]

The A/E should be selected on a competitive basis at least for a public project. [8]

In 1997, FIDIC published its document on quality-based selection, setting out to help all those responsible for the selection of consultants to achieve that objective on the basis of qualification, experience, ability and integrity.

The appointed A/E is expected to provide his service in a strictly fiduciary manner, which is essential not only to the whole relationship between the employer and the architect/consulting engineer but also to that between the employer and the contractor.

The principles of behaviour set by FIDIC in the codes of ethics[8] for consulting engineers are summarized as "competence", "integrity", "impartiality", "fairness to others" and "anti-corruption".

The quality and cost based selection process, an alternative to the quality based selection referred to above, can be divided into: [14]

(1) preparation of the terms for reference (TOR),
(2) preparation of cost estimate and the budget,
(3) advertising,
(4) preparation of the short list of consultants,
(5) preparation and issuance of the request for proposals (RFP) which should include: the letter of invitation, instructions to consultants, the TOR and the proposed draft contract,
(6) receipt of proposals,
(7) evaluation of technical proposals, consideration of quality,
(8) public opening of financial proposals,
(9) evaluation of financial proposals,
(10) final evaluation of quality and cost, and
(11) negotiations and being award of the contract to the selected firm.

1.5.3 Pre-investment studies

Pre-investment studies is perhaps a most critical task for the A/E because the results may determine the employer or promoter's investment policy as well as the feasibility and basic features of individual projects. [8]

However, it is the pre-investment study that determines whether or not this project is feasible, viable, acceptable, workable, and finally sustainable and at what cost.

The knowledge required of the A/E is extremely wide and the team may include engineers, planners, economists, sociologists, ecologists and other specialists. Various alternative solutions are normally presented to the particular problem under study with commentary on advantages and disadvantages of each option. It is clear that such a study has to be totally independent and free from any bias or influence, be commercial, technical, political, financial or any other kind.

1.5.3.1 The employer's needs

The services the employer wants the A/E to provide vary widely, such as advising on identification of project and resources; quantification of objectives; a particular case or situation; assessment of the project requirements and to devise alternative solutions for his consideration, although occasionally he may have a general knowledge of what he needs and wants to have the details; etc. Sometimes he may want the A/E to provide the information on a particular phase or aspect in order to decide whether or not to proceed with a project, or to raise funds or obtain other authority for its implementation. In other cases he may wish to ascertain the effect a project may have on the environment.

Such services may involve an individual consulting engineer, or a small group of specialists qualified to deal with the particular task.

1.5.3.2 Importance and extent

Once appointed the first tasks for the A/E are to acquaint himself with and to review the employer's requirements, and then to agree a brief for the investigations. Normally the A/E will inspect the site and consider all physical, financial and political constraints, devise possible solutions and compare them. Then he will submit a report to the employer summarizing his findings and conclusions and recommending a preferred solution with an estimate of its capital and likely maintenance costs.

The report should recommend the basis for the economic design and/or engineering of a project. It should therefore cover the main design choices and the major problems that may have different cost, time and future revenue implications. These should therefore have been investigated thoroughly by this stage. Investigations and reports cost only a small fraction of the total cost of implementation of the project, so it is false economy to restrict the scope or thoroughness of the A/E's investigations and design/engineering studies.

1.5.4 The A/E's TOR[7][8][15]

1.5.4.1 Statement of work

It is usually the employer who initiates the brief for investigation but the final version should be agreed between him and the A/E in order to incorporate the latter's specialist knowledge and experience of similar studies. The A/E's duties in respect of investigation and reports should be defined in a statement of work (SOW) as a part of the brief. In developing the brief the employer must state his long-term and short-term objectives and indicate if he is aware of all the constraints. SOW should also indicate the nature and details of the information to be presented in the A/E's report and supporting documents, available information and the time available for preparation of the report, and whether interim or progress reports are required. Site conditions and their possible exploitation often cause the employer to incur additional expenditure which he has not realized. In such cases the employer's basic needs for exploration and advice shall be met by the A/E before the brief is finalized.

1.5.4.2 Flexibility and alternatives

SOW should afford flexibility so that the A/E can consider new developments that may occur while the investigation is in progress. The extent to which the A/E may consider alternative solutions should be agreed. It is generally advisable for the SOW not to indicate a preferred alternative, as the value of a report can be seriously diminished if it can be inferred that any matter has been prejudged. This is particularly true when environmental issues are involved.

1.5.4.3 Type of investigation or study

The extent, nature and depth of the A/E's study depend on the value and complexity of the project, the number of solutions to be considered, and the nature and number of the decision-making processes involved before one of the alternatives can be chosen. For some projects, a single study with the preferred alternative and its cost estimate will suffice; others may require a series of studies with increasing details. To develop a preferred alternative for a major project some or all of the following studies may be required:

☐　assessing the requirements of a particular sector of an economy or industry so that individual projects to meet them can be identified;

☐　pre-feasibility study of likely viable demand for the product or service of the project to be studied, availability of the resources required within acceptable cost and time limits, and justifiability of a feasibility study;

☐　feasibility (or pre-investment) study, including preliminary surveys for investigating technical and economic viability, an estimate of capital and operating costs, and other information to enable the employer to decide if he should raise funds for the project; but not including detailed design. Some outline design may be done, however;

☐　preparation of a master plan as a long-term development program, generally indicating how construction and expenditure can be phased;

☐　examination of the effect which the proposed development would have on the environment;

☐　geotechnical study of the nature of the ground where works may be constructed and assessment of possible problems;

☐　developing the preferred alternative in detail, especially in technical terms, which is sometimes referred to as the final design study;

☐　identification of possible sources of finance and assistance which the employer may arrange for the provision and repayment of the funds needed, in which cases a special report on the financing arrangements will be needed.

1.5.4.4　Technical investigations

They are used to resolve the engineering aspects of a project. Further investigations will be required when a project is likely to affect the environment; economic, social or environmental studies may be needed in such cases.

1.5.4.5　Economic and social investigations

The economic, demographic or other studies may be needed for certain project and its employer. Although the studies shall generally be done under direction of the A/E, a specialist in another discipline usually performs them. The team for any large study may include several non-engineering specialists.

1.5.4.6　Use of existing information and plans

The existing data, topographical maps and marine charts where these exist shall be made available. It may be necessary to carry out additional studies to supplement the available information, and in the case of topographic surveys to make possible the preparation of accurate plans of the sites considered for the construction of the works.

1.5.4.7 Site exploration

Thorough site exploration is fundamental to the economic design, efficient planning and execution of any project. Its importance cannot be over-emphasized. Existing topographical and geological maps and data should be studied and thorough investigations carried out under the direction of the A/E. Special investigations, soil investigation and so on should be undertaken by specialists who should provide bore hole logs, samples and laboratory test results. Their reports should include an interpretation of results by a geotechnical expert, but the A/E must make the final assessment of the results in so far as they may affect the design and specification of the works.

1.5.4.8 Physical data

The access to the site both for permanent and temporary use during construction, statutory or local authority restrictions on the site or the special rights of adjacent landowners and the availability of services such as drainage, sewerage, water, electricity and gas may require investigation. The source and transportation of materials for construction and operation and disposal of waste material should be studied as well. In addition, the data on the climate and the incidence of storms and flooding shall also be collected. The existing ambient noise levels may have to be measured and studies undertaken to predict and restrict as necessary the noise levels during and after construction.

1.5.4.9 Models

There are four principal types of modes. Experimental and mathematical models are used in complex problems to generate information needed to assess the effects of various design options. Physical and computer graphic models are valuable in setting the layout of projects, studying the use of space and deciding on the methods and sequence of work in construction. Physical models are valuable for discussing proposals with the employer's staff and also for training those who will take over the operating activities of the completed project.

1.5.4.10 Public participation

The execution of civil engineering works affects the public in some way and there usually are some people who suffer. The suffering people may reasonably have objections to a proposed scheme. The public shall be invited to take part in the decision-making process at an early stage and encouraged to express their views and preferences which shall be duly taken into account during the feasibility and planning stages. Involvement of the public in planning stage of a project can reduce objections and eliminate the need for a public inquiry. It can also produce constructive ideas to the benefit of the project.

1.5.4.11 Outline program

In formulating an outline program for the detailed design and construction of the works the A/E must take account of the availability of finance, labour, materials and feasible methods of construction and temporary works. The outline program must be realistic and should not impose unnecessary restrictions on tenderers or the contractor.

1.5.4.12 Cost estimate

The A/E is usually required to prepare estimates of the capital cost of proposed alternatives, of the land, buildings, plants and machinery involved, and that of the operating and maintenance costs of the completed facility. Some of the estimates may have to be obtained from or confirmed by outside sources. Capital and annual cost estimates shall be drawn up in a manner that the economic viability of alternative schemes or sites can be assessed and the costs to the employer of implementing the chosen alternative.

1.5.4.13 A/E's report[7]

When the investigations are completed, the A/E should present to the employer a report summarizing his investigations and conclusions. Normally he will outline the alternative best suited to the requirements, in his opinion (which will usually, but not always, be the least costly of the feasible options), and give forecasts of costs and financial viability to help the employer visualize the project as a whole.

1.5.5 Proceeding with project[7]

If convinced by the A/E's report the employer decides to proceed with the project, he must take steps to obtain the legal authority to start the building or works. It is always necessary to obtain planning permission or its equivalent and seek comments from adjacent landowners or other people who may be affected by the project. Having done this the employer can instruct the A/E to proceed with the design and/or engineering and the preparation of tender documents. This work will usually proceed concurrently with the procurement of finance and the purchase of the necessary land.

1.5.6 Design and preparation of documents[8][15]

1.5.6.1 The A/E's services

The A/E may be involved in engineering and detailed design, preparation of tender and/or contract documents and contract administration services during implementation phase. Engineering and detailed design services should normally be followed by contract administration services during the implementation phase. The same A/E should normally provide both services. It is a mistake to employ different professionals for these two interrelated services. Most important reason for this conclusion is perhaps the avoidance of any doubt the responsibility for the engineering and other services provided.

To enable the A/E to do engineering and detailed design and to provide adequate information on which contractors can be invited to tender, further site investigations are usually required. Laboratory or model tests may have also to be undertaken, and the approval of outside authorities obtained for certain aspects of the proposed works. The A/E must pay attention to the aesthetic and environmental aspects of his design, and should be free to seek other advice as he wishes on these and other subjects. Consultation with contractors during the design stage is strongly nowadays advocated by some employers wishing to make sure that the design is suitable for safe and economic construction.

1.5.6.2 Specialist design and development services

The specialist services may be original designs of certain processes or items of machinery or equipment or construction artifacts. Services may extend to research and development of innovative work and assistance in registration of patents for inventions.

Certain specialist work may be delegated to specialists who may later, with the employer's approval, become nominated subcontractors for the detailed design and construction stages of that part. Equitable arrangements for payment for specialist design work must be made between the A/E and specialist consultant or contractor. The A/E should retain ultimate responsibility for all specialist design work.

1.5.6.3 Outline designs

In some cases outline designs, giving more information than was contained in the A/E's initial investigation report but less detailed than tender designs, may be required for approval by the employer. The need for such designs should be stated in the agreement between the A/E and the employer.

1.5.6.4 Tender designs

Ideally, detailed designs of the works should be completed before tenders for construction are invited. In some cases this is not practicable because of the urgency of the works, and the drawings on which tenders are invited are supplemented by a series of further drawing issued by the A/E during construction.

The more complete the drawings, specifications and bills of quantities are, at the time of tender, the better the tenderers will understand what is required. The more accurate will be their prices, the smoother the subsequent execution of the works will be and the lower the cost. Certain information can be known only when construction is under way, and in such cases redesign or supplementary design work during the construction phase may be unavoidable.

1.5.6.5 Time

The A/E should advise the employer of the time needed to prepare designs and tender documents. Shortage of time allowed to this stage is highly likely to lead to delays and additional expenditure. An early decision should be taken as to the type of contract to be used for construction, as this will affect the form of the documents the A/E has to prepare.

1.5.6.6 Contract documentation

In remeasurement, lump sum or cost reimbursement contracts, the tender documents are prepared by the A/E on behalf of the employer. They are used first to obtain tenders from contractors and subsequently in the administration of the contract.

1.5.7 Project management[16]

The "project management" is defined by the PMI as "application of knowledge, skills, tools and techniques to project activities to meet project requirements."

In addition to managing engineering, procurement, construction and commissioning, the project management team will also be required to cover other significant elements of a project, such as securing project financing, maintaining public awareness and obtaining government approvals.

1.5.8 Programme manager

This service differs from project management in that it involves elements other than those relating to construction, such as financing and operation.

Words, Phrases and Expressions 1.5

[1] parties to a contract 合同当事人,合同双方
[2] building and civil engineering procedure 建筑与土木工程程序
[3] direct labour 直接人工,自建工程 (As an alternative to execution by contract the employer may resort to the method known as direct labor. Under this system the employer uses labor already available to him or recruited specifically for the project. He undertakes all the responsibilities of management and risks of construction, and pays all wages, costs and expenses as they are incurred. There is no contract as such. 业主可不将工程发包他人,而是以名为自营工程的办法完成。按此办法,业主使用已有或专门为本工程召募的人力。业主承担管理的全部责任和全部施工风险,并支付所有发生的工资、成本和开销。没有同任何人签订的合同。)
[4] contractual relationship 合同关系,契约关系 (These refer to the various contractual relationships established among the contractual and non-contractual parties involved in a contract. Usually, the employer and the contractor are the contractual parties, while the engineer, consultant engineers, and design firms involved in the design and supervision of the works are not the parties to the contract between the employer and the contractor and they are non-contractual parties. Each of them has their own contract of employment with the employer. 合同关系指参与合同的契约方和非契约方之间的各种契约关系。一般业主和承包商是契约方,而参与工程设计和监督的工程师、咨询公司和设计单位不是业主和承包商之间所签合同的当事人,因此是非契约方。他们每一方都同业主签有自己的雇佣合同。)
[5] purpose-built civil engineering projects 专门建造的土木工程项目
[6] internal bidding procedure 内部招标程序
[7] to keep a check on sth. 控制
[8] provide vt. 形成,产生 (produce a useful result) A *prudently run direct labour*

organization will operate some kind of internal bidding procedure to keep a check on prices and provide a basis for cost control. 谨慎而周密计划自建工程的组织，愿意设立某种内部招标程序，以便控制价钱，为控制费用打下基础。

[9]　initiation/promotion 发起

[10]　legal authority 法定权限，合法权限，正式权限

[11]　statutory regulations 法律规定

[12]　appreciate *vt.* 重视（put a high value on）*Lack of appreciation of the time required for proper investigation.* 不重视正常调查需要的时间。

[13]　define the functions that the project is to perform 明确该项目应具备的功能

[14]　brief *n.*（业主要求的）简要说明，任务书

[15]　brief *vt.* 就计划、程序等指示某人；向……提供重要情况（instruct about（plans, procedures, etc.）; give essential information to *sb.*）

[16]　constraints *n.* 制约因素，制约条件

[17]　solutions *n.* 解决办法

[18]　terms of reference 职权范围，工作职责说明书（The terms of reference of the Committee include: (a) to coordinate activities of all departments…该委员会的职权范围包括：(a)协调所有部门的活动……）

[19]　work statement 或者 statement of work（SOW）工作说明书（A narrative description of *products, services, or results* to be supplied. 对应提供的产品、服务或成果的文字说明。）

[20]　interim or progress reports 中间或进展报告

[21]　viable *adj.* 财务上持久的，财务上可行的（financially sustainable）

[22]　capital and operating costs 初始投资与使用费

[23]　a master plan 总计划

[24]　long-term development program 长期发展项目组合

[25]　a geotechnical expert 岩土技术专家，岩土工程专家

[26]　services *n.* 公用设施（facilities supplying some public demand such as telephone, etc.）*Availability of services such as drainage, sewerage, water, electricity and gas may require investigation.* 是否有排水、污水管道、自来水、电力和燃气等公共设施可供使用可能要进行调查。

[27]　outline program 初步计划

[28]　the purpose for which the project was commissioned 当初批准该项目的目的

[29]　proceed with 动手，继续（已停顿之事）（to go forward especially after stopping）*If, having considered the A/E's report, the employer decides to proceed with the project he must take steps to obtain the legal authority to construct the works.* 在考虑建筑师/工程师的报告后，雇主如果决定将这一项目继续下去，就应采取步骤，取得政府对工程施工的正式批准。

[30]　tender documents 招标文件

［31］ outline design 方案设计
［32］ detailed designs 详细设计
［33］ remeasurement contract 计量合同
［34］ lump sum contract 总价合同
［35］ cost reimbursement contracts 成本加成合同

Notes 1.5

［1］ Certain information can be known only when construction is under way, and in such cases redesign or supplementary design work during the construction phase may be unavoidable.

【译文】 某些资料只有在施工进行过程中才能获得，在这种情况下，在施工阶段进行过程中进行重新设计或补充设计的工作就是难以避免的。

［2］ it is false economy to restrict the scope or thoroughness of the A/E's investigations and design studies.

【译文】 为省钱而限制建筑师/工程师调查与设计研究的范围或深度实际上并不能节省成本。

［3］ Brief A/E

【译文】 向 A/E 提供情况，提出要求。

1.6 Context of construction projects

Construction projects are initiated and carried out in a context that is much broader than the construction projects themselves. Context refers to activities, events and conditions which occur or exist external and internal to the construction projects. The conditions include the institutions where economic, social and other activities take place and are kept in order, such as modes of production, business cycle and legal systems. The contextual conditions external to a construction project include physical, economic, social, political and cultural conditions; legal systems; and the processes and changes therein; and the societal values and attitude. Contextual activities, events and conditions internal to a construction project include division of responsibilities and organization structure of the employer, the A/E, and contractor, information, financial, material, and human resources available for the project and their technical and managerial competence. It is important for the members of management team for the construction project to consider the construction project team in its context, and recognize and understand how the context influences a construction project and its management.

1.6.1 Economic environment

1.6.1.1 Business cycle

History shows that market economies seem to grow in spurts. Periods of growth and expansion are followed by periods of contraction and decline; periods of near full employment are followed by periods of high unemployment, and periods of stable prices are followed by periods of inflation.

The periodic ups and downs in the economy are collectively referred to as "the business cycle" which tends to cause damage to the economy, society and certainly the construction projects.

We need to be aware of impacts which the business cycles may have on construction projects and the contract administration.

When output is declines, fewer inputs, such as labour, are needed, the unemployment rate rises, and a smaller percentage of the capital stock is used, more of plants and equipment are running at less than full capacity.

A depression is a severe and prolonged recession. The U. S. economy experienced a depression between 1929 and the late 1930s. The most severe recession since the 1930s took place during the period from 1980 to 1982.

The social consequences of the depression of the 1930s are perhaps the hardest to comprehend. In addition to lost output and serious, immediate social consequences, recessions may lead to less output in the future. Recession may decrease the demand for construction projects.

Common sense tell us, however, that declining real output tends to lower the rate of investment and capital production. Investment is a key to future economic growth and progress.

Recessions may generate some benefits. They are likely to slow down the rate of inflation therefore, recessionary policies can be used to counteract inflation or prevent future inflations.

Ups in the business cycle, as opposed to downs, often but not always, tends to encourage inflation.

A growth in inflation during the life cycle of a construction project may lead to an unexpected rise in the cost of raw materials and labour costs thus increasing the estimated debt of the project.

On the other hand, if the inflation surprises the participants in a construction project, it can hurt them. It can be said that inflation higher than expected benefits debtors and inflation lower than expected benefits creditors.

Even when anticipated, inflation will cause some costs or losses. If the participants in a construction project are not fully informed, or do not understand what is happening to prices in general, they may make mistakes in their tendering or acceptance of a tender.

When inflation is unanticipated, the degree of risk associated with contracts in the economy increases. Increases in uncertainty may make the employers reluctant to invest in capital projects. To the extent that the level of investment falls, the prospects for long-term economic growth are dimmed.

No matter what the real economic cost of inflation, it makes people uneasy and unhappy. That is why a provision is made for adjustment for changes in cost in most standard forms of contract, which will be described in chapter 7.

1.6.1.2 Commercial environments

Nowadays the construction projects have been increasingly linked with the domestic and international financial markets which have become more speculative and sensible to any event likely to occur in any corner of the globe.

The problems with them may lead to disasters. A recent typical example is the crisis which led to a collapse of the financial and property markets in Asian countries in 1997. It is this crisis that brought the prosperous emerging economies in the region to a sudden halt. The factors responsible for its happening have not been fully identified and understood. The parties to a construction project should be alerted of the likely negative impacts whenever the project has any connection with the markets.

A fluctuation in currency exchange rates can also have detrimental effects on the promoters' efforts to pay off its debt.

A devaluation of the currency of the host country would cause a decrease in value of the revenue which a capital project may generate.

Economic globalization has made the economies in the world more international trade oriented, more closely connected with and dependent on each other than ever before. Recession or inflation in one economy can quickly trigger that in another. That is why we as participants in a construction project, especially that has strong international background, should keep an eye on the performance of other economies.

1.6.2 Social and cultural environment

The parties to a construction project need to understand the way in which the project and its management interact with people. This may require an understanding of aspects of the social, demographic, educational, ethical, ethnic, religious, and other characteristics of the people whom the construction project affects or who may have an interest therein.

1.6.3 Political and international environment

Some members of a project team may need to be familiar with applicable international, national, regional, and local laws and customs, as well as the political climate that could affect the construction project. Other international factors to consider are time-zone differences, national and regional holidays, travel requirements for face-to-face meetings, and the logistics of teleconferencing.

1.6.3.1 Political environment

A political change can lead to all sorts of problems. For example, when put under political pressure from the public, a government promoting a public facility may be forced to hold a public inquiry, which is sure to cause major delays in the construction of the facility.

Politically sensitive issues to be addressed may include the length of the concession period, the toll or tariff, the promoter being forced to cap the toll level of the facility. It may also see the principal being forced to increase the taxes and royalties of the profits from the facility.

Change of government may result in an overall change in fiscal policies or even force the government to sell it's assets.

Total change to the political circumstances in a country may lead to nationalization of all industries. This would be totally disastrous if the promoter had made a significant start on the construction of the facility.

1.6.3.2 International environment

☐ Trade barriers

The barriers to any traded goods and/or services play a significant part in making successful a contract, especially the one international.

In recent years, some industrialized countries have raised their duties on the goods and/or services from the developing countries, which made and will make international construction contracts difficult to enter into and then administer.

☐ Licensing requirements

The licensing requirements must be met in a manner by which the contract can be operational smoothly. Licensing is sometimes required to meet international obligations or on the grounds of safety, health or the environment. These are sometimes in support of anti-smuggling efforts, or to monitor movement of highest-technology products on behalf of other governments.

1.6.4 Legal context

1.6.4.1 Legal system

A basic consideration for international construction project may be that the legal framework in a country may not be sufficiently developed to address all the issues relevant to the implementation of the construction project.

Changes to the law may result in delays. If, for example, the project is to build a toll road crossing a number of regions, the promoter may have to negotiate with two separate bodies to agree on the same issue.

The legal aspects to be considered include company law, standards and specifications, commercial law, liabilities and ownership, royal decrees, statutory enactments, resolution of disputes and the type of concession agreements.

It has been internationally accepted that rule of law and mature and trusted legal systems are critical for the contracts formed for businesses, including procurement of works and services. A legal system well-developed for a successful economic and social development shall provide the society with:

☐ Contract rights

Under such a legal system, documents, certificates, contracts, and rights and obligations shall be securely protected.

☐ Property rights

One of the most critical economic freedoms is the right to protection of private property. That private property is efficiently and effectively protected is the most important pre-requisite for a contract to be successful.

☐ Arbitration

Disputes have been resolved through mediation and arbitration in a number of international urban centers for many years, with the process governed by the arbitration rules set by some national, regional or international organizations, such as the United Nations, International Chamber of Commerce, etc..

On 10 June 1958 the New York Convention on the Recognition and Enforcement of Foreign Arbitral Awards was done at New York. China is one of the signatories. The administrator of a contract, especially an international contract shall be aware that the Convention enables the courts in a country or jurisdiction to enforce domestic awards and those made in non-convention through the courts of other signatory countries.

1.6.4.2 Common law system[13]

The common law comes both from the previous customs and practices and from past decisions of courts in the United Kingdom and, therefore, is also known as case law or judge-made law. The common law system can be categorized by subject into substantive law and procedural law, on one hand.

Substantive law refers to all areas which define and deal with the rights and duties of individuals, groups, organisations and the state. Contract law comes from this category.

Procedural law deals with the legal rules through which the process of law is set in motion to enforce some substantive right or remedy. These rules are often complex and numerous.

On the other hand, sources of common law other than the constitution in written form may be listed as follows:

(a) judicial decisions when they are recognized as precedents in the common law,

(b) equity,

(c) legislation or statute law,

(d) regulations and delegated or subordinate legislation,
(e) international treaties, and
(f) custom.

Equity forms part of the substantive law developed by the decisions of the Court of Chancery in the United Kingdom. There are certain general principal on which the court exercised its jurisdiction, many of which are embodied in what called the twelve maxims of equity. A number of these maxims have an effect on construction contracts. An example of an application of one of these maxims is the area of penalties. The maxim concerned provides that "equity looks to the intent rather than to the form".

Tort is a legal term used in both the common law and the Romano-Germanic systems of law to describe various wrongs which may give rise to civil proceedings mainly in the form of an action for damages.

A tort must be distinguished from a breach of contract in that the obligations and rights of the parties to a contract emanate from the terms of that contract whereas in tort they arise from the operation of the law without any need for consent from the parties. Thus, unlike the situation under contract, a person has no choice in the obligations placed upon him by the law in his behaviour towards others.

The same facts of a particular situation may give rise to an action in contract and in tort under the common law system. Concurrent liability in contract and in tort is not, however, universally recognised.

The law of torts regulates a wide variety of unlawful behaviour, of which nuisance, slander, libel, trespass and negligence are related to construction.

1.6.5 Physical environment

If a construction project affects its physical surroundings, its participants should be knowledgeable about the local ecology, physical geography, geology and hydrology that could affect the construction project or be affected thereby.

All parties should be aware of the sensitivity that environmental issues can arouse. Existing environmental laws can also result in the project being redefined in order to meet set standards and specifications. The parties involved in the project should be aware of any future changes in the environmental standards and specifications that may have an adverse effect on the project during its operation.

Furthermore, external environmental factors such as a change in weather conditions should be addressed.

Words, Phrases and Expressions 1.6

[1] context of construction projects 建设项目及其与环境的联系
[2] modes of production 生产方式

[3] business cycle 经济周期
[4] common law 普通法
[5] sources of common law 普通法渊源
[6] constitution in written form 书面宪法,成文宪法
[7] judicial decisions 司法判决
[8] precedent *n*. 先例,前例(earlier happening, decision, etc., taken as an example or rule for what comes later)
[9] equity *n*. 衡平法,平衡法(the principles of justice forming part of English law and used to correct laws when these would apply unfairly in special circumstances.)
[10] legislation or statute law 立法或成文法
[11] regulations 法规
[12] delegated or subordinate legislation 从属立法
[13] international treaties 国际条约
[14] substantive law 实体法
[15] procedural law 程序法
[16] jurisdiction 裁决权
[17] twelve maxims of equity 衡平法十二准则
[18] law of torts 侵权法
[19] Romano-Germanic systems of law 罗马日耳曼法系
[20] damages 损失赔偿费
[21] breach of contract 违约解除
[22] terms 条款
[23] concurrent liability 共存责任

Chapter 2　Procurement of works

2.1　Procurement of works

2.1.1　Procurement of works or services

As described in 1.2.1 (*"Construction project and its product"*), a construction project is often a part of a private sector initiated capital or a public purpose project initiators of which usually meet their needs for buildings or engineering works by acquiring the *construction and related engineering services and/or architectural and engineering services* from outside their organizations. The acquisition involves a number of processes which are collectively referred to as "*procurement of works*" or "*procurement of consultant services*".

2.1.2　Procurement VS contracts

The procurement involves contracts that are legal documents between a buyer and a seller. A contract is a mutually binding agreement which obligates the seller to provide the specified products, services, or results, and obligates the buyer to provide monetary or other valuable consideration[16]. In the case of a construction contract (the contract entered into or to be entered into for construction activities), this is an agreement under which the builder or constructor, commonly referred to as the contractor, undertakes for reward to carry out building or civil engineering works for the employer.

A construction contract is a service contract and not a contract for goods.[17] While the contractor will use goods to build the structure, those goods will not be moveable when the performance is rendered.

A contract may involve both goods and services in which case one must decide which is the dominant part of the transaction. A contract to change oil and lubricate a car is probably a service contract because the goods (oil and grease) are incidental. A contract for a piano in which the seller is to deliver it and tune it after it is moved is a transaction in goods because the services involved are incidental to the sale of the goods. A contract in which an artist is to paint a portrait is probably a service contract. The artist is to deliver the completed painting which could be viewed as specially produced goods. However, the dominant nature of the contract is probably the service of the artist painting the picture.

It is worthwhile to point out that sometimes in this book the term "contract" means the construction project or the objective of the contract, i.e. the works to be carried out by the contractor. Therefore, the readers are strongly suggested to make it clear what the term really means in the context where it is used.

2.1.3 Procurement agent and contract administrator

Some employers set a management team for their procurement of the works and the related architectural and/or engineering services. However, it is difficult for the employer who is not familiar with the contracting to form such a team of his own personnel because of its increasing complexity. It is common, therefore, that the employer engages a firm as his procurement agent or then the contract administrator.

In some circumstances, it may be advantageous for the employer to engage or continue with a specific firm. These circumstances include choosing a firm who would continue from the feasibility study stage to project preparation and from project preparation to the execution stage. If a firm has carried out pre-investment studies for a project and is technically qualified to undertake the preparation services, the advantages of continuity will be a consistency in basic technical approach and a commitment to the project cost estimate on which the investment decision was based. If a different firm were retained for detailed engineering, it would normally wish to make an in-depth review or even to repeat the preliminary design work and cost estimate done by another firm.

It is normally advisable that the same firm which undertook the previous preparation work of the project should implement and supervise the project. This ensures that contract documents are interpreted properly during the implementation stage and that modifications in design, if found necessary in the course of implementation, suit the basic concepts. In certain types of projects, such as large dams, power stations and industrial projects, the design consultants prepare working drawings only after contract award and as implementation proceeds. Hence, for these types of projects, the design consultants should normally be appointed to ensure that the contractor works satisfactorily.[8]

There may be circumstances where continuing with the same consulting firm may not be in the interests of the project. These include cases in which preliminary design and engineering is found to be unsuitable or where relations between an employer and the consulting firm have deteriorated to such an extent as to have a detrimental effect on the project. It is normal, therefore, to enter into separate contracts for the pre-investment and preparation assignments. When a contract covers preparation and implementation (e. g., detailed engineering and supervision), such a contract should allow for a detailed review of implementation requirements and revision or termination of the contract, if necessary, near the end of the preparation phase. If a change of firms between phases is unavoidable, and such a change may have a bearing on the legal liabilities of the original and the new firm, the firm taking over should be given an opportunity to check and comment on the previous consulting work and should be required to accept appropriate responsibility for the complete project design.[8]

Further circumstances where it might be advantageous to approach a single firm rather than conduct a selection process would be where the firm:

☐ has a close association with a project similar to the employer's project;

☐ has expertise not widely available; or

☐ has undertaken similar assignments for, and has a good working relationship with, the employer.

Words, Phrases and Expressions 2.1

[1] management *n*. 管理部门（层、人员、职能），管理学，管理（过程、方法、活动、方式）(1. The term management is used in a variety of ways. It can refer to the members of an organization or a project team who make key decisions. It can also be used to refer to a discipline of knowledge that has accumulated over the years through applications of scientific research and observation of managers in practice. Hence, management can also refer to the collective wisdom that has evolved from scientific study that can then be applied to specific managerial situations. In most cases, however, management is defined as a process of planning, organizing and staffing, directing, and controlling activities in an organization or of a project in a systematic way in order to achieve a common goal. 管理这个术语有多种用法。可以指组织或项目班子成员中作出重要决策者。有时还可以指一门通过管理人员经多年积累在实践中应用科学研究和观察结果而形成的学科，因此，该术语又可以指从科学研究中演变出来，然后可应用于具体的管理局面中的集体智慧。但是，在大多数情况下，管理定义为对组织内或项目的活动进行系统的规划、组织、配备人员、指导和控制，以实现一共同目标的过程。2. The management functions include: (1) relating to an organization's internal and external environments, (2) planning, (3) organizing, (4) staffing and human resource management, (5) leadership and interpersonal influence (directing, or actuating), and (6) control. Collectively the functions comprise what is termed the management process. In summary, we can say that management is the process of achieving desired results through efficient utilization of human and material resources. 管理职能包括：(1) 处理组织同内外环境的关系，(2) 规划，(3) 组织，(4) 配备人力资源并进行管理，(5) 领导并利用人际关系施加影响（指导或激励调动）以及 (6) 控制。这些职能综合起来就构成了所谓的管理过程。概括起来，可以把管理理解成有效地利用人力和物质资源，争取理想结果的过程。)

[2] binding 有约束力的

[3] agreement 协议，协议书；取得一致，同意

[4] provide *v. t.* 给予（afford）*provide monetary or other valuable consideration*. 给予货币或其他有价值的报酬

[5]　consideration *n*. 报酬，对价，约因（Value, generally in the form of goods, money, or promises and services and so on involved in a contract. 合同中涉及的价值，一般表现为物品、金钱或许诺与服务等/In UK contracts 'Valuable consideration' means payment. 在英国合同中，'有价值的报酬'指支付款项。）

[6]　service contract 服务合同

[7]　contract for goods 购货合同

[8]　procurement agent 采购代理人，招标代理人

[9]　contracting 发包（*The procurement agent or contract administrator may seek support early from specialists in the disciplines of contracting, purchasing, and law.* 采购代理人或合同管理者会尽早谋求发包、采购和法律专家的帮助。*Disciplines* 是"专业"的意思，不一定译出来。）

[10]　project preparation 项目准备

[11]　pre-investment studies 投资前研究

[12]　commitment to the project cost estimate 专心致志估算项目费用

[13]　detailed engineering 详细设计

[14]　preliminary design 初步设计

[15]　working drawings 施工图纸

[16]　contract award 定标，授标

[17]　preliminary design and engineering 初步建筑与工程设计

[18]　enter into separate contracts for the pre-investment and preparation assignment 为投资前和准备工作单独签合同

[19]　have a bearing on 同……有关（联）系，对……有影响

Notes 2.1

[1]　词语解释

1) allow *vt*. 允给，让……得到(give; let(*sb*. or *sth*.) have; agree to give)

例1　Allow *sb*. a monthly subsidy. 按月发给某人津贴。

2) allow for 为……在时间、资金或人力等方面做准备或留有余地(assign as a share or suitable amount(as of time or money))，考虑到，顾及，体谅

例1　Allow an hour for lunch. 规定一个小时的午饭时间。

例2　When a contract covers preparation and implementation (e.g., detailed engineering and supervision), such a contract should allow for a detailed review of implementation requirements and revision or termination of the contract, if necessary, near the end of the preparation phase. 当合同涉及准备和实施（例如详细设计和准备）工作时，这种合同就应考虑到详细审查各项实施要求，以及修改该合同或在必要时在准备阶段接近结束时终止该合同的工作量。

例3　It takes about an hour to get there, allowing for possible traffic delays. 把路上可能的耽搁算进去，大约一个小时可到那里。

例 4 We must allow for his inexperience. 我们必须考虑到他的缺乏经验。

3) allow 酌加，酌减(reckon as a deduction or an addition)

例 1 Allow one percent for leakage 少算百分之一作为漏损。

4) cost n. 成本、费用

可以将 cost 译为"费用"或"成本"。但"费用"或"成本"两者含义不同。

费用——《现代汉语词典》、《新华词典》：花费的钱；开支。

成本——《现代汉语词典》：生产一种产品所需的全部费用。《新华词典》：也叫生产成本、成本价格。产品价值中属于投入的生产要素价值的那一部分。

不难看出，费用含义要比成本广。经济学中的定义是：成本就是生产过程所消耗的生产要素的价值，也称要素成本；而费用是生产和流通过程从开始，直到将产品交付使用所消耗的生产要素的价值，不限于生产作业。

项目与项目管理遍及国民经济、社会、政治、文化各个方面，不能用"成本"一词将项目与项目管理限制于生产活动。所以 project cost management 最好译成"项目费用管理"，而非"项目成本管理"。

[2] 疑难词语解释

1) a commitment to the project cost estimate

【译文】为编制项目费用估算而进行的思考，付出的努力，开展的工作等。

2) It is normal, therefore, to enter into separate contracts for the pre-investment and preparation assignments. When a contract covers preparation and implementation (e.g., detailed engineering and supervision), such a contract should allow for a detailed review of implementation requirements and revision or termination of the contract, if necessary, near the end of the preparation phase.

【译文】所以，一般单独签订投资前和准备工作的合同，而一份合同经常规定准备和实施（例如，详细设计和施工监督）两项工作，这样的合同应当规定（allow for）详细地审查（业主对）实施（提出的）要求，而且在必要时还应有（allow for）关于在准备阶段接近结束时修改或终止合同的规定。

3) If a change of firms between phases is unavoidable, and such a change may have a bearing on the legal liabilities of the original and the new firm, the firm taking over should be given an opportunity to check and comment on the previous consulting work and should be required to accept appropriate responsibility for the complete project design. a bearing on the legal liabilities of the original and the new firm.

【译文】如果（项目）各阶段之间难免更换公司，并且更换影响到原来与新公司的责任，就应让接替者有机会核对与评价已经进行过的咨询工作，但应要求新公司承担项目设计的有关责任。

2.2 Procurement management

Procurement management includes following processes:[16]

(1) Procurement planning-to determine what, when and how to procure.

(2) Contracting planning-to document works or services requirements and to identify potential sellers. This is a pre-tender process.

(3) Request seller responses-to obtain tenders, proposals, quotations or information, as appropriate, which is, in construction industry, referred to as tendering and described in Chapter 3.

(4) Selection of sellers-to evaluate tenders or proposals, to review quotations, offers, or information, to choose among potential sellers, and to negotiate a written contract with each seller. This process is also described in Chapter 3. The pre-contract arrangements needed to form a contract are described in Chapter 4.

(5) Contract administration-to manage the contract and relationship between the buyer and seller, to review and document how a seller is performing or whether or not he has performed to establish required corrective actions, to provide a basis for future relationships with the seller, manage contract-related changes and, when appropriate, to manage the contractual relationship with the outside buyer of the project. Chapter 4 describes in detail the general concepts of contract administration, while the chapters following Chapter 4 describe the particular aspects of the A/E's contract administration from commencement throughout the expiration of the defects notification period.

(6) Contract closure-to complete and settle each contract, including the resolution of any open items, and to close each contract applicable to the project or a project phase.

These processes interact with each other and with the processes needed for the project team to manage the project in terms of the scope, time, cost, quality, human resources, communications, and risk as well.

Words, Phrases and Expressions 2.2

[1] processes of procurement management 采购管理的过程
[2] procurement planning 采购规划
[3] contracting planning 发包规划
[4] document *vt*. 将……写入书面文件；将要求、建议或设想用书面文件提出，并说明根据或理由(furnish documentary evidence of; furnish with documents; provide with factual or substantial support for statements made or a hypothesis proposed, especially to equip with exact references to authoritative supporting information) *document works, services or results requirements* 将对(待采购)的工程、服务或结果的要求写成书面文件。
[5] identify *vt*. 识别(establish the identity of) *Identify potential sellers and prepare procurement documents*. 识别潜在的卖主并编写采购文件。
[6] tenders, proposals, quotations or information 标书(用于施工服务采购)、建议书(用于咨询或设计服务采购)、报价(用于物品或货物采购)或信息(用于咨询或信息采购)

[7]　seller selection 卖方选择
[8]　evaluate tenders or proposals 评标
[9]　review quotations, offers or information 审查与评价报价、报盘或信息
[10]　pre-contract arrangements 签约前安排,签约前工作
[11]　form a contract 订立合同
[12]　contract closure 合同结尾
[13]　defects notification period 缺陷通知期

Notes 2.2

[1]　形式动词

这是笔者的一个杜撰。英文常用名词表示动词的"过程"或"结果"。如用 payment、arrangement、building 和 performance 分别表示 pay、arrange、build 和 perform 的"过程"或"结果"。

使用这些名词的主动句有：The contractor has made payments to the workers in his employment; The host has made arrangements for the conference; The contractor is carrying out the building of the bridge under the Engineer's instructions, 等等。

使用这些名词的被动式句子有：Payment has been made to the workers; Arrangements have been made for the conference; Building is carried out under the Architect's instructions 和 Performance is rendered, 等等。

读者一定注意到，当 pay、arrange、build 和 perform 变成名词 payment、arrangement、building 和 performance 时，就不再是语法上的动词，为了弥补它们失去的动词功能，必须另外找一个动词。上述句子中的 make、carry out 和 render 就是起这种弥补作用的动词，它们本身没有"支付"、"安排"、"建造"或"完成"的意思，这些意思只能由 payment、arrangement、building 和 performance 表达。这就是笔者将 make、carry out 和 render 称为"形式动词"的理由。

[2]　疑难句解释

(1) While the contractor will use goods to build the structure, those goods will not be moveable when the performance is rendered.

【译文】"虽然承包商用物品建造构筑物，但是在使用这些物品将构筑物建造起来之后，这些物品就不能动了。"在这句话中，performance 是概括 use 和 build 这两个动词的，而 render 是形式动词。

(2) It is the project management team's responsibility to help tailor the contract to the specific needs of the project. In construction industry, contracts can be called an agreement or subcontract.

【译文】　设法使合同适合于项目的具体需要乃是项目经理部之责。

(3) It is worthwhile to point out that sometimes in this book the term contract means the objective of the contract, i.e. the works to be carried out by the contractor. Therefore, the readers are strongly suggested to make it clear what the term really means in the context

where it is used.

【译文】 值得注意的是,本书中 contract 这一术语有时指其标的,即应由承包商完成的工程。因此,希望读者在遇到这个术语时,务必要弄清它的具体含义。

2.3 Procurement planning[16]

Procurement planning is to determine how to best meet the employer's needs for the works and/or related engineering services, by procuring from outside his organization or by using his own workforce. If only can the procurement meet his needs then the works and/or services to be procured shall be identified. This process includes consideration of potential sellers, particularly if the buyer wishes to exercise some degree of influence or control over contracting decisions. The procurement planning also involves reviewing the risks inherent in the procurement decisions and the types of contract planned to be used with respect to mitigating risks and transferring risks to the seller.

2.3.1 Direct labour[7][18]

In some cases, an employer can use his own personnel and equipment to erect the building, civil engineering works or other facilities to meet his needs, which is called "force account" in the United States or "direct labour" in the United Kingdom, the commonwealth countries and Hong Kong. The use of direct labour may be justified where:

☐ quantities of work involved cannot be defined in advance;

☐ works are small and scattered or in remote locations for which qualified construction firms are unlikely to tender at reasonable prices;

☐ work is required to be carried out without disrupting ongoing operations;

☐ risks of unavoidable work interruptions are better borne by the employer than by a contractor; and

☐ there are emergencies needing prompt attention.

Under this system the employer uses labour already available within his own organization or recruited specifically for the project. He undertakes all the responsibilities of management and risks of construction, and pays all wages and expenses as they are incurred. There may be no contract in the normal sense, unless the employer requires the work to be undertaken in competition with contractors.

In most cases, however, it is impossible for an employer to have the needed works erected by using force account. The only way for him to obtain the needed works is to procure it from outside of his organization.

2.3.2 Procurement strategy[16]

If the works and/or related architectural and engineering services shall be procured externally the options available should be reviewed at the inception stage of a capital or public purpose project, and a decision made as to the most appropriate option. The appropriate procurement strategy shall be established at this stage, with decisions being reached on the following matters:

(1) the works to be executed under each contract (often called "contract packaging"); and,

(2) for each contract

☐ the extent of design to be provided to, or to be carried out by, the contractor, and

☐ lump-sum, measure-and-value, cost-plus or other basis for determining the final contract price.

For a large capital or public project, decisions on the number and scope of contracts may be critical to the eventual success of the project. Having a large number of contracts may give the employer more control than under a single contract for the entire project, and may be more economic by maximizing competitive pricing. However, these advantages may be offset by a greater extent of coordination risk borne by the employer.

For each contract, the party responsible for the design will develop it during the detailed design stage. If the design is carried out by (or on behalf of) the employer, he will have a greater control over the details. However, problems may on occasions arise from the division of responsibility between designer and builder/constructor.

If the contractor is responsible for the design, he will wish to develop it in his own interests, subject to any constraints in the contract. The employer will have less control over the design than he would have if he was responsible for providing it. Under contractor-design, the contract price would typically have been tendered on a lump-sum basis, so any change in cost (increase or decrease) to the contractor, resulting from design development, would not be passed on to the employer.

2.3.3 Procurement decisions

As previously stated, procurement should commence with strategic decisions on contract packaging and, for each contract, on the allocation of design responsibility and on the basis for determining the contract price. It is only after these above matters have been considered and decisions reached, that a decision should be made on the appropriate form of contract for the particular works.

The procurement decisions shall be documented as a procurement management plan. The plan describes how the procurement processes will be managed from developing procurement documentation through contract closure and may cover:

☐ types of contracts to be used;

☐ who will prepare independent estimates and if they are needed as evaluation criteria;

☐ the actions the project management team can take on its own, if the employer's organization has a procurement, contracting, or purchasing department;

- [ ] standardized procurement documents, if they are needed;
- [ ] managing multiple providers;
- [ ] coordinating procurement with other project aspects, such as scheduling and performance reporting;
- [ ] constraints and assumptions that could affect planned purchases and acquisitions;
- [ ] handling the lead times required to purchase or acquire items from sellers and coordinating them with the project schedule development;
- [ ] setting the scheduled dates in each contract for the contract deliverables and coordinating with the schedule development and control processes;
- [ ] identifying performance bonds or insurance contracts to mitigate some forms of project risk;
- [ ] establishing the direction to be provided to the sellers on developing and maintaining a contract work breakdown structure;
- [ ] establishing the form and format to be used for the contract statement of work;
- [ ] Identifying pre-qualified selected sellers, if any, to be used; and
- [ ] procurement metrics to be used to manage contracts and evaluate sellers.

Words, Phrases and Expressions 2.3

[1] force account 自营工程(美国常用)
[2] direct labour 直接劳务,自营工程(英国和英联邦国家常用)
[3] contracting decisions 发包决策
[4] documentation n. 书面凭据(documentary evidence)
[5] evaluation criteria 评价标准

Notes 2.3

疑难句解释

[1] There may be no contract in the common sense, unless the employer requires the work to be undertaken in competition with contractors.
【译文】 除非业主要求(本单位施工力量)要同(外面的)承包商竞争才能承担这一工程,否则就可能没有通常的合同。

[2] This process also includes reviewing the risks involved in each make-or-buy decision.
【译文】 这一过程还要审查各个自制购买决策中难以避免的风险。

[3] The output of procurement planning shall be a procurement management plan.
【译文】 采购规划的成果应当是一份采购管理计划。

[4] This process involves consideration of whether, how, what, how much, and when to acquire.
【译文】 这一过程必须考虑是否要采购,如何采购,采购何物,采购多少以及何时采购。

2.4 Procurement methods

2.4.1 Variety of procurement methods

In addition to the long lasting design-tender-construct method, many new procurement methods or systems evolved during the 1980s and 1990s, giving an employer greater choice and flexibility. The main requisites of each method or system are briefly described as below.

(1) Design-build

The employer purchases the completed building or engineering works from a contractor who undertakes both design and construction.

(2) Management contracting

The employer appoints design and cost consultants and a contractor or consultant to manage the construction for a fee. The appointed firm or the joint venture of firms is referred to as management contractor. Specialist contractors are appointed to undertake the construction work by negotiation or in competition. The management contractor takes some contractual risks in delivering the completed building(s) or civil engineering works at an agreed price and on time and employs specialist contractors as subcontractors. The employer retains some time and price risks. The contractual and organi zational relationships under this system are shown in Figure 2.1。

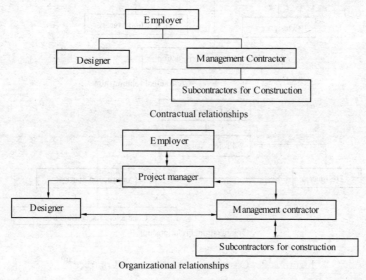

Figure 2.1 Management Contract[19]

(3) Construction management

A professional firm is paid a fee to provide the management service and the specialist contractors enter into direct contracts with the employer, who retains the time and price risks. The appointed professional firm is referred to as construction manager. The contractual and organizational relationships under this system are shown in Figure 2.2。

(4) Design and management

The employer appoints a single firm to design and deliver the completed building(s) or engineering works, although specialist contractors are appointed to carry out the construction work by negotiation or in competition. There are two variants:

□ Contractor: the design and manage firm takes a contractual risk in delivering the completed building(s) or civil engineering works to an agreed price and on time and employs specialist contractors and sometimes designers as subcontractors.

□ Consultant: a designer/manager is employed as the employer's agent and specialist contractors enter into direct contracts with the employer, who retains time and price risks.

(5) EPC (Engineer, Procure, Construct)

If an employer does not wish to be involved in the day-to-day progress of the work, provided the end result meets the performance criteria he has specified and the parties concerned (e.g. sponsors, lenders and the employer) are willing to see the contractor paid more for the construction of the facility in return for the contractor bearing the extra risks associated with enhanced certainty of final price and time, then the employer wishes the contractor to take total responsibility for the design and construction of the process or power facility and hand it over ready to operate "at the turn of a key".

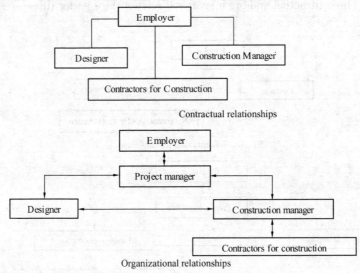

Figure 2.2 Construction Management Contract[19]

(6) Concession arrangement

It is more often than not a concession arrangement would be negotiated to own or operate or own and operate the facility to be built for a period of time for the benefit of all the parties involved before handing over the facility to the employer (owner). "BOT", that is, Build-Operate-Transfer agreement, is such a concession agreement.

In some cases, the financier may dictate the terms of contract for the project to be financed, such as in a case where the World Bank is involved, certain amendments to the provisions of the standard form of contract used are mandatory.

2.4.2 BOT concept

The concept of BOT has become increasingly popular in recent years in China. It requires the presence of three elements: a feasible and viable project; a willing government to grant a concession agreement empowering a concessionaire to operate and benefit from the constructed project by that concession; and financiers willing to take the financial risk of undertaking the project. On the expiry of the concession period, the project reverts to the authority that had granted the concession. The contractual relationships with a BOT project in China are shown in Figure 2.3.

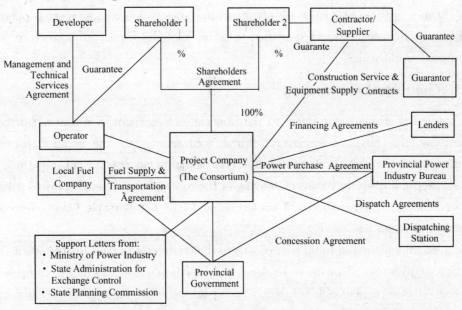

Figure 2.3 Contractual Structure of BOT Power Plant Project in China[39]

The concept is not new although the acronym is. The first construction project financed through such an agreement was the Suez Canal. The works was completed and opened for international navigation in 1869. The concession agreement in that case was for a period of 99 years. A similar approach was later adopted for the concession of the Panama Canal project.

Although a period of 99 years proved to be too long in the case of the Suez Canal and it was cut short in 1956 due to a political motivation, the project financing arrangements were successful. A more reasonable period nowadays for a concession agreement would be 10 to 40 years. However, despite the success of these arrangements, the traditional methods of financing infrastructure projects around the world remained, until recently, unchanged. These traditional methods entailed using either fiscal resources of the respective countries or sovereign borrowings from leading agencies, such as the World Bank, the International Monetary Fund, the African Development Bank and similar organizations.

In recent years, however, there have been major global shortcomings in funding the ever-growing need for infrastructure projects around the world. This led many governments to reconsider their policy of self-financing such projects. Many resurrected the concept of BOT rather than seeking to impose higher taxes. This approach was encouraged by the fact that in many countries around the world, the national economic growth and the annual income per capita have increased to a level that permitted the participation of the general public in such financial activities for the benefit of the whole society.

In summary, although the traditional methods of financing a project, i.e. self-financing and borrowing, represent the main methods used today, the concept of involving private finance in the construction of infrastructure projects is becoming more popular. The use of private finance is expected to rise to an estimated 15% of infrastructure projects implemented world-wide.

2.4.3 Contractual arrangement[20]

Contractual arrangement refers to the financial arrangement of a contract or basis of payment by the buyer (employer, prime contractor) to the seller (contractor, subcontractors or suppliers). Traditionally, contractual arrangements fall into three broad categories depending on the means of arriving at the contract sum (contract price); these are lump sum contracts, remeasurement contracts, and cost reimbursable (also referred to as cost plus) contracts.

In remeasurement and lump sum contracts the contractor is paid for the work done in accordance with rates and/or prices tendered by him beforehand, whether in competition or otherwise. In cost reimbursable contracts, he is paid the actual ascertained costs incurred by him under predefined heads, plus a fee to cover his administration and profit. In some cases a bonus and penalty based on the time for completion of the works may also be introduced and similarly based on the actual cost of the works, as in target contracts.

(1) Lump sum contracts

A lump sum contract is one in which the contractor undertakes to carry out all the work specified and shown on the drawings for a tender sum of money. The contractor is responsible for the assessment of all the costs he will incur in fulfilling the specified requirements. The contract will not usually provide for any adjustment of the quoted lump sum unless the requirements specified at the time of tender are altered. Payment is not made as a single sum other than for very small jobs, but is made in interim stage payments related to the amount of work done or at stated intervals during the progress of the works.

Tenderers for such contracts may be asked to quote rates for items of work listed in a schedule for use by the A/E for the fixing of prices for new or varied work ordered during the execution of the project.

This category of contract can also include incentives for meeting or exceeding selected project objectives, such as schedule targets.

(2) Cost reimbursable contracts

This category of contract involves payment (reimbursement) to the seller for seller's actual costs, plus a fee typically representing seller profit. Costs are usually classified as direct costs or indirect costs. Direct costs are costs incurred for the exclusive benefit of the project (e. g. salaries of full-time project staff). Indirect costs, also called overhead and general and administrative costs, are costs allocated to the project by the project team as a cost of doing business (e. g. salaries of management indirectly involved in the project, cost of electric utilities for the office). Indirect costs are usually calculated as a percentage of direct costs. Cost-reimbursable contracts often include incentive clauses where if the seller meets or exceeds selected project objectives, such as schedule targets or total cost, then the seller receives an incentive or bonus payment.

(3) Remeasurement contracts

They are also referred to as admeasurement, "measure and value", unit price or rate contracts, and a more general term "time and material (T&M)" can be used to refer to this category. Remeasurement contracts are a hybrid of cost reimbursable and lump sum type arrangements. These types of contracts resemble cost reimbursable type arrangements in that they are open ended. The buyer does not define the full contract value and the exact quantities at the time of the contract award. Thus, remeasurement contracts can grow in contract value as if they were cost reimbursable type arrangements. Conversely, they can also resemble lump sum arrangements. For example, the buyer can preset unit rates and the seller when both parties agree on the rates for a specific resource category. For procurement of works measurement contracts can be:

☐ Bill of quantities contracts

The bill of quantities contract is based on a detailed bill of quantities (computed by the A/E's organization from drawings) which includes brief descriptions of the work to be undertaken, against each item of which the contractors tendering enter a unit rate or price. The tender price is the aggregated amount of the various quantities priced at the quoted rates. During the performance of the work the actual quantity executed under each item is measured and valued at the quoted rate. In the *FIDIC Conditions of Contract* provision is made for the adjustment of rates for varied or additional work and for the fixing of new rates by the A/E.

☐ Schedule of rates contracts

The usual form of the schedule of rates contract is a list of work items covering the building or engineering works which an employer may want done. No quantities are given. Sometimes tenderers are invited to affix rates to the items; sometimes they are invited to quote a percentage to be added to or deducted from rates previously entered by the A/E. The contract usually relates to a defined location and to a stated period of time. The contractor may be called on to undertake any of the work items anywhere in that location during the term of the contract.

Another form of this type of contract is that applied to one specific project where it is essential to commence the works without delay. In this case a short list of the most important items of work is prepared, complete with representative quantities. The tenderers price the items of work and, if the employer or procurement agent accepts, these prices become the basis for negotiating prices for subsequent items of work as they become necessary.

The mutual trust and confidence between the parties involved in a construction project is necessary for the contract to run effectively; this is so particularly for schedule of rates contracts.

2.4.4 Selection of procurement methods

It is vital that an employer makes the correct choice of procurement method in an increasingly complex situation, with a wide range of criteria and systems, as described in this chapter. The choice between the procurement methods and the types of contract thus resolves itself into a complex process of balancing technical, economic and contractual factors. In this regard a wise employer will rely largely on the advice of the A/E or procurement agent.

A/E's have an important role to perform in selection among procurement methods. The employers range from being experienced that they may have their own professional team and an effective procurement policy to being inexperienced with little or no knowledge of procurement methods and who requires extensive professional advice. Furthermore, the employer's needs vary considerably with regard to certainty of price, cost limits, time requirements, complexity of design and many other factors.

2.4.4.1 Criteria for selection of procurement methods[18][21]

Prior to 1980s an employer tended to prioritize time, cost and quality of the works or services to be procured when selecting sellers. This approach is too simplistic as these important criteria form only part of what the employer needs to examine. A set of criteria have been ever since formed based on a number of diverse and valuable case studies, which examine the processes to select sellers, the decisions made and the subsequent review of the completed projects. Usually the criteria involve a number of questions representing the matters which the employer generally has to consider before a most appropriate procurement method can be chosen.

☐ Design input: does the employer want to influence the design and, if so, to what extent?

☐ Control over the project: how hands-on does the employer wish to be in the management of the project?

☐ Cost certainty: what level of cost certainty does the employer require before signing the contract and on completion of the works?

☐ Risk taking: is the employer prepared to accept the risk by direct management or does he wish to transfer it to third parties?

☐ Flexibility: to what extent is the employer's brief likely to be changed during the execution of the works?

☐ Market conditions: how are market conditions likely to change during the course of the project with possible consequences for design or construction?

☐ Programme security: how crucial is the final completion date?

☐ Value for money: does the employer want to contribute to and take benefit from value management and value engineering, and how will any resulting savings be shared?

2.4.4.2 Factors affecting choice of type of contract [7]

In weighing the merits of the different types of contract, a wise employer will wish to know as accurately as possible before his works is contracted out the total expenditure he will incur and the time required for completion of the works. He will, in most cases, also wish to ensure that the works is done at the minimum cost compatible with satisfactory materials, workmanship and time. The extent to which these objectives are attained will depend largely on the quality and scope of the information which he can make available or furnish to the tenderers. The better the information, the more precise will be the tenders and the more accurate will be the forecast of the final cost.

☐ Exceptional urgency

If the design-tender-construct method is used the contractors are expected to compete in tendering for the contract by planning well to execute and complete the works efficiently and thus at minimum cost. However, sometimes the value of time to the employer may be so much great as to justify a commencement on site before the detailed design is completed. If it is not possible to obtain competitive tenders, a cost reimbursement contract may have to be used. Such an arrangement gives the A/E more discretion to amend design as construction proceeds, which is often valuable for the projects which use rapidly changing technology or which involve long lead times for the delivery of essential items of plant and equipment.

☐ Contingencies and risks

Contingency sums are sometimes specified to be included in tenders for likely but unknown detailed items of work. The sums are to be expended at the A/E's instruction. There are also unforeseen risks against which insurance cover must be provided. However much care and time is devoted to preliminary investigations, the contingent and unforeseen risks are usually potentially costly in civil engineering work. The uncertainty of occurrence or magnitude of risks invites insurance, and protection is usually best secured by the selection of a contractor who, by reason of his experience and reputation, can obtain cover against the risks at competitive rates in the insurance market.

The extent and nature of the liabilities imposed under a remeasurement contract will have a direct bearing on the contract price. If a particular risk is disproportionately large it is more economical for the employer to carry it. If the employer has reason to secure protection through the contract against every conceivable construction risk this will inevitably result in higher tender prices. The A/E can assess, as the employer's adviser or procurement agent, the effects of contractual liabilities and judge how much financial value their imposition or relaxation will be to the employer. As a matter of fact, a sound sense of engineering judgment based on experience may be far more telling than mathematically correct economics.

□ Financing a project

The way in which a project is financed may affect the choice of type of contract. The employer should keep the A/E fully informed of his intentions in this respect, so that the latter can weigh this factor against all the others to be considered. In some cases, the contractor is required to raise at least part of the finance for the project.

2.4.4.3 Selection process

A very effective process for selection of a procurement method is described below:

(1) When the procurement criteria/objectives are agreed the employer is asked to determine the relative importance of each and to rank them in order of merit. For example, cost certainty might be paramount and could receive a ranking of nine or ten on a scale of 1 to 10, while the balance of risk might be less important with a ranking of three to four.

(2) Each procurement option is then discussed to ensure that the managerial and contractual arrangements between the employer and the rest of all the project's stakeholders are fully understood. The principal working methods of each option, with their advantages and disadvantages, are discussed. This normally results in the elimination of one or two procurement methods, leaving a shortlist of preferred options.

(3) The extent to which each short-listed method meets the employer's list of specific objectives is then considered. Each procurement method is discussed in turn and a weighting assigned to each objective according to how well that method meets the objective. Thus under construction management, cost certainty would be highly weighted and might score eight or nine, whereas under a two-stage contract, it might only merit a five.

(4) The final task is to determine the most appropriate procurement method for the project. Taking each method and each item on the employer's objectives list in turn, the importance ranking is multiplied by the weighting assigned to it for each short-listed procurement option. After totaling the tallies, the best option is identified as the one with the highest score.

Words, Phrases and Expressions 2.4

[1] procurement methods (procurement systems) 采购办法

[2]　the long lasting design-tender-construct method 历史悠久的先设计，再向承包商招标，由其施工的采购办法

[3]　design-build 设计—建造，设计—施工

[4]　management contracting 管理承包

[5]　construction management 施工管理

[6]　design and manage 设计与管理承包

[7]　EPC (Engineer, Procure, Construct) 设计—采购—施工

[8]　involve *vt*. 使……参与（engage as a participant）

[9]　provided *conj*. 倘若，如果（on condition (that); with understanding）

[10]　concession 特许权（right given by owner (s) of land, or by a government, to do *sth*.）

[11]　concession arrangement 特许权安排

[12]　Build-Operate-Transfer agreement 建造—运营—转让协议

[13]　concessionaire 特许权接受者，特许权受让者（a holder of a concession）

[14]　contractual arrangement 合同安排

[15]　contract sum (contract price) 合同价（格）

[16]　lump sum contracts 总价合同

[17]　remeasurement *n*.（重新）计量（calculation of the actual quantities of work executed by the contractor as required in the contract for payment to him. It is also referred to as "measure and value".）

[18]　remeasurement contracts 计量合同

[19]　admeasurement contracts 计量合同

[20]　measure and value contracts 计量合同

[21]　unit price or rate contracts 单价合同

[22]　reimburse *vt*. 报销（pay back to *sb*.; repay; make payment of an equivalent to）

[23]　cost reimbursable contracts 成本加成合同

[24]　tender *vt*. 向买方提出……的价格，向买方报……的价格，报价（present for acceptance; offer）*The contractor is paid for the work done in accordance with rates and/or tendered by him beforehand.* 就已经完成的工程按照承包商事先提出的单价或价钱向其支付款项。

[25]　represent *vt*. 是，表示，表现，象征，体现（be, give, make a picture, sign, symbol, or example of）*This category of contract involves payment (reimbursement) to the seller for seller's actual costs, plus a fee typically representing seller profit.* 这类合同要求向卖方支付（为卖方报销）各项实际费用，再加一笔酬金，表示卖方利润。

[26]　direct costs 直接费（用）

[27]　indirect costs 间接费（用）

[28]　costs incurred for the exclusive benefit of the project 仅为本项目开销的费用

[29]　overhead 管理费，摊销费

[30]　electric utilities 用电设施
[31]　time and material (T&M) contracts 工料合同
[32]　bill of quantities contracts 工程量清单合同
[33]　rate *n*. 单价（price per unit quantity of an item of work. Not used in this book to mean the speed of work unless otherwise stated.）
[34]　unit rate 单价
[35]　tender price 标价
[36]　conditions of contract 合同条件
[37]　schedule of rates contracts 单价表合同
[38]　criteria for selection of procurement methods 选择采购办法的准则
[39]　stakeholders *n*. 利害关系者（Persons and *organizations* such as *customers*, *sponsors*, *performing organization* and the public, that are actively involved in the *project*, or whose interests may be positively or negatively affected by execution or completion of the project. They may also exert influence over the project and its *deliverables*. 积极参与项目或者利益可能受到项目实施或完成的积极或消极影响的个人或组织，例如顾客、发起人、实施组织和公众。他们亦可能对项目及其可交付结果施加影响。）
[40]　shortlist 决选名单（list of candidates (for a position, etc.) that has been reduced to a small number from which a final selection is to be made.）
[41]　contingency sums 不可预见费，应急费用
[42]　invite *vt*. 增加了……可能性（increase the likelihood of）*The uncertainty of occurrence or magnitude of risks invites insurance* 风险发生与大小的不确定性提高了购买保险的可能性。
[43]　tend *vt*. 往往，经常，容易，动辄（be inclined to move; have a direction）*Prior to 1980s an employer when selecting seller tended to prioritize time, cost and quality of the works or services to be procured*. 20 世纪 80 年代之前，业主选择卖方的优先顺序经常是待采购工程或服务的时间、费用和质量。
[44]　design input 对设计提出的要求
[45]　programme 施工计划
[46]　programme security 施工计划的稳妥性
[47]　value for money 物有所值
[48]　value management and value engineering 价值管理与价值工程
[49]　hands-on 直接插手

Notes 2.4

[1]　疑难句解释

If an employer does not wish to be involved in the day-to-day progress of the work, provided the end result meets the performance criteria he has specified and the parties concerned (e. g. sponsors, lenders and the employer) are willing to see the contractor paid more for the construction of the facility in return for the contractor bearing the extra risks associated with enhanced certainty of final price and time, then the employer wishes the contractor to take total responsibility for the design and construction of the process or power facility and hand it over ready to operate "at the turn of a key".

【译文】 若最终结果满足业主事先为工程规定的性能标准，有关各方（例如项目发起人、贷款人和业主）也愿意在承包商为提高最后价格和时间的可靠程度而承担额外风险时增加向其支付的款项，这时如果业主不想参与工程日常实际工作，则业主就希望承包商承担该加工或发电设施的设计和施工任务并在完成时交给业主，使其能够"用钥匙一开"就能使用。

[2] 词语解释

1) cover *vt*. 够用，足数，（钱）用于……((of money) enough for)

例 1　$1200 will cover my needs for the journey. 这1200美元够我此次行程之需。

例 2　We have only just covered our expenses, made enough for our expenses, but no profit. 敝号所得仅抵开销，所入足抵所出，但无利可图。

例 3　The tender, rates and prices tendered by the contractor shall cover all his obligations under the contract and all matters and things necessary for the proper execution and completion of the works and the remedying of any defects therein. 承包商所报标价、单价和价钱应足以支付他为履行所有合同义务，以及正常施工、竣工并弥补工程缺陷所需之所有事物的开销。

2) cover *vt*. 支付费用，负担开支 (make provision for (a demand or charge) by means of a reserve or deposit)

例 1　Some piling activities are very appropriately covered by method-related charges. 某些打桩工序的费用最好由与方法有关的费用负担。

例 2　These items will be measured in m^3 to cover the extra cost involved. 这些分项工程按立方米计量，用以支付额外必要之费。

例 3　An allowance shall be made to cover oncosts and profit. 应考虑现场费用和利润的数额。

例 4　These items rarely cover the actual costs incurred. 这些事项难以支付实际开销的费用。

例 5　Such amounts have not already been covered by payments made to the contractor. 已经支付给承包商的款项中还没有计入上述款项。

例 6　The contractor is paid the actual ascertained costs incurred by him under predefined heads, plus a fee to cover his administration and profit. 向承包商支付他在事先确定的各名目下经过核实的实际开销，再加一笔酬金，用以弥补其行政管理费和利润。

3) equipment 施工设备，施工机具

Equipment may refer to the mechanical, electrical or other items being supplied under a contract, or it may refer to the items used by a contractor to do his work. The definition in a contract should state which. Whichever is meant, plant is often the word used to refer to the other. The equipment used by a contractor is, in the FIDIC Conditions of Contract, referred to as the contractor's equipment meaning all appliances and things of whatsoever nature (other than temporary works) required for the execution and completion of in executing the works and the remedying of any defects therein, but does not include Plant, materials or other things intended to form or forming part of the permanent works. Occasionally, 'equipment' in this book means 'contractor's equipment'.

【译文】 equipment 可能指根据合同供应的机械、电气或者其他设备，也可指承包商用以施工的设备。合同定义中应说明 equipment 为何含义。不管含义为何，都用 plant 指未定义的其它设备。在 FIDIC 合同条件中，承包商使用的设备叫做 contractor's equipment，包含为实施和完成建筑或土木工程并修补其中任何缺陷所需要的所有用具和任何性质的物品（临时工程除外），但不包括准备构成或正在构成永久工程一部分的设备、材料或其它物品。本书有时候用 equipment 代替 contractor's equipment。

4) plant 设备

In the FIDIC Conditions of Contract, plant means machinery, apparatus and the like intended to form or forming part of the permanent works.

【译文】 在 FIDIC 合同条件中，plant 指准备构成或已构成永久工程一部分的机器、用具以及其他等。

5) essential items of plant and equipment 由于 plant 和 equipment 均为不可数名词，故在表达 plant 或 equipment 数目时，须借助可数名词 item 或 piece。

6) thereafter, thereby, therein, thereinafter, thereof, thereto, thereunder, thereupon, therewith 等在合同与其他一些文书中常用。其中 there 指上文提到之事物。这种用法使句子精炼。下面即为一例。

The Bidder and any of its personnel or agents will be granted permission by the Employer to enter upon its premises and lands for the purpose of such visit, but only upon the express condition that the Bidder, its personnel, and agents will release and indemnify the Employer and its personnel and agents from and against all liability in respect *thereof*,

【译文】 雇主将允许投标人，及其下属人员或代理人进入雇主管辖之所和土地察看现场，但是明确的先决条件是投标人，及其下属人员或代理人将保障雇主投标人，及其下属人员或代理人不承担与此有关的所有责任。

【解释】 ***thereof*** 中的 ***there*** 代表 *The Bidder and any of its personnel or agents enter upon its premises and lands for the purpose of such visit*。这句话若如下所写，则嫌啰嗦：The Bidder and any of its personnel or agents will be granted permission by the Employer to enter upon its premises and lands for the purpose of such visit, but only upon the express condition that the Bidder, its personnel, and agents will release and indemnify the Employer and its personnel and agents from and against all liability in respect of *the Bidder and any of its personnel or agents entering upon its premises and lands for the purpose of such visit*.

实际上，这种省略汉语中很多。如：小王买了两只鸡，三头鹅，然后捆上这两只鸡和三头鹅的翅膀，并蒙上这两只鸡和三头鹅的眼睛。显然，可简写为：小王买了两只鸡，三头鹅，然后捆其翅膀，蒙其眼睛。"其"译成英文，就是 thereof。还可再简单：小王买了两只鸡，三头鹅，捆翅膀，蒙眼睛。

再如，小王买了两只鸡，三头鹅，然后为这两只鸡，三头鹅垒了窝，并为这两只鸡，三头鹅找来米糠。简单点，应这样写：小王买了两只鸡，三头鹅，为其垒了窝，找来米糠。"其"译成英文，就是 therefor。自然，还可再简单点：小王买了两只鸡，三头鹅，垒了窝，找来米糠。

又如，小王买了两只鸡，三头鹅，并向卖主了解这两只鸡，三头鹅的情况。简单点，应这样写：小王买了两只鸡，三头鹅，并向卖主了解其情况。"其"译成英文，就是 in connection therewith, in relation thereto 或 in respect thereof。

又再如，小王买了两只鸡，三头鹅，以后吃蛋就得靠这两只鸡，三头鹅。可简写为：小王买了两只鸡，三头鹅，以后吃蛋就得靠它们。"它们"译成英文，就是 rely thereon 或 rely thereupon，如此等等。希望读者自己在阅读这种句子时，细细体会。

另外，如下结构也常见：hereafter, hereby, herein, hereinafter, hereof, hereto, hereunder, hereupon, herewith, 等。不过，其中的 here 是指下文或本文提到的事物。

7) Bill of quantities contracts 和 Schedule of rates contracts 的差别

【解释】 Bill of quantities contracts 有工程量，而 Schedule of rates contracts 没有工程量。Bill of quantities contracts 适合于招标时能够估算出工程量的情况，而 Schedule of rates contracts 适合于招标时无法估算出工程量的情况。

2.5 Contracting planning

The contracting planning is a process to prepare the documents needed to support the "request seller responses" process and select seller process.[16]

2.5.1 Standard forms

Major works, whether they are predominantly building, civil engineering, chemical engineering, electrical engineering, mechanical engineering, or any combination thereof, are frequently complex, which has made the conditions of contract increasingly complex. It is becoming increasingly important for them to be based upon a standardized form of contract with which the contracting parties and financial institutions are familiar.[8]

In most cases, the parties to a contract will react favourably to such a standardized form of contract, which should lessen the likelihood of unsatisfactory performance, increased costs and disputes. If the contract is to be based on standard forms of contract, tenderers should not need to make financial provision for unfamiliar contract conditions. The widespread use of standard forms also facilitates the training of personnel in contract administration, reducing the need for them having to work with ever-changing forms of contract.

Standard forms of contract include conditions of contract, specification (descriptions of items to be procured), forms of tender and appendixes to tender, forms of agreements, non-disclosure agreements, forms of tender security, forms of performance security, tender or proposal evaluation criteria checklists, or standardized versions of all parts of the needed tender documents. The employer or other organizations that perform substantial amounts of procurement can have many of these documents standardized. Buyer and seller organizations performing intellectual property transactions ensure that non-disclosure agreements are approved and accepted before disclosing any project specific intellectual property information to the other party.

When the contracting planning process ends, the employer's project team shall produce a set of procurement documents.

2.5.2 Procurement documents[16]

Procurement documents are used to seek proposals from prospective sellers. A term such as tender, bid, or quotation is generally used when the seller selection decision will be based on price (as when buying commercial or standard items), while a term such as proposal is generally used when other considerations, such as technical skills or technical approach, are paramount. However, the terms are often used interchangeably and care is taken not to make unwarranted assumptions about the implications of the term used. The term "tender documents" is used for procurement of works in place of "procurement documents".

Sometimes procurement documents are simply the invitation for tender, request for proposal, request for quotation, tender notice, invitation for negotiation, and contractor initial response.

If the works to be procured are complex the procurement documents may refer to prequalification documents and the tender documents.

Experience has shown that, particularly for contracts which include contractor-design, prequalification of tenderers is particularly desirable. It enables the employer to establish the competence of a known number of companies and joint ventures who are subsequently invited to tender. Restricting tendering to a pre-determined number encourages the better qualified entities to tender in the knowledge that they have a reasonable chance of success.

Procedures for the prequalification of prospective tenderers may be imposed by the applicable laws, or by the requirements of the financial institutions which will be providing funds for the project. In particular, they may not permit any limit on the number of prequalified tenderers.

Typically, the employer initiates the prequalification stage of the project, by publishing advertisements which either (i) contain all the necessary information on the project and on how applicants should apply for prequalification, or (ii) describe how to obtain a document which contains all this prequalification information and which should include: information on the prequalification procedure, including qualification criteria and any relevant policies which the employer may have; instructions on the language and content of each application for prequalification and on the time and place for its submission; and information on the contract. Other information on the contract may also be included.

Tender documents must be drafted with care, particularly in respect of quality, tests and performance criteria for contractor-design contracts. If tender documents are deficient, the employer may pay an exorbitant price for unacceptable works. He must therefore ensure that adequate resources are allocated to the skilled tasks of drafting the technical and commercial aspects of the tender documents, and of analysing the tenderers' proposals.

The buyer or employer structures procurement documents to facilitate an accurate and complete response from each prospective seller and to facilitate easy evaluation of the tenders. These documents include a description of the desired form of the response, the relevant contract statement of work (e. g. drawings, specification, and bill of quantities, etc.) and any required contractual provisions (e. g. a copy of a model contract, non-disclosure provisions). With government contracting, some or all of the content and structure of procurement documents can be defined by regulation.

The complexity and level of detail of the procurement documents should be consistent with the value of, and risk associated with, the planned purchase or acquisition. Procurement documents are rigorous enough to ensure consistent, comparable responses, but flexible enough to allow consideration of seller suggestions for better ways to satisfy the requirements. Inviting the sellers to submit a proposal that is wholly responsive to the request for tender and to provide a proposed alternative solution in a separate proposal can do this.

Issuing a request to potential sellers to submit a proposal or tender is done formally in accordance with the policies of the buyer's organization, which can include publication of the request in public newspapers, in magazines, in public registries, or on the Internet. The procedure will be described in Chapter 3.

2.5.3 Evaluation criteria[16][22]

Evaluation criteria are developed and used to rate or score proposals. They can be objective or subjective. Evaluation criteria are often included as a part in the procurement documents. and can be limited to tender price if the works to be procured is an ordinary building or civil engineering works from a number of acceptable contractors.

Other selection criteria can be identified and documented to support an assessment for a more complex works or services. For example:

☐ Understanding of need. How well does the tender or proposal address the statement of work?

☐ Overall or life-cycle cost. Will the selected seller produce the lowest total cost (purchase cost plus operating cost)?

☐ Technical capability. Does the seller have, or can the seller be reasonably expected to acquire, the technical skills and knowledge needed?

☐ Management approach. Does the seller have, or can the seller be reasonably expected to develop management processes and procedures to ensure a successful project?

☐ Technical approach. Do the seller's proposed technical methodologies, techniques, solutions, and services meet the procurement documentation requirements or are they likely to provide more than the expected results?

☐ Financial capacity. Does the seller have, or can the seller reasonably be expected to obtain the necessary financial resources?

☐ Production capacity and interest. Does the seller have the capacity and interest to meet potential future requirements?

☐ Business size and type. Does the seller's enterprise meet a specific type or size of business, such as small business, women-owned, or disadvantaged small business, as defined by the buyer or established by governmental agency and set as a condition of being award a contract?

☐ References. Can the seller provide references from prior customers verifying the seller's work experience and compliance with contractual requirements?

☐ Intellectual property rights. Does the seller assert intellectual property rights in the work processes or services they will use or in the products they will produce for the project?

☐ Proprietary rights. Does the seller assert proprietary rights in the work processes or services they will use or in the products they will produce for the project?

To evaluate a bid, as recommended by the World Bank for use with prequalification, the employer shall consider the following:

1. the tender price, excluding provisional sums and the provision, if any, for contingencies in the summary bill of quantities, but including daywork items, where priced competitively;
2. price adjustment for correction of arithmetic errors as follow:

If the tender is substantially responsive, the Employer shall correct arithmetical errors on the following basis:

(1) if there is a discrepancy between the unit rate and the total price that is obtained by multiplying the unit price and quantity, the unit rate shall prevail and the total price shall be corrected, unless in the opinion of the Employer there is an obvious misplacement of the decimal point in the unit rate, in which case the total price as quoted shall govern and the unit rate shall be corrected;

(2) if there is an error in a total corresponding to the addition or subtraction of subtotals, the subtotals shall prevail and the total shall be corrected; and

(3) if there is a discrepancy between words and figures, the amount in words shall prevail, unless the amount expressed in words is related to an arithmetic error, in which case the amount in figures shall prevail subject to (1) and (2) above.

3. price adjustment due to discounts offered in a manner described below:

The tenderer shall quote any unconditional discounts and the methodology for their application in the Letter of Tender. The Letter of Tender and other schedules, including the Bill of Quantities, shall be prepared using the relevant forms furnished in the tender document entitled Bidding Forms. The forms must be completed without any alterations to the text, and no substitutes shall be accepted. All blank spaces shall be filled in with the information requested.

4. converting the amount resulting from applying 1 to 3 above, if relevant, to a single currency.

5. adjustment for nonconformities as follow:

Provided that a tender is substantially responsive, the Employer shall rectify nonmaterial nonconformities related to the Tender Price. To this effect, the Tender Price shall be adjusted, for comparison purposes only, to reflect the price of a missing or non-conforming item or component. The adjustment shall be made using the method indicated in the following 6~8.

6. assessment, as indicated, of adequacy of Technical proposal with requirements.

7. evaluation, as indicated, of alternative times for completion if permitted under the tender documents.

8. evaluation, as indicated, of technical alternatives if permitted under the tender documents.

Words, Phrases and Expressions 2.5

[1]　standard forms 标准格式
[2]　non-disclosure agreements 保密协议
[3]　tender security 投标保证，投标保证书，投标保证金
[4]　performance security 履约保证，履约保证金，履约保证书
[5]　checklists 核对表
[6]　perform *vt*. 执行，履行（do (a piece of work, *sth*. that one is ordered or has promised to do)) *organizations that perform substantial amounts of procurement can have many of these documents standardized.* 进行大量采购的组织可能已经统一了许多此类文件的格式。
[7]　produce *vt*. 提出，完成（compose, create, or bring out by intellectual or physical effort）*When the solicitation process ends, the employer's project team shall produce following outputs*：当询价过程结束时，业主的项目班子应当提出下列成果：
[8]　procurement documents 采购文件
[9]　tender 标书（statement of the price at which one offers to supply goods or services, or to do *sth*.; an offer or proposal for acceptance）

[10]　warrant *vt*. 保证，担保（guarantee）*care is taken not to make unwarranted assumptions about the implications of the term used*. 对于所用的这一术语，注意不要有不可靠的先入之见。

[11]　invitation for tender 投标邀请（书、信、函、广告、电子邮件……）

[12]　request for proposal 征求建议书（信、函、广告、电子邮件……）

[13]　request for quotation 询价函（信、广告、电子邮件……）

[14]　tender notice 招标通知

[15]　invitation for negotiation 议标邀请信（书、函）

[16]　contractor initial response 承包商初步回复

[17]　response *n*. 回复

[18]　prequalification 资格预审

[19]　structure *vt*. 编排、组织、安排（form into a structure or form according to a structure; give a structure to）*The buyer structures procurement documents to facilitate an accurate and complete response from each prospective seller and to facilitate easy evaluation of the tenders*. 买方在编排采购文件时既要便于所有有意的卖主作出准确而又完整的回复，亦应便于评标。

[20]　address *vt*. 处理（deal with; treat）*How well does the tenderer's tender or proposal address the contract statement of work?* 投标者的标书或建议书对于合同工作说明书中说明的问题处理得如何？

[21]　develop *vt*. 酝酿，提出（cause to unfold gradually）*Evaluation criteria are developed and used to rate or score proposals*. 提出评价标准并将其用于评价建议书的优劣或为其评定分数。

[22]　reference *n*. （关于人品或能力的）证明信（书）、推荐信（书）、介绍信；（关于人品或能力的）证明人、推荐人、介绍人（statement about a person's character or abilities; person willing to make a statement about a person's character or abilities）*Can the seller provide references from prior customers verifying the seller's work experience and compliance with contractual requirements?* 卖主能否提供从前顾客介绍其工作经验和遵守合同要求的证明？

[23]　intellectual property rights 知识产权

[24]　proprietary rights 专有（技术）权

Notes 2.5

疑难词语解释

[1]　However, the terms are often used interchangeably and care is taken not to make unwarranted assumptions about the implications of the term used.

【译文】　然而，这些术语经常混用，因此，务必不要想当然地从字面上理解所用术语的真实含义，否则他人不会承担因此而造成的后果责任（unwarranted）。

Chapter 3 Tendering Procedure

3.1 Request for seller responses[16]

This is a process to obtain responses such as tenders or proposals from prospective sellers (contractor, subcontractors or suppliers) on how the buyer's (employer, contractor) requirements can be met. The process is referred to as inviting for tenders or to invite to tender if an employer or his procurement agent procures works, i. e. *construction and related engineering services*. A tender is an offer submitted in response to a formal invitation for tenders.

The employer, his procurement agent or other organizations involved in the project may maintain lists of prospective and previously qualified tenderers, or more generally called sellers, who can be asked to tender for, propose, or quote on the works to be procured. These lists generally have information on relevant past experience and other characteristics of the prospective sellers.

3.1.1 Preliminary inquiry for invitation to tender[23]

Sometimes, a preliminary inquiry shall be made to the prospective tenderers as to whether they wish to be invited to tender for a particular project. This will give them necessary information on which to base their decision. Figure 3.1 is a sample letter of preliminary inquiry.

Architect's address
12 August 2006

Contractor's address

Dear Mr. Lu Ban,

<u>Re: Inquiry on willingness to tender for Good Luck Plaza</u>

We are writing you to seek your intention to tender for our Good Luck Plaza project. Enclosed you will find a set of plans and specifications covering the construction of the project.

All interested tenderers will be invited formally to tender for our project. Would you please indicate your willingness to tender by 25 August 2006?

We look forward to your early response.

Yours faithfully,

Wang Fazhan
Project manager

Figure 3.1   Letter of Preliminary Inquiry

3.1.2   Indicating willingness to tender[23]

The recipient of the letter of preliminary inquiry shall respond to the inquiry. A sample letter indicating his willingness to tender is shown in Figure 3.2. Figure 3.3 illustrates a possible letter of tactful refusal.

Contractor's address
23 August 2006

Architect's address

Dear Mr. Wang,

Re: Our willingness to tender for Good Luck Plaza

Thank you for your letter of 19 August 2006 asking us whether we wish to be invited to submit a tender for this project in accordance with the condition which you set out.

We shall be pleased to accept such an invitation if extended and we look forward to hearing from you further.

Yours faithfully,

Lu Ban
Marketing manager

Figure 3.2   Letter of Willingness to Tender

Contractor's address
23 August 2006

Architect's address

Dear Sir,

Re: Inquiry on willingness to tender for Good Luck Plaza

We very much regret that our commitments at the present time are such that we are reluctantly compelled to decline. However, we hope that we may be given the opportunity to tender for any future projects with which you may be concerned.

Yours faithfully,

Lu Ban
Marketing manager

Figure 3.3　Letter of Refusal

3.1.3　Formal invitation to tender

Once a contractor has confirmed his willingness to tender, the next stage is the employer's formal invitation to tender which will be accompanied by the tender documents. A formal invitation to tender could be in the form shown in Figure 3.4.

The form that an invitation to tender for the contracts financed by the World Bank in whole or in part depends on the presence of prequalification.

The *Standard Bidding Documents for Procurement of Works and User's Guide*[22] published by the World Bank in May 2005 sets out:

"The Invitation for Tenders for contracts, subject to prequalification, is sent only to firms determined by the Borrower to be qualified in accordance with the Borrower's prequalification procedure. This procedure must be reviewed and commented on by IBRD if the potential contract is to be eligible for IBRD financing.

Ideally, the Letter of Invitation for Tenders is sent to the qualified Bidders at the time that the prequalification results are announced.

For major works, prequalification shall be used. If, exceptionally, prequalification is not used, the appropriate Specific Procurement Notice/Invitation for Bids form (see below) shall be used.

If tenders are invited openly from contractors without using a prequalification procedure, the Invitation for Tenders should be issued directly to the public as:

(a) a *General Procurement Notice* (for procurement by ICB) in *UN Development Business online* and in the *Development Gateway's dgMarket*;

(b) an advertisement in at least one newspaper of national circulation in the Employer's country (or in the official gazette, or in an electronic portal with free access);

(c) an advertisement in *Development Business* and/or well-known technical magazines for large or important contracts; and

(d) a letter addressed to contractors who, following the publication of the *General Procurement Notice*, have expressed interest in tendering for the Works.

Its purpose is to supply information to enable potential bidders to decide on their participation. Apart from the essential items listed in the standard documents, the Invitation for Bids should also indicate any important tender evaluation criteria (for example, the application of a margin of preference in bid evaluation).

The Specific Procurement Notice/Invitation for Bids form should be incorporated in the tender documents and should be consistent with the information contained in the Bidding Data."

From Architect
30 August 2006

Contractor's address
Dear Mr. Lu,

Re: Invitation to tender for Good Luck Plaza

Following your acceptance of the invitation to tender for the above, I now have pleasure in enclosing the following:
1. Two copies of the Bills of Quantities;
2. Two copies of the general arrangement drawings indicating the general character, shape and disposition of the works;
3. Two copies of the form of tender;
4. Addressed envelopes for the return of the tender and instructions relating thereto; and
5. Two copies of the Instructions to Tenderer.

Will you please also note:
1. Drawings and details may be inspected at the Room 104, ground floor of our office at the above address, 08:00-11:30 am, 3-9 September 2006;
2. The site may be inspected by arrangement with the employer;
3. Tendering procedure can be found in the Instructions to Tenderer.

Will you please acknowledge receipt of this letter and enclosures and confirm that you are able to submit a tender in accordance with these instructions?

Yours faithfully,

Guo Fengzhan
Project manager

Figure 3.4　Formal Invitation to Tender[23]

The *Standard Bidding Documents for Procurement of Works* published by the World Bank in May 2000 recommends the following form of invitation to tender (as shown in Figure 3.5) to be used in the procurement of works for a project funded by the loan or credit from the World Bank.

Form of Invitation for Bids[24]
[*letterhead paper of the Employer*]
_____[*date*]

To: [*name of Contractor*]
　　[*address*]
Reference: [*Insert IBDR Loan No. or IDA Credit No.*]

Contract Name, and Identification No. _____/_____.

Dear Sirs,

We hereby inform you that you are pre-qualified for tendering the above cited contract. A list of pre-qualified and conditionally pre-qualified Applicants is attached to this invitation.

On the basis of information submitted in your application, you would [not] (*insert if appropriate*) appear eligible for application of the domestic bidder price preference in tender evaluation. Eligibility is subject to confirmation at bid evaluation.

We now invite you and other pre-qualified Applicants to submit sealed bids for the execution and completion of the cited contract.

You may obtain further information from, and inspect or acquire the tendering documents at, our offices at [*mailing address, street address, and cable/telex/facsimile numbers*].

A complete set of bidding documents may be purchased by you at the above office, on or after [*time and date*] and upon payment of a nonrefundable fee of [*insert amount and currency*].

All bids must be accompanied by a security in the form and amount specified in the bidding documents, and must be delivered to [*address and exact location*] at or before [*time and date*]. Bids will be opened immediately thereafter in the presence of bidders' representatives who choose to attend.

Please confirm receipt of this letter immediately in writing by cable, fax, or telex. If you do not intend to bid, we would appreciate being so notified also in writing at your earliest opportunity.

Yours truly,

Authorized signature _____.
Named and title _____.
Employer _____.

Figure 3.5 Form of Invitation to Tender that was
Used for a Contract Financed by World Bank

Words, Phrases and Expressions 3.1

[1] eligible *adj*. 符合……要求（条件）的；适于选中的；合格的（fit, suitable to be chosen; having the right qualifications：~ *for promotion* (*a position, a pension, membership in a society*) 符合提拔（任职、领取养老金、入会）的要求）*This procedure must be reviewed and commented on by IBRD if the potential contract is to be eligible for IBRD financing.* 如果这一可能签署的合同要符合由国际复兴开发银行提供资金的要求，该程序务必要由该行审查并提出意见。

[2] major *adj.* 大的，*major works* 大工程，*major cities* 大城市
[3] *General Procurement Notice* 一般采购通知
[4] ICB 国际竞争性招标，国际公开招标（International Competitive Bidding）
[5] procurement by ICB（以）国际竞争性招标（方式）采购，（以）国际公开招标（方式）采购
[6] *UN Development Business online* 联合国发展事业网
[7] *Development Gateway's dgMarket* 发展之路上的 *dgMarket*（发展之路市场）栏目
[8] national circulation 全国发行
[9] gazette 公报（governmental periodical with legal notices, news of appointments, promotions, etc. of officers and officials）
[10] official gazette 政府公报
[11] electronic portal with free access 免费浏览的电子网站
[12] letter addressed to contractors 按地址寄给承包商的信函
[13] purpose *n.* 用途（that which one means to be）*Its purpose is to supply information to enable potential tenderers to decide on their participation.* 其用途是提供信息，使有意的投标人能够决定自己是否参加。
[14] incorporate *vt.* 使……成为一部分（make, become, united in one body or group）*The Specific Procurement Notice/Invitation for Tenders form should be incorporated in the tendering documents.* 具体的采购通知/投标邀请格式应当列入招标文件之中。
[15] acknowledge *vt.* 致函或宣布收到，函谢（make known the receipt of）*Will you please acknowledge receipt of this letter and enclosures and confirm that you are able to submit a tender in accordance with these instructions?* 收到本函及附件后请回音，并说明贵方能否提交符合本须知的标书。
[16] reference *n.* 附注；援引注；编号；索引号（direction, note, etc. telling where certain information may be found）
[17] on the basis of 根据……
[18] domestic bidder price preference （给予）本地（国内）投标人（的）价格优惠
[19] subject to *adj.* 须经……的；以……为条件的；听候；依照（conditional (ly) upon）

Notes 3.1

疑难词语解释

[1] The Invitation for Tenders for contracts, subject to prequalification, is sent only to firms determined by the Borrower to be qualified in accordance with the Borrower's prequalification procedure.

【译文】 须资格预审时，仅邀请由借款国断定符合借款国资格预审程序的公司投标。

[2] The project team must have access to an appropriate set of technical skills and knowledge.

【译文】 项目班子必须能够取得一套适当的技能和知识。

[3] We shall be pleased to accept such an invitation if extended and we look forward to hearing from you further.

【解释】 if extended 是 if invitation is extended to us 的简略。extended = made = given,是一种文雅说法。

3.2 Tender documents

For the sake of uniformity the use of capital letters in such words as contract, letter of tender, accepted contract amount, contract price, clause, sub-clause, contractor, architect, engineer, employer, works, payment certificate, etc. have been omitted in the majority of text of this book.

The tender documents prepared during contracting planning, as described in chapter 2, shall be used to invite for tenders from prospective contractors.

Tender documents form the basis on which a tenderer, that is the potential building or civil engineering contractor, will prepare his tender. When the employer has accepted a successful tenderer as the contractor, the most part of the tender documents will become important components of the contract documents and the basis for the contractor to execute, complete the contract works and remedy defects therein.

It is essential that the documents shall collectively detail all the requirements of the project in a comprehensive and unambiguous way.

The tender documents should give tenderers all available information that is likely to influence their tenders. These documents should therefore convey the clearest possible picture of the character and the quantity of the work, the site and subsoil conditions, the responsibilities of the contractor, the terms of payment and any other conditions, technical or commercial, which may affect the execution of the work. Such information is vital to the tenderer to enable him to work out a viable plan of operation, make provision for the necessary temporary works, estimate the cost of the work, assess the finance required and evaluate the risks involved. A lack of adequate and reliable information at the time of tendering will lead to uneconomic tender prices and claims for additional payments during the running of a contract.

The tender documents in connection with a remeasurement contract (also referred to as unit rate contract, unit price contract or schedule of rates contract) generally consist of:

☐ Instructions to tenderers, which do not usually form part of the contract documents;

☐ Form of tender (with which may be associated the requirements for a tender security or tender bond, details of experience and financial resources);

☐ Conditions of contract;

☐ Specification;

☐ Drawings;

☐ Bill of quantities;

☐ Data affecting the execution of the works, including such things as a site investigation report, details of access, limitations of working hours, local conditions and other work being

carried out on site, machinery, services and supplies which will be provided by the employer for the contractor's use; hydrographic surveys and special safety measures;
- ☐ Form of agreement; and
- ☐ Performance bond or performance security (if required).

In lump sum, cost reimbursable and all-in contracts, some of these documents may not be required, depending on the circumstances. In the case of direct labour, although there may be no contract, the specification and drawings and maybe bill of quantities will nevertheless be required.

The number of sets of tender documents made available to the tenderers is normally not less than two, and one copy of the priced bill of quantities is often required to be returned with the tender.

Drawings, including bar bending schedules where available, should always be issued with the tender documents. Notes on drawings are added to amplify the design details; they should be legible and certainly not conflict with the specification.

3.2.1 Instructions to tenderers

Their purpose is to direct and assist the tenderers in the preparation of their tenders, to ensure that they are presented in the form required by the employer. These instructions will vary from project to project. Some of the more usual and important items frequently included are:

(a) the documents to be submitted with a tender, including a programme and a general description of the arrangements and methods of construction which the contractor proposes for carrying out the works, together with particulars of the legal and financial status and technical experience of the tenderer;

(b) the place, date and time for the delivery of tenders;

(c) the instructions on visiting the site;

(d) the instructions on whether or not tenders on alternative designs will be considered, and if so the conditions under which they may be submitted;

(e) the notes drawing attention to any special conditions of contract, materials and methods of construction to be used and unusual site conditions; and

(f) the instructions on completion of the bill of quantities and/or schedule of rates, and production of tender and/or performance bonds.

In the case of selective tendering the particulars of tenderers' legal and financial status and technical experience are required at a pre-qualification stage.

In order to simplify the preparation of tender documents, the *Standard Bidding Documents for Procurement of Works and User's Guide* (SBDW), published in May 2005 by the World Bank, groups the provisions that shall remain unchanged for all procurements in the "*Instructions to Bidders*" and the provisions specific to a particular procurement in the "*Bidding Data*".

In Hong Kong, instructions to tenderers are called "tender conditions" or "conditions of tender", while bidding data called 'special conditions of tender'.

3.2.2 Definition of tender

The word *"tender"* means a number of things or matters, and is very confusing in general, and particularly to a layman.

In the context of procurement of works, as an intransitive verb, *"tender"* means *making an offer (to carry out work, supply goods, etc.) at a stated price*. For example, *tender for the construction of a new motorway*.

As a noun, *"tender"* means *a statement of the price at which one offers to supply goods or services, or to do something*. For example, *to invite (call) tenders for a new bridge; to put in (make, send in) a tender for something; to accept the lowest tender*.

The FIDIC *Conditions of Contract for Works of Civil Engineering Construction*, Fourth Edition 1987, amended 1992 (herein after referred to as the *Red Book of Fourth Edition*) defined *"tender"* as "the contractor's priced offer to the employer for the execution and completion of the works and the remedying of any defects therein in accordance with the provisions of the contract, as accepted by the letter of acceptance". The definition is confusing in that it not only treats the "contractor's priced offer" as same as the amount "as accepted by the letter of acceptance", but also confuses it with the "evaluated tender amount", on one hand. On the other hand, the *Red Book of Fourth Edition* by *"tender"* means "letter of tender" as well.

Fortunately, the *Red Book* (see 3.2.5.5) has eliminated the embarrassing situation in favour of its users.

The *FIDIC Contracts Guide*, including a modified version of the *Red Book*, as noted hereinafter, uses "letter of tender" in place of *"tender"* to mean the *statement*.

The *Example Form of Instructions to Tenderers* included in the *FIDIC Contracts Guide* by "tender amount" instead of *"tender"* means the sum offered or the amount entered in the letter of tender (*stated price*), while the *Instructions to Bidders* in the SBDW by "bid price (s)" instead of *"bid"* mean the prices quoted by the bidder in the letter of bid and in the bill of quantities.

The *Red Book* included in both the *FIDIC Contracts Guide* and the SBDW, by *"tender"* means the "letter of tender" and all other documents which the tenderer submitted with the "letter of tender".

In addition, the *FIDIC Contracts Guide* has differentiated the "tender amount" not only from the "evaluated tender amount" but also from the "accepted tender amount".

"Evaluated tender amount" is determined by making any correction for errors in the submitted amount; making an appropriate adjustment for any acceptable variations, deviations, discounts or other alternative offers not reflected in the submitted amount and making an allowance for any acceptable varied times for completion offered in the alternative tenders.

The "accepted contract amount" is defined as "the amount accepted in the letter of acceptance for the execution and completion of the works and the remedying of any defects".

3.2.3 Letter of tender

The letter of tender is the tenderer's written offer to execute the work in accordance with the other contract documents, and for such sum as may be ascertained in accordance with the conditions of contract and within the time for completion. Normally the tenderer must submit a letter of tender which complies fully with the specification, but in addition he may also submit an alternative offer. A sample letter of tender and appendix thereto can be found in Figure 3.6 and Figure 3.7, respectively.

Letter of Tender[22]

NAME OF CONTRACT:
TO:

We have examined the Conditions of Contract, Specification, Drawings, Bill of Quantities, the other Schedules, the attached Appendix and Addenda Nos. _____ for the execution of the above-named Works. We offer to execute and complete the Works and remedy any defects therein in conformity with this Tender which includes all these documents, for the sum of (in currencies of payment)

_____

or such other sum as may be determined in accordance with the Conditions of Contract.

We accept your suggestions for the appointment of DAB, as set out in Schedule.

[*We have completed the Schedule by adding our suggestions for the other Member of the DAB but these suggestions are not conditions of this offer*].

We agree to abide by this Tender until _____ and it shall remain binding upon us and may be accepted at any time before that date. We acknowledge that the Appendix forms part of this Letter of Tender.

If this offer is accepted, we will provide the specified Performance Security, commence the Works as soon as is reasonably practicable after the Commencement Date and complete the whole of the Works in accordance with the above-named documents within the Time for Completion.

Unless and until a formal Agreement is prepared and executed this Letter of Tender, together with your written acceptance thereof, shall constitute a binding contract between us.

We understand that you are not bound to accept the lowest or any tender you may receive.

Signature _____ in the capacity of _____

duly authorized to sign tenders for and on behalf of _____
Address:_____
Date:_____

Figure 3.6 Letter of Tender

Appendix to Tender[22]

[Note: with the exception of the items for which the Employer's requirements have been inserted, the following information must be completed before the tender is submitted]

Item	Sub-Clause	Data
Employer's name and address	1.1.2.2&1.3	
Contractor's name and address	1.1.2.3&1.3	
Engineer's name and address	1.1.2.4&1.3	
Time for Completion of the Works	1.1.3.3	Days
Defects Notification Period	1.1.3.7	365 days
Electronic transmission systems	1.3	
Governing Law	1.4	
Ruling language	1.4	
Language for communications	1.4	
Time for access to the Site	2.1	Days after commencement Date
Amount of Performance Security	4.2	% of the Accepted Contract Amount, in the currencies and proportions in which the Contract Price is payable
Normal working hours	6.5	
Delay damages for the Works	8.7 & 14.15(b)	% of the final Contract Price per day, in the currencies and proportions in which the Contract Price is payable
Maximum amount of delay damages	8.7	% of the final Contract Price
If there are Provisional Sums: Percentage for adjustment of Provisional Sums	13.5(b)	%
If Sub-Clause 13.8 *applies*: Adjustments for Changes in Cost:	13.8	for payments each month[Year] [currency]

Table(s) of adjustment data

Coefficient:	Country of origin:	Source of index	Value on stated date(s)*	
scope of index	Currency of index	Title/definition	Value	Date
$a=0.1$ Fixed				
$b=$ Labour				
$c=$				
$d=$				
$e=$				

* These values and dates confirm the definition of each index, but do not define Base Date indices.

Total advance payment	14.2	% of the Accepted Contract Amount
Number and timing of instalments	14.2	
Currencies and proportions	14.2	% in

Start repayment of advance payment	14.2(a)	When payments are	% of the Accepted Contract Amount less Provisional Sums
Repayment amortization of advance payment	14.2(b)		%
Percentage of retention	14.3		%
Unit of Retention Money	14.3		% of the Accepted Contract Amount
If Sub-Clause 14.5 applies:	14.5(b)		[list]
Plant and Materials for payment when shipped on route to the Site			[list]
Plant and Materials for payment when delivered to the Site	14.5(c)		[list] [list]
Minimum amount of Interim Payment Certificates	14.6		% of the Accepted Contract Amount

If payments are only to be made in a currency/currencies named on the first page of the Letter of Tender:

Currency/currencies of payment	14.15		as named in the Letter of Tender

If some payments are to be made in a currency/currencies not named on the first page of the Letter of Tender:

Currency/currencies of payment   14.15

Currency Unit   Percentage payable in the Currency   Rate of exchange: number of Local per unit of Foreign

Local: [name]
Foreign: [name]

Periods for submission of insurance:

(a) evidence of insurance	18.1		Days
(b) relevant policies	18.1		Days
Maximum amount of deductibles for Insurance of the Employer's risks	18.2(d)		
Minimum amount of third party insurance	18.3		
Date by which DAB shall be appointed	20.2	*Either*: One sole member/adjudicator *Or*: A DAB of three members	
Appointment(if not agreed) to be made by	20.3	The President of FIDIC or a person Appointed by the President	

If there are Sections:

Definition of Sections:

Description	Time for Completion	Delay Damages
(Sub-Clause 1.1.5.6)	(Sub-Clause 1.1.3.3)	(Sub-Clause 8.7)

[*In the above Appendix, the text shown in italics is intended to assist the drafter of a particular contract by providing guidance on which provisions are relevant to the particular contract. This italicized text should not be included in the tender documents, as it will generally appear inappropriate to tenderers.*]

Initials of signatory of Tender

Figure 3.7   Appendix to Tender

3.2.4　Form of tender security

The tender security is a document, as usually required in the tender documents, accompanying a tender or proposal as assurance that the tenderer (1) will not withdraw the tender within the period specified therein for acceptance, and (2) will execute a written contract and furnish such securities as may be required within the period specified in the tender (unless a longer period is allowed) after receipt of the specified forms.

If the contractor fails to discharge his obligations under the tender documents satisfactorily, a bank, insurance company or other acceptable guarantor undertakes to pay a specified sum to the employer.

The sum is specified in the tender documents, affording the employer reasonable protection against irresponsible tenders, but it shall not be set so high as to discourage tenderers. The tender security, at the tenderer's option, shall be in the form of a certified check, a letter of credit or a bank guarantee from a reputable bank. Tenderers shall be allowed to submit bank guarantee directly issued by a bank of their choice. Tender security shall remain valid for a specified period beyond the validity period for the tenders, in order to provide reasonable time for the employer to act if the security is to be called. Tender security shall be released to the unsuccessful tenderers once it is determined that they will not be awarded a contract. The amount of the security is usually 10 per cent of the tender price and the contractor is almost certain to include the cost of providing the security in his tender. The tender security usually reads like in Figure 3.8.

A Form of Tender Security[22]

Brief description of Contract
Name and address of Beneficiary
(whom the tender documents define as the Employer).

We have been informed that　　(hereinafter called the "Principal") is submitting an offer for such Contract in response to your invitation, and that the conditions of your invitation (the conditions of invitation, which are set out in a document entitled Instruction to Tenderers) require his offer to be supported by a tender security.

At the request of the Principal, we (*name of bank*)　　hereby irrevocably undertake to pay you, the Beneficiary/Employer, any sum or sums not exceeding in total the amount of(say:　　)upon receipt by us of your demand in writing and your written statement (in the demand) stating that:

(a) the Principal has, without your agreement, withdrawn his offer after the latest time specified for its submission and before the expiry of its period of validity, or
(b) the Principal has refused to accept the correction of errors in his offer in accordance with such conditions of invitation, or
(c) you awarded the Contract to the Principal and he has failed to comply with sub-clause 1.6 of the conditions of the Contract, or

(d) you awarded the Contract to the Principal and he has failed to comply with sub-clause 4.2 of the conditions of the Contract.

Any demand for payment must contain your signature(s) which must be authenticated by your bankers or by a notary public. The authenticated demand and statement must be received by us at this office on or before (*the date 35 days after the expiry of the validity of the Letter of Tender*), when this guarantee shall expire and shall be returned to us.

This guarantee is subject to the Uniform Rules for Demand Guarantees, published as number 458 by the International Chamber of Commerce, except as stated above.

Date _____ Signature(s)

Figure 3.8 A Form of Tender Security

3.2.5 Conditions of contract

Contracts consist of various statements, promises, and stipulations, written or oral, which are grouped together under the word "terms". It is the terms of a contract that specify the extent of each party's duties, obligations, and rights. The conditions of contract define the terms under which the work is to be carried out, the relationship between the employer and the contractor, the powers of the A/E and the terms of payment.

3.2.5.1 Express and implicit terms[20]

(1) Express terms

Statements made by each of the parties, which are intended to be incorporated into the contract, are known as express terms. It may be necessary, in the event of a dispute, for the court to decide what was said or written by the parties.

Where the contract is wholly oral and the terms are in dispute, it will be a matter to be decided by the DAB (the person or three persons appointed under sub-clause 20.2 or sub-clause 20.3 of the *Red Book*), the arbitrators, or the court from evidence presented to it. However, problems may arise where evidence is conflicting and difficult to substantiate. Therefore, in the case of building contracts, where there are generally many terms, it is essential to have everything in writing to prevent disputes of this nature arising. Where the contract is in writing it is usually obvious what the parties have written, although there may be problems of interpretation, arising from ambiguity, which the court has to resolve. Also, in most cases where the terms of a contract are in writing, the court may refuse to allow oral evidence to be admitted, if it has the effect of adding to, varying, or contradicting the written agreement.

(2) Implicit terms

In addition to express terms inserted by the parties a contract may contain and be subject to implicit terms. Such terms originate from custom or statute. Also, a term may be implied by the court, where it is necessary to achieve the result which the parties obviously intended the contract to have.

3.2.5.2 Terms, condition and warranty of contract[20]

Not all the express terms created by a contract are of equal importance, so various words have been used to classify the terms of a contract.

(1) Condition

"Condition" is the word used to denote a vital term that goes to the root of the contract. A condition is so essential to the nature of the contract that its non-performance may fairly be considered by the other party as a substantial failure to perform the contract at all. So if certain conditions are not carried out in a contract, the injured party may repudiate (refuse to accept the contract) and sue for damages. For example, in the case of a construction contract, if the employer refuses to pay the contractor for work that has been properly executed (carried out), the contractor can take the employer to a court of law, sue for damages and repudiate the contract.

(2) Warranty

In contrast, however, warranty is a term considered to be subsidiary to the main purpose of the contract, an obligation which, although it must be performed, is not so vital as is a condition. A warranty may be defined as a term, the breach of which may give rise to a right to claim damages but not a right to reject the contract and treat it as repudiated. For example, in the case of construction contracts, the contractor may fail to provide the A/E with a site office. In such a case the employer cannot break the contract, but will be entitled to recover the cost of providing a site office from the contractor.

The imposition of conditions of contract which are biased in favour of either party can be uneconomical.

3.2.5.3 Standard forms of contract

For a number of reasons it is now common practice to use standard forms of contract in procurement of building and/or engineering works worldwide:[20]

(1) All the parties to a contract become familiar with their terms as a result of common usage, and therefore gain a greater understanding of the contract and there is less likelihood of disputes arising over the interpretation of contract clauses.

(2) The parties to a contract are clearly made aware of their duties, obligations and rights, as they are all written in one set of well-referenced documents. The employer is fully aware of the arrangement for payment to the contractor on certification by the A/E. The A/E has extensive powers of control and direction of the works as described in the conditions of contract, but he must exercise his duties reasonably. The duties and obligations of the contractor are well documented. It is unwise to make modifications to the standard conditions of contract as this can lead to uncertainties and disputes that are better avoided. Using a set of standard documents avoids the waste of time would result from having to

individually draft the many terms necessary for each new contract.

(3) In the event of disputes, standard forms of contract provide a means of settlement, without the need to refer the matter to a court of law. An arbitration agreement clause in the documents allows the parties to refer their disputes to the decision of an arbitrator.

(4) Standard forms of contract ensure comprehensiveness of coverage, better balance of rights or obligations between employer and contractor, general acceptability of its provisions, and savings in time and cost for tender preparation and review, leading to more economical prices.

On the other hand, however, there are some arguments against the use of standard forms of contract.

(1) Standard forms do not cover every event that might lead to a dispute for the reason that they do not take into account all the particulars of each individual contract.

(2) It is argued that the standard forms available give A/E too much power, especially in regard to variations.

3.2.5.4 General and particular conditions of contract

In each contract, the conditions of contract governing the rights and obligations of the parties will comprise the appropriate general conditions together with particular (special) conditions.

The general conditions of contract must be retained intact to facilitate its reading and interpretation. It will be necessary to prepare the particular conditions for a particular contract, taking account of the circumstances relevant to the particular contract. Any amendments and additions to the general conditions of contract, specific to the contract in hand, should be introduced in the conditions of particular application, or in the Appendix to Tender.

The conditions of particular application take precedence over the general conditions of contract.

3.2.5.5 FIDIC conditions of contract[13]

FIDIC has long been renowned for its standard forms of contract, especially *the Red Book*, for use between employers and contractors.

The first edition of the *Red Book*, i.e. the *Conditions of Contract (International) for Works of Civil Engineering Construction* was published in August 1957. Perhaps because of its long title, in a very short time it became popularly known as the 'Red Book' (its cover was printed in red). It was prepared by the FIDIC and the International Federation of Building and Public Works, now known as the FIEC (International European Construction Federation).

The second edition of the *Red Book was* published in July 1969, when the document was approved and ratified by the International Federation of Asian and Western Pacific Contractors' Associations. A supplementary section containing *Conditions of Particular Application to Dredging and Reclamation Work* was then added as Part III. The second edition, however, included no changes in the text. A reprint of the second edition in 1973 added the approval and ratification by the Associated General Contractors of America and the Inter-American Federation of the Construction Industry.

However, the publication of the fifth edition of the ICE (Institution of Civil Engineers) Form in June 1973 in the United Kingdom provided an impetus for a further revision of the *Red Book*. This fifth edition of the ICE was criticised for the style, language and lack of clarity of the document, which obviously provided food for thought for those responsible for the *Red Book*, and so it was in March 1977 that the third edition of the *Red Book* was published, incorporating some significant changes.

The third edition remained unaltered and no amendments were issued until the fourth edition was published in September 1987, when major revisions were made which extended even to the title of the document.

The word "international" was deleted, inviting parties from all over the world to use the *Red Book* not only in international contracts but also in domestic contracts.

Part II of the *Red Book* which is referred to as the "*Conditions of Particular Application*" was expanded and produced in a separate booklet. It is linked to Part I by the corresponding numbering of the clauses, so that Parts I and II together comprise the conditions governing the rights and obligations of the parties. Part II must be specifically drafted to suit each individual contract. To assist in the preparation of Part II, explanatory material and example clauses are included providing the parties with options for their use where appropriate.

In 1988, the fourth edition was reprinted with a number of editorial amendments which were identified at the end of the document. These amendments were of a very minor nature and did not affect the meaning of the relevant clauses but simply clarified their intention.

Later, in 1992, further amendments were introduced in a reprint of the fourth edition of the *Red Book*.

In November 1996, FIDIC published a document entitled "*Supplement to Fourth Edition* 1987 - *Conditions of Contract for Works of Civil Engineering Construction-Reprinted* 1992 *with Further Amendments*". It was intended to provide the user with alternative arrangements in three controversial areas of the *Red Book*, thus giving him a choice in the method to be used for: settlement of disputes; payment; and preventing delay in certification for the purpose of payments.

In addition to the *Red Book* FIDIC had also published:

Conditions of Contract for Electrical and Mechanical Works including Erection on Site (1987) (the *Yellow Book*);

Conditions of Contract for Design-Build and Turnkey (1995) (the *Orange Book*);

Conditions of Subcontract for Works of Civil Engineering Construction (1994) (the *Purple Book*);

Client/Consultant Model Services Agreement (1998) (the *White Book*).

During the use of the rainbow (red, yellow, orange, white) books, FIDIC noted that certain projects had fallen outside of the scope of the Books. Accordingly FIDIC had not only updated the standard forms but expanded the range, and published in September 1999 a suite of four new standard forms of contract to suit for the great majority of construction and plant installation projects around the world. This new suite comprises:

(a) *Conditions of Contract for Construction* for building and engineering works designed by the employer, referred to as the *Construction Contract* or simply "CONS" (referred to as the *Red Book* hereinafter in this book);

(b) *Conditions of Contract for Plant and Design-Build* for electrical and mechanical plant and for building and engineering works designed by the contractor, referred to as the *Plant and Design/Build Contract* or simply "P&DB" (the *Yellow Book*);

(c) *Conditions of Contract for EPC/Turnkey Projects*, referred to as the *EPC/Turnkey Contract* or simply "EPCT" (the *Silver Book*), and

(d) *Short Form of Contract*, referred to as the *Short Form* (the *Green Book*).

The *Red Book* is recommended for works designed by the employer or by his representative, the engineer. Under the usual arrangements for this type of contract, the contractor constructs the works in accordance with a design provided by the employer. However, the works may include some elements of contractor-designed civil, mechanical, electrical and/or construction works.

P&DB is recommended for the provision of electrical and/or mechanical plant, and for the design and execution of works. Under the usual arrangements for this type of contract, the contractor designs and provides, in accordance with the employer's requirements, plant and/or other works; which may include any combination of civil, mechanical, electrical and/or construction works.

EPCT is recommended where one entity takes total responsibility for the design and execution of an engineering project. Under the usual arrangements for this type of contract, the entity carries out all the Engineering, Procurement and Construction: providing a fully-equipped facility, ready for operation (at the "turn of the key"). This type of contract is usually negotiated between the parties.

Short Form is recommended for works of relatively small capital value. Depending on the type of work and the circumstances, this form may also be suitable for contracts of greater value, particularly for relatively simple or repetitive work or work of short duration. Under the usual arrangements for this type of contract, the contractor constructs the works in accordance with a design provided by the employer or by his representative (if any), but this form may also be suitable for a contract which includes, or wholly comprises, contractor-designed civil, mechanical, electrical and/or construction works.

It is inadvisable to make modifications to the FIDIC conditions of contract, as the legal result may often be different from what is intended. Where alterations to meet the needs of a particular project are unavoidable, the addition of special conditions is preferable to the modification of standard clauses; attention should always be drawn specifically to such amendments and additions. The standard conditions should not be altered or added to without appropriate legal or expert advice.

It is very interesting that the third edition of the *Red Book* is still in use in many parts of the world, such as Hong Kong, Singapore, Middle East countries including the United Arab Emirates (UAE), etc. . When the writer of this book visited Abu Dhabi in the UAE in August 2006 he found that the third edition of the *Red Book is incorporated by reference into* the tender documents for a reclamation project as follow:

Conditions of Contract

The applicable Conditions of Contract shall comprise the following:

1. Part I: General Conditions

The General Conditions of Contract shall be:

The *Conditions of Contract (International) for Works of Civil Engineering Construction*, Third Edition, 1977 prepared by FIDIC (The *Fédération Internationale des Ingénieurs-Conseils*)

P. O. Box 86

CH-1000 Lausane 12, Switzerland

Tel: 0041 21 654 4425

Fax: 0041 21 654 4417

www.fidic.org

from whom prints of the Conditions may be obtained.

2. Part II: Conditions of Particular Application

The Conditions of Particular Application which follow amplify and amend selected clauses of the General Conditions and include additional clauses. Such amplified, amended and additional clauses shall take precedence over the General Conditions.

The reluctance of the employers and/or contractors in the areas to use the subsequent editions of the *Red Book* may be explained in part by the familiarity of the employers, contractors and the courts with the third edition and its adequacy to cover their obligations and rights.

The writer of this book believes that the relationships created by a form of contract not only have legal nature but also economic and social nature. There is a complicated interaction between the organization of the construction industry and the contractual relationships which have significant impacts on the efficiency and effectiveness of the former, positively or negatively.

The contractual relationships, as a matter of fact, are determined by and formed in the economic, social and cultural contexts where the construction activities are carrying on. Like in the other sectors of economy it is the continuously improved division of labour that has been ever increasing the construction productivity. The improved division of labour in turn has been recognized by and formulated in the constantly updated forms of contract.

On the other hand, however, if the framework of responsibilities created by a form of contract is suitable for and can be accepted by the parties involved in the construction activities there will not be any necessity for change because changes are not free.

3.2.5.6 Common law basis of FIDIC forms of contract[13]

The *Red Book of Fourth Edition* originated through the ACE (the Association of Consulting Engineers in the United Kingdom) Form from the ICE (the Institution of Civil Engineers in the United Kingdom) Form which is based on legal concepts rooted in the common law system.

The legal framework of the *Red Book of Fourth Edition* is basically set out in its clauses 5, 26 and 70.

Clause 5 of the *Red Book of Fourth Edition* provides for the law of the contract (the applicable law of the contract) to be specified in Part II of the conditions. Where that law follows the common law system there would be little, if any, conflict of laws between the concepts of the *Red Book of Fourth Edition* and those under which the contract might have to be construed because of the fact that the *Red Book of Fourth Edition* is based on the common law concepts. In fact, in the early years after the *Red Book* was first introduced, it used to stipulate English law to govern these contracts whenever the engineer or the contractor was of British nationality.

It has become more usual and appropriate in recent years to specify as the applicable law of the contract the law of the country where the works is constructed. In such cases or where the contract follows a system of law other than the common law, it is inevitable that some conflict would occur between the applicable law of the contract and some of the legal concepts incorporated in the *Red Book of Fourth Edition*.

Clause 26 of the *Red Book of Fourth Edition* requires the contractor to conform in all respects with all national statutes, ordinances, laws, regulations and by-laws of any local or other duly constituted authority in connection with the execution and completion of the works and the remedying of any defects therein. It therefore requires conformity not only with the laws of the country where the works is located but also with the laws of any other country where part of the works is executed. It means that if the works or the supply of materials for such works is carried out in more than one country, then the laws of all these countries involved would have to be observed.

Clause 70 of the *Red Book of Fourth Edition* provides for the possibility of either introducing new, or making changes to already existing, national or state statutes, ordinances, decrees or other laws or any regulation or by-law of any local or other duly constituted authority in the country in which the works are being, or are to be, executed.

If such laws are introduced or if such changes are implemented in already existing laws causing additional or reduced costs to the contractor, then such additional or reduced costs must be determined by the engineer and accordingly, an adjustment should be made to the contract price. The matters covered by this clause are referred to in Part II of the Form.

In this connection, reference should also be made to clauses 71 and 72 which deal with changes imposed by the government of the country where the works are being or are to be executed in respect to currency restrictions or the rates of currency exchange.

3.2.5.7 Standard forms of contract for indigenous building and civil engineering works in China

There are other standard forms of contract used for building or civil engineering work in China. For example:

Conditions of Contract for works of building construction (GF-1999-0201) was published by MOC and the State Industrial and Commercial Administration Bureau on 24 December 1999, which comprises three parts, namely, Part I -Form of Agreement, Part II - General Conditions and Part III -Conditions of Particular Application. Part III shall prevail over Part II and comprise amendments, additions and details of Part II.

Conditions of Contract for works of hydraulic and water conservancy engineering construction (GF-2000-0208) published by the Ministry of Water Conservancy, the Ministry of Electricity and the State Industrial and Commercial Administration Bureau in September 1993 and subsequently revised twice in 1997 and 2000, respectively.

Model Tender documents for Domestic Procurement of Works of Highway Construction prepared and put into use by the Ministry of Communications in 1994 and revised in 1999. The conditions of contract included in the tender documents are notably similar to the *Red Book of Fourth Edition*.

3.2.6 Specification

The specification describes in detail the work to be executed, the character and quality of the materials and workmanship, and any special responsibilities of the contractor that are not covered by the conditions of contract. It may also lay down the order in which various portions of the work are to be executed, the methods to be adopted, and particulars of any facilities to be afforded to other contractors.

It is usually divided into two parts: the one dealing with general information and the general duties and obligations of the contractor not specifically covered by the conditions of contract, and the other specifying materials and workmanship requirements.

Precise and clear specifications are a prerequisite for tenderers to respond realistically and competitively to the requirements of the employer without qualifying or conditioning their tenders. In the context of international competitive tendering, the specifications must be drafted to permit the widest possible competition and, at the same time, present a clear statement of the required standards of materials, plant, other supplies, and workmanship to be provided. Only if this is done will the objectives of economy, efficiency, and equality in procurement be realized, responsiveness of tenders be ensured, and the subsequent task of tender evaluation facilitated. The specifications should require that all materials, plant, other supplies to be incorporated in the works are new, unused, of the most recent or current models, and incorporate all recent improvements in design and materials unless provided otherwise in the contract.

A clause setting out the scope of the works is often included at the beginning of the specifications, and it is customary to give a list of the drawings, description of access to the site, supply of electricity and water, provision of offices and mess facilities and statements regarding suspension of works during frost and bad weather, damage to existing services, details of borings, water levels and similar clauses.

Where the contractor is responsible for the design of any part of the permanent works, the extent of his obligations must be stated.

Most specifications are normally written specially by the employer or A/E to suit the contract for works in hand. There are no standard specifications for universal application in all sectors in all countries, but there are established principles and practices.

It is considerably advantageous to standardize "general specifications" for repetitive works in recognized public sectors, such as highways, ports, railways, urban housing, irrigation, and water supply, in the same country or region where similar conditions prevail. The general specification should cover all classes of workmanship, materials, and equipment commonly involved in construction, although not necessarily to be used in a particular works contract. Deletions or addenda should then adapt the general specifications to the particular works.

Care must be taken in drafting specifications to ensure that to avoid conflict with any of the provisions of the conditions of contract and they are not restrictive. In the specification of standards for materials, plant, other supplies, and workmanship, recognized international standards should be used as much as possible. Where other particular standards are used, whether national standards or other standards, the specification should state that materials, plant, other supplies, and workmanship meeting other authoritative standards, and which ensure substantially equal performance as the standards mentioned, will also be acceptable.

3.2.7 Drawings

The drawings are prepared by the A/E. They should clearly depict the nature and scope and give details of the contract works. For many reasons this is not always practicable but they should be as comprehensive and detailed as possible and tenderers must be given sufficient information to enable them to understand what is required so that they can submit properly considered tenders.

All drawings should contain ample descriptive and explanatory notes that should be legible and free of abbreviations, and be suitably referenced.

All available information should be provided relating to site conditions and in particular data regarding soils and groundwater. Drawings may incorporate or be accompanied by schedules, such as those recording details of steel reinforcement and manholes.

During the progress of the works, the A/E or his/her representative often finds it necessary to issue further drawings in amplification of those issued at the tender stage. Some may give details of components which were the subject of provisional sums or prime cost in the bill of quantities.

Provisional sum means a sum included in the contract and so designated in the bill of quantities for the work or services that can be identified but is of uncertain scope or cannot be detailed at the time the tender documents are issued. The sum may be used by, in whole or in part, or not at all, on the instructions of the A/E for the execution by the main contractor or a nominated subcontractor of any part of the works or for the supply of goods, materials, plant or services, or for contingencies. The contractor or the subcontractor shall be entitled to only such amounts in respect of the work, supply or contingencies to which such provisional sums relate as the A/E shall determine in accordance with the relevant clauses in the contract.

The work for which a provisional sum can be used for can be either "defined" or "undefined". "Defined work" is work which is not completely designed at the time the tender documents are issued but for which certain specified information can be given. "Undefined work" is work for which such specific information cannot be given.

Sometimes "provisional quantities" are given for sections or individual parts of the works where only the quantity is in doubt and it is desired to obtain competitive rates for valuing the work on measurement. Provisional quantities differ from provisional sums in that there is no contractual power for the A/E to nominate a subcontractor to undertake the work except by agreement with the contractor. However, the practice is not entirely satisfactory in that if the actual quantity is radically different from the provisional quantity, the rate may be inappropriate and unless it is treated as a variation there is no mechanism for substituting a new rate.

A prime cost sum is a sum of money provided in the bill of quantities for payment of the nominated subcontractors and nominated suppliers and is to be expended in favor of such persons as the A/E shall instruct. Normally, prime cost sums are incorporated into tender as cover for the cost of nominated work. In so doing the employer does not have to enter into a series of special contracts. After the initial nomination these specialists become subcontractors and suppliers to the contractor. In the adjustments, the prime cost sums would be deducted from the contract price and substituted by the final accounts of the specialist subcontractors and suppliers so nominated.

The prime cost covered by the provisional sums when the *Red Book* is used. Subclause 13.5 of the Red Book states: "…For each provisional sum, the Engineer may instruct: (b) plant, materials or services to be purchased by the contractor, from a nominated subcontractor or otherwise; and for which there shall be included in the contract price…"

3.2.8 Bill of quantities

This is a list of items giving brief identifying descriptions and estimated quantities of the work comprised in a contract. In conjunction with the other tender documents, it forms the basis on which tenders are obtained. When priced it affords a means of comparing tenders. When the contract has been entered into, the rates in the priced bill of quantities are applied to assess the value of the actual quantities of work carried out. The bill of quantities should be prepared and measurements made in accordance with the procedure set out in the method of measurement specified in tender documents.

Its objectives are:
(a) to provide sufficient information on the quantities of the works to be performed to enable tenders to be prepared efficiently and accurately and
(b) when a contract has been entered into, to provide a priced bill of quantities for use in the periodic valuation of works executed.

A bill of quantities contains the following components.

☐ A list of principal quantities enables contractors tendering for the work to make a quick assessment of the scope of the contract.

☐ A preamble to the bill of quantities draws the attention of contractors to the method of measurement used in preparing the bill of quantities, together with any amendments that have been made to the method of measurement.

☐ Where work cannot be adequately defined at the bill preparation stage, the best method is to value it on a daywork basis. One approach is to list craft operatives and labourers and the principal items of materials and plant, with a provisional quantity inserted against each of them. Each item is then priced by the contractor and he will be paid at the rates in the schedule for the actual work done and plant and materials used.

☐ It is usual for a bill of quantities to be subdivided into separate parts or bills relating to the various sections of work. For example,

Bill No. 1- Preliminaries/General items

Bill No. 2- Preambles

Bill No. 3- Excavator

Bill No. 4- Concretor

Bill No. 5- Bricklayer

Bill No. 6- Drainlayer

Bill No. 7- Asphalter and Roofer
Bill No. 8- Capenter and Joiner
Bill No. 9- Ironmonger
Bill No. 10- Steel and Metalworker
Bill No. 11- Plasterer
Bill No. 12- Plumber
Bill No. 13- Glazier
Bill No. 14- Painter
Bill No. 15- External Works
Bill No. 16- Prime cost and Provisional Sums

☐ A grand summary at the end of the bill of quantities tabulates the totals of each bill. In addition, provision is normally made for a contingency item to be added to the amounts brought forward from various bills. Provision is also made for an adjustment item that follows the contingency item, whereby the contractor can adjust his tender figure if he wishes to do so, after full consideration of the work involved, the tendering climate and any other relevant factors. The total of these items constitutes the contractor's tender for the contract. A sample bill of quantities is given in Figure 3.9~Figure 3.17.

Bill No. 1: General Items

Item No.	Description	Unit	Quantity	Rate	Amount
101	Performance Bond/Guarantee	Sum	Item		
102	Insurance of the Works	Sum	Item		
103	Insurance of Contractor's Equipment	Sum	Item		
104	Third-Party insurance	Sum	Item		
105	Allow for maintenance of Works for 12 months after completion	Month	12		
106	etc.				
112	Provide and equip Engineer's offices	Nr	2		
113	Maintain Engineer's offices for 24 months, including services	Month	24		
114	etc.				
121	Provide diversion road	Sum	Item		
122	Provide for traffic control and maintenance of diversion road	month	24		
123	etc.				
132	Provide for cleaning the Site on completion, etc.	sum	Item		

Total for Bill No. 1
(carried forward to Summary, p_____)

Figure 3.9 Bill No. 1: General Items[22]

Bill No. 2: Earthworks

Item No.	Description	Unit	Quantity	Rate	Amount
201	Excavate top soil to maximum depth 25cm and stockpile for reuse, maximum haul distance 1km	m^3	95,000		
202	Excavate topsoil to maximum depth 25cm~50cm, and dispose	m^3	15,000		
203	etc.				
206	Excavate fill material from cuttings or approved borrow pits, haul up to 1km, deposit, shape, and compact to fill	m^3	258,000		
207	Excavate rock in cuttings and dispose, any depth	m^3	25,000		
208	etc.				

Total for Bill No. 2
(carried forward to Summary, p _____)

Figure 3.10 Bill No. 2: Earthworks[22]

Bill No. 3: Culverts and Bridges

Item No.	Description	Unit	Quantity	Rate	Amount
301	Excavate in all materials other than rock from ground level to underside of foundations, maximum depth 5m, and dispose	m^3	18,500		
302	Excavate in all material other than rock, depth 5m to 7.5m	m^3	2,500		
303	Provisional Item As Item 302, depth 7.5m to 10m	m^3	500		
304	etc.				
311	Concrete class B in abutments	m^3	18,500		
312	etc.				
318	Mild steel reinforcement in abutments and piers up to 20 mm diameter	t	370		
319	etc.				

Total for Bill No. 3
(carried forward to Summary, p _____)

Figure 3.11 Bill No. 3: Culverts and Bridges[22]

Schedule of Daywork Rates: 1. Labor

Item No.	Description	Unit	Nominal quantity	Rate	Extended amount
D100	Ganger	Hour	500		
D101	laborer	Hour	5,000		
D102	Bricklayer	Hour	500		
D103	Mason	Hour	500		
D104	Carpenter	Hour	500		
D105	Steel work Erector	Hour	500		
D106	etc.	Hour			
D113	Driver for vehicle up to 10 tons	Hour	1,000		
D114	Operator for excavator, dragline, Shovel, or crane	Hour	300		
D115	Operator for tractor with dozer blade or ripper	Hour	500		
D116	etc.	Hour			

Subtotal

| D122 | Allow _____ percent of Subtotal for Contractor's overhead, Profit, etc., in accordance with paragraph 3 (b) above |

Total for Daywork: Labor
(carried forward to Daywork Summary, p _____)

Figure 3.12 Schedule of Daywork Rates: 1. Labor[22]

Schedule of Daywork Rates: 2. Materials

Item No.	Description	Unit	Nominal quantity	Rate	Extended amount
D201	Cement, ordinary Portland, or equivalent in bags	t	200		
D202	Mild steel reinforcing bar up to 16 mm diameter to BS 4449 or equivalent	t	100		
D203	Fine aggregate for concrete as specified in Clause	m^3	1,000		
D204	etc.				
D222	Gelignite (Nobel Special Gelatine 60%, or equivalent) including caps, fuse, wire, and requisite accessories	t	10		

Subtotal

| D122 | Allow _____ percent of Subtotal for Contractor's overhead, Profit, etc., in accordance with paragraph 3 (b) above |

Total for Daywork: Materials
(carried forward to Daywork Summary, p _____)

Figure 3.13 Schedule of Daywork Rates: 2. Materials[22]

Schedule of Daywork Rates: 3. Contractor's Equipment

Item No.	Description	Nominal quantity (hours)	Basic hourly rental rate	Extended amount
D301	Excavator, face, shovel, or dragline:			
1	Up to and including 1 m^3	500		
2	Over 1 m^3 to 2 m^3	400		
3	Over 2 m^3	100		
D302	Tractor, including bull or angle dozer:			
1	Up to and including 150 kW	500		
2	Over 150 kW to 200 kW	400		
3	Over 200 kW to 250 kW	200		
D303	Tractor with ripper:			
1	Up to and including 200 kW	400		
2	Over 200 kW to 250 kW	200		
D304	etc.			

Total for Daywork: Contractor's Equipment
(carried forward to Daywork Summary, p _____)

Figure 3.14 Schedule of Daywork Rates: 3. Contractor's Equipment[22]

Daywork Summary

	Amount ( )	% Foreign
1. Total for Daywork: Labor		
2. Total for Daywork: Materials		
3. Total for Daywork: Contractor's Equipment		
Total for Daywork (Provisional Sum) (carried forward to Tender Summary, p _____)		

Figure 3.15 Daywork Summary[22]

Summary of Specified Provisional Sums

Bill no.	Item no.	Description	Amount
1			
2	2.8	Supply and install equipment in pumping station	1,250,000
3			
4	4.32	Provide for ventilation system in subway tunnel	3,500,000
Etc.			

Total for Specified Provisional Sums 4,750,000
(carried forward to Grand Summary (B), p)

Figure 3.16 Summary of Specified Provisional Sums[22]

Grand Summary

General Summary	Page	Amount
Bill No. 1: Preliminary Items		
Bill No. 2: Earthworks		
Bill No. 3: Drainage Structures		
etc.		
Total for Daywork (Provisional Sum)		
Subtotal of Bills	(A)	
Specified Provisional Sums in included in subtotal of bills	(B)	4,750,000
Total of Bills Less Specified Provisional Sums (A-B)	(C)	
Add Provisional Sum^a for Contingency Allowance	(D)	[sum]
Tender Price (A+D) (Carried forward to Form of Tender)	(E)	

Figure 3.17　Grand Summary[22]

3.2.9　Form of agreement

The form of agreement, if any and when completed, is a legal undertaking entered into between the employer and the contractor for the execution of the work in accordance with the other contract documents. It is not essential to have a form of agreement if there is a written acceptance by the employer of the tender submitted by the contractor. A sample form of agreement is shown in Figure 3.18.

Contract Agreement

This Agreement made the _____ day of _____ 20

Between _____ of _____ (hereinafter called "the Employer") of the one part

and _____ of _____ (hereinafter called "the Contractor") of the other part

　　Whereas the Employer is desirous that the Works known as _____ should be executed by the Contractor, and has accepted a Tender by the Contractor for the execution and completion of these Works and the remedying of any defects therein

The Employer and the Contractor agree as follows:

1. In this Agreement words and expressions shall have the same meanings as are respectively assigned to them in the Conditions of Contract hereinafter referred to.
2. The following documents shall be deemed to form and be read and construed as part of this Agreement:
(a) The Letter of Acceptance dated
(b) The Letter of Tender dated
(c) The Addenda nos

(d) The Conditions of Contract
(e) The Specification
(f) The Drawings, and
(g) The completed Schedules.

3. In consideration of the payments to be made by the Employer to the Contractor as hereinafter mentioned the Contractor hereby covenants with the Employer to execute and complete the Works and remedy any defects therein, in conformity with the provisions of the Contract.

4. The Employer hereby covenants to pay the Contractor in consideration of the execution and completion of the works and the remedying of defects therein, the Contract Price at the times and in the manner prescribed by the Contract.

In Witness whereof the parties hereto have caused this Agreement to be executed the day and year first before written in accordance with their respective laws.

SIGNED by:	SIGNED by:
For and on behalf of the Employer in the presence of	For and on behalf of the contractor in the presence of
Witness:	Witness:
Name:	Name:
Address:	Address:
Date:	Date:

Figure 3.18　Conthelt. Agreement[22]

3.2.10　Performance security/bond

The performance security/bond is a document whereby a bank, insurance company or other acceptable guarantor undertakes to pay a specified sum if the contractor fails to discharge his obligations satisfactorily. It is not always required by the employer. Figure 3.19 shows a sample performance security.

A Form of Performance Security-Demand Guarantee

Brief description of Contract _____

Name and address of Beneficiary _____

(whom the Contract defines as the Employer).

We have been informed that _____ (hereinafter called the "Principal") is your contractor under such Contract, which requires him to obtain a performance security.

At the request of the Principal, we (*name of bank*) hereby irrevocably undertake to pay you, the Beneficiary/Employer, any sum or sums not exceeding in total the amount of (the "guaranteed amount", say:____) upon receipt by us of your demand in writing and your written statement stating:

(a)that the Principal is in breach of his obligation(s) under the Contract, and

(b)the respect in which the Principal is in breach.

[Following the receipt by us of an authenticated copy of the taking-over certificate for the whole of the works under clause 10 of the conditions of the Contract, such guaranteed amount shall be reduced by ____ % and we shall promptly notify you that we have received such certificate and have reduced the guaranteed amount accordingly.][1]

Any demand for payment must contain your [minister's/directors'] signature(s) which must be authenticated by your bankers or by a notary public. The authenticated demand and statement must be received by us at this office on or before (*the date 70 days after the expected expiry of the Defects Notification Period for the Works*) (the "expiry date"), when this guarantee shall expire and shall be returned to us.

We have been informed that the Beneficiary may require the Principal to extend this guarantee if the performance certificate under the Contract has not been issued by the date 28 days prior to such expiry date. We undertake to pay you such guaranteed amount upon receipt by us, within such period of 28 days, of your demand in writing and your written statement that the performance certificate has not been issued, for reasons attributable to the Principal, and that this guarantee has not been extended.

This guarantee shall be governed by the laws of ____ and shall be subject to the Uniform Rules for Demand Guarantees, published as number 458 by the International Chamber of Commerce, except as stated above.

Date _____ Signature(s)

[1] *When writing the tender documents, the writer should ascertain whether to include the optional text shown in parentheses[].*

Figure 3.19 Form of Performance Security-Demand Guarantee[22]

Words, Phrases and Expressions 3.2

[1] estimate *vt*. 估算(An assessment of the likely quantitative result. Usually applied to project costs and duration and should always include some indication of accuracy (e.g., +or -15%). Usually used with a modifier (e.g., preliminary, conceptual, feasibility). Some application areas have specific modifiers that imply pre-set accuracy ranges (e.g., order of magnitude estimate, budget estimate, and definitive estimate in construction). 对可能的结果所做的一种估计。通常用于项目费用和持续时间，且总是指出估价的准确度大小（例如+15%或-15%）。使用时一般都带有修饰语（例如，初步（设计）估算，概念（设计）估算，可行性研究估算）。有些应用领域有具体的修饰语，指明事先确定的准确度范围（例如，数量级估算、预算估算、施工预算）。

[2] variation *n*. 变更

[3] discharge *vt*. 履行(perform (a duty, obligation))

[4]　successful tenderer 中标的投标人,中标人,中标者
[5]　form of agreement 协议书格式
[6]　performance bond 履约保函
[7]　execute *vt*. 办理(签字盖章)手续,使……产生法律约束力 (make legally binding)
[8]　bar bending schedules 钢筋(加工)表
[9]　general description of the arrangements 施工总说明
[10]　methods of construction 施工方法
[11]　produce *vt*. 提交,呈递,出具（put or bring forward to be looked at or examined）
　　　production of tender and/or performance bonds 提交标书和履约保函
[12]　selective tendering 选择性招标,邀请招标,邀标
[13]　context *n*. 总体关系,(因果)背景,(因果)环境,局势,内外联系,形势(the whole situation, background, or environment relevant to a particular event, personality, etc.)
[14]　Letter of Acceptance 中标通知书,中标函
[15]　Evaluated Tender Amount 评标价(纠正标书中计算错误、将不同货币换算成投标货币等之后的标价)
[16]　Letter of Tender 投标书,投标函
[17]　Tender Amount 标价
[18]　Accepted Tender Amount(由招标人)接受的标价
[19]　Accepted Contract Amount(由招标人)接受的合同价,中标合同价
[20]　Appendix to Tender 标书附录
[21]　Turnkey 转动钥匙
[22]　Turnkey contract 交钥匙合同(The term sometimes used to describe a contract where the contractor agree to engineer, procure and construct (EPC) a facility and complete it to the point of readiness for operation or occupancy. Alternatively such contracts are called "Package deal" contracts. They have become more accepted in the industrial projects. The term has no precise legal meaning and its use is best avoided. The alleged advantages of such contracts are project cost, coordination and speed. Against this must be set the substantial disadvantage that the employer is sometimes deprived of an impartial third-party check. "Package deal" contracts are most suitable for specialist engineering fields where companies possessing highly developed expertise may offer such proposals as the only access to that expertise. The terms "design-build", "turnkey" and "EPC" are often used interchangeably although there are some minor differences among the three systems of delivery of services. For example, turnkey often is used in situations where the contractor is not only responsible for design and construction but also for financing the facility. 有些时候这一术语用于说明承包商承担设施的设计与施工,以及采购任务,交付之时就能使用或占用的合同。此类合同也可以称为"成套交易"合同。工业项目中这类合同开

始多起来。但这一术语法律意义不清楚，因此最好避免使用。据说，这类合同的好处是省钱、便于协调、速度快。反对者自然列举其致命缺点：有些时候，剥夺了业主请第三方进行公正检查的机会。"成套交易"合同最适合于专业工程行业。这些行业的公司掌握非常成熟的专业知识，要想获得这种专业知识，唯一的办法就是看这些公司是否愿意提供之。"设计—建造"、"交钥匙"与"设计—采购—施工"这三个术语经常混用。这三种服务方式之间有些差别。例如，当承包商不仅负责设施的设计与施工，还负责筹集资金时，就用"交钥匙"这个术语。）

[23] Employer's requirements 业主要求说明书
[24] electronic transmission systems 电子传送办法
[25] Governing Law 适用（于合同的）法（律）
[26] Ruling language 裁决语言
[27] Delay damages for the works 工程误期赔偿费
[28] provisional sums 暂定金额
[29] base date 基准日期（"Base Date" means the date 28 days prior to the deadline for submission of tenders.）
[30] advance (payment)（业主在开工时暂时借给承包商的）预付款
[31] retention (money) 保留金（generally a percentage of the measured value of the work up to a stated limit）
[32] interim payment certificates 期中付款证书
[33] percentage payable in the currency（应当）用该种货币支付的百分比，此处 payable 是 percentage 的后置定语。payable 的这种用法在合同条件中经常使用。
[34] rate of exchange 汇率
[35] third party insurance 第三方（责任）保险
[36] certified check 保兑支票
[37] letter of credit 信用证（A document issued by a bank, usually in connection with international trade, whereby the bank replaces the buyer as the paying party. The exporter is basing his risk of getting paid on the bank rather than on the importer. The bank will have to be reimbursed by the importer. 银行开出的证件，一般用于国际贸易。根据该文件银行代替买方作为付款人。出口方将从该行，而不是从进口的买方取得货款。这样，开出信用证的银行就得从进口方索回相应的款项。）
[38] bank guarantee 银行保证书
[39] principal n. 委托人，本人（person for whom another acts a agent in business）
[40] validity n. 有效 (state of being valid)
[41] period of validity 有效期
[42] notary public 公证
[43] terms of payment 支付条件
[44] condition （合同中的）基本条款
[45] warranty （合同中的）保证条款

【46】 one set of well-referenced documents 彼此配合良好的一套文件
【47】 general conditions 一般条件,通用条件
【48】 particular (special) conditions 具体条件,专用条件
【49】 prime cost sums 指定费用额
【50】 provisional quantities 暂定(工作)量
【51】 time for completion 竣工时间
【52】 formulate *v.t.* (formulated, formulating) (经过深思熟虑之后,正式)提出 (to devise or develop, as a method, etc. and then put into a systematized statement, i.e. express clearly and exactly)
【53】 nominated subcontractor 指定分包商
【54】 alternative tenders 可供(招标人)选择的标书,投标人在招标文件要求之外另外提出供招标人选择的标书
【55】 preamble (工程量清单的)前言
【56】 daywork 计日工作,零星工作,点工
【57】 preliminaries 开办费,开工准备费,开工动员费
【58】 general items 一般事项费用,待摊费用
【59】 summary 汇总表
【60】 grand summary 最终汇总表
【61】 contingency item 不可预见事项
【62】 allow for maintenance of works for 12 months after completion 为工程竣工后维护十二个月准备的费用额
【63】 provide *vt.* 设置 (make available (*sth.* needed, wanted)) *provide diversion road* 修筑临时道路
【64】 provide (for) *vi.* 预备,准备 (take measures with due foresight) *provide for traffic control and maintenance of diversion road* 为临时道路的保养与交通管制做准备
【65】 performance certificate 完工证书
【66】 taking-over certificate 接收证书
【67】 call *vt.* 要求支付 (demand payment of a sum of money)
【68】 In consideration of 为了报答……
【69】 be subject to 服从……,受制于……,受……制约 (owing obedience (to))

Notes 3.2

[1] 疑难词语解释

1) When the employer has accepted a successful tenderer as the contractor, the most part of the tender documents will become important components of the contract documents and the basis for the contractor to execute, complete the contract works and remedy defects therein.

【译文】 当业主将中标人选定为承包商时,招标文件的大部分就变成了合同文件的重要部

分，承包商将据此施工、完成该合同工程并弥补其中所有缺陷。

2) Tender documents form the basis on which a tenderer, that is the potential building or civil engineering contractor, will prepare his tender.

【译文】 招标文件将成为投标人，即有意的建筑或土木工程承包商编制标书的根据。

3) If the contractor fails to discharge his obligations under the tender documents satisfactorily, a bank, insurance company or other acceptable guarantor undertakes to pay a specified sum to the employer.

【译文】 若承包商未令人满意地履行招标文件规定的义务，银行，保险公司，或其他（业主所）认可的保证人承诺向业主支付一笔规定的数额。

【解释】 注意，当 discharge 的直接宾语是 contract 时，discharge 就是"解除合同，不再履行"（release from performing a contract）的意思，详细情况见第 8 章。另外，在用 discharge 表达"解除某人的责任、义务（relieve *sb*. of a duty, obligation)"时，必须加用介词 from。例如，*The contractor shall be discharged from the contract when he has satisfactorily met all the requirements set out in it*. 在承包商令人满意地满足了合同中载明的所有要求之后，就应解除之。

4) Unless and until a formal Agreement is prepared and executed this Letter of Tender, together with your written acceptance thereof, shall constitute a binding contract between us.

【译文】 在正式的协议书拟定并签字盖章生效之前，本投标函连同贵方对此的书面接受应成为约束我们双方的合同。

5) The tenderer will execute a written contract and furnish such securities as may be required within the period specified in the tender after receipt of the specified forms.

【译文】 投标人愿意签署一份书面合同并在标书规定的收到规定格式后的时间内提交标书可能要求的上述保证书。

6) In the context of procurement of works, as an intransitive verb, "tender" means making an offer (to carry out work, supply goods, etc.) at a stated price.

【译文】 就工程采购而言，*tender* 当及物动词用时，意思是"主动表示按照说明的价格完成工程，供应货物等"。

7) Tender security shall remain valid for a specified period beyond the validity period for the tenders, in order to provide reasonable time for the employer to act if the security is to be called.

【译文】 为了在提出支付投标保证书中规定的金额时让业主有合理的准备时间，投标保证应当在标书有效期满之后一段规定的时间内继续有效。

8) The employer is fully aware of the arrangement for payment to the contractor on certification by the A/E.

【译文】 业主完全知道接到建筑师/工程师签发的支付证书后要安排向承包商支付款项。

9) In consideration of the payments to be made by the Employer to the Contractor as hereinafter mentioned the Contractor hereby covenants with the Employer to execute and complete the Works and remedy any defects therein, in conformity with the provisions of

103

the Contract. The Employer hereby covenants to pay the Contractor in consideration of the execution and completion of the works and the remedying of defects therein, the Contract Price at the times and in the manner prescribed by the Contract.

【译文】 作为如下文所提及的雇主支付给承包商的各款项的报酬,承包商特此与雇主签约,保证按照合同的诸项规定,实施、完成本工程并修补其中任何缺陷。雇主在此立约保证按照合同规定的时间和方式,向承包商支付合同价,作为本工程实施、完成并修补其中任何缺陷的报酬。

10) This guarantee is subject to the Uniform Rules for Demand Guarantees, published as number 458 by the International Chamber of Commerce, except as stated above.

【译文】 除了上述情况以外,本保证书遵守国际商会公布的458号文件"见索即付保证书统一规则"。

[2] general items 与 preliminaries 的区别

【解释】 preliminaries 比 general items 具体,主要指开工准备工作需要的费用,而 general items 泛指需要分摊到各个具体工作上去的费用。另外,preliminaries 这个词对于外行人过于生癖,所以近年来开始用 general items 代替 preliminaries。

3.3 Inviting for tenders

The employer who decides to have work executed by contract may select the contractor by open tendering or selective tendering. Occasionally he may prefer to negotiate a contract with a particular contractor. This procedure does not have the benefits of competition, but it may be appropriate in cases of urgency, or where the selected contractor's expertise is unique and the firm is well known to the employer.

Occasionally tendering takes place in two stages. During the first stage selected tenderers submit tenders based on an outline design or, in the case of an all-in contract, on design proposals. The second stage comprises tenders on a preferred design, or the development of a detailed design and negotiation of a contract with a tenderer.

3.3.1 Open tendering

Competitive tendering is another name of open tendering. Under open tendering, tenders are invited by public advertisement. This system may appear to provide maximum competition, but it has serious disadvantages and is generally not recommended.

The A/E may have the problem of deciding whether or not to recommend a tender that is not the lowest, and this may conflict with the standing orders of the employer. A large number of tenders may be submitted, including some from firms of inadequate experience or doubtful financial standing. The system also involves too many contractors in abortive tendering and hence wasted resources.

3.3.2 Selective tendering

The practice of inviting for tenders from selected contractors eliminates the undesirable factors connected with open tendering. Tenderers may be selected from a standing list or on an *ad hoc* basis for each project. Employers having a continuing programme of works frequently keep standing lists of approved contractors for various types and values of work. These lists are normally reviewed periodically to allow for changing circumstances. An *ad hoc* list may be compiled after public advertisement or from knowledge of suitable firms. The preparation of either standing or *ad hoc* lists is usually by means of a pre-qualification procedure, in which contractors invited to pre-qualify are asked to submit details of their relevant experience and other information. They are then assessed on such factors as their financial standing, their technical and organizational ability, their performance and their health and safety records. Pre-qualification may not be necessary for firms already well known to the employer or the A/E.

The aim is to identify only those firms that are capable of carrying out the contract satisfactorily. Normally the number of contractors invited to tender is not fewer than four nor more than eight. Major employers in developed countries and international bodies such as the World Bank have standard procedures for selecting tenderers.

Words, phrases and Expressions 3.3

[1] open tendering 公开招标
[2] standing *adj*. 既定的,不变的(established and permanent)
[3] standing orders 现行规则
[4] doubtful financial standing 令人生疑的财务状况
[5] on an *ad hoc* basis for each project 为每一个项目专门…

3.4 Contractor's planning

3.4.1 Contractor's project

A building or engineering works is the product of a multi-phase construction project and needs a variety of inputs from diverse contributors. Inputs to a phase of the project are different from those to the next. The participants in a portion or an aspect of the project are usually different from those in another portion or aspect. The participants in a portion, aspect or phase usually treat as the portion, aspect or phase as a project. As a result the procurement by the employer of works or architectural and/or engineering services creates numerous projects for the other participants, such as the design project for the A/E, the management contract for the consulting engineer, the construction contract and subcontracts for the contractor, the subcontractors, and the suppliers, etc.

The participants set their own project management teams. For example, if the contractor accepts an invitation from the employer to tender he will set up a team to make pre-tender planning and then prepare and submit a tender. When he wins a contract he will have his personnel make a pre-contract planning. When a contract is formed between him and his employer he will appoint a competent person as his site representative who will in turn set up a team to manage the project defined by the contract. The effort of the contractor and his personnel makes up a project.

3.4.2 Contractor's project management

Project management is the application of knowledge, skills, tools and techniques to project activities to meet project requirements. It is done by applying and integrating five process groups. The process groups are associated with each other by their performance and used to initiate, plan, execute, monitor and control, and close a project. They may also interact and overlap in relation to scope, cost, schedule, etc.

In order for his project to be successful, the contractor's team must:

(1) establish clear and achievable objectives and select appropriate processes to meet them;

(2) identify and meet the employer's requirements set in the tender documents and then those set in contract documents;

(3) adapting the specification, drawings, bill of quantities and approach to the different concerns and expectations of the various stakeholders; and

(4) balance the competing demands of scope, time, cost, quality, resources, and risk to produce a quality product (a tender, proposals, works, etc.).[16]

3.4.2.1 Initiating of a project

The decision as to whether or not a contractor should accept the invitation to tender involves him in a careful planning process. He can accept it only when he is satisfied that he has adequate technical and financial resources available to complete the contract on schedule as required by the employer.

When the contractor's management decides to tender for a contract and appoints a team to prepare a tender a new project is "initiated". When the contractor has won a contract and appointed a contractor's representative then the execution of works project is "initiated", etc. It is noticeable that both the tendering as a project and the execution of works as another are initiated by the contractor's management instead of either the tendering team or the contractor's representative.

The initiator shall clearly identify the objectives of a project, including the reasons for tendering for or entering into a specific contract. The documentation for this decision also contains a basic description of the scope, duration, and a forecast of the resources which maybe available for the project. The relationship of the project to the contractor's strategic plan identifies the management responsibilities within his organization.

The initial scope description and the resources which the contractor is willing to put in the project need further refining during the initiation process. The contractor's representative as the project manager will be selected. All these together with initial assumptions and constraints shall be documented in a "project charter".[16] The contractor's management by the project charter authorizes the project or a phase of it. The charter documents the contractor's business needs and the employer's needs for works and/or engineering services.

The charter covers majority of the letter of power specifying the duties and authority of the contractor's project manager for specific construction contracts in China.

3.4.2.2 Project planning

Project planning is the most important prerequisite for a successful project. The planning process group is used to define, integrate, and coordinate all subsidiary plans into a project management plan.[16]

The subsidiary plans include but are not limited to scope and variation management plan, time programme (schedule) management plan, cost management plan, quality management plan, claim management plan, risk management plan, record and correspondence management plan, contract administration plan, staffing management plan, etc.

The project management plan defines how the project is executed, monitored and controlled, and closed.

The term "programme" which has been ever used in the construction industry means to a great extent the same thing as the term "project management plan" does. "To a great extent" means that the "project management plan" covers much more aspects of a project and its management than the "programme" which shall be updated to reflect the new concepts, tools, techiques and their more satisfactory results.

The project management plan documents the collection of outputs of the processes to produce the subsidiary plans and other components and shall, at least for the purpose of compliance with contractual requirements, include:

(1) the order in which the contractor intends to carry out the works, including the anticipated timing of each stage of design (if any), contractor's documents, procurement, manufacture of plant, delivery to site, construction, erection and testing;

(2) each of these stages for work by each nominated subcontractor;

(3) the sequence and timing of inspections and tests specified in the contract; and

(4) a supporting report which includes:

(i) a general description of the methods which the contractor intends to adopt, and of the major stages, in the execution of the works, and

(ii) details showing the contractor's reasonable estimate of the number of each class of contractor's personnel and of each type of contractor's equipment, required on the site for each major stage.

The contractor's project management plan may describe:

(1) how to select and apply the processes best suitable for managing the specific project, including the dependencies and interactions among those processes, and the essential inputs and outputs;

(2) how to execute the works to meet the project objectives;

(3) how to monitor and control the changes;

(4) how to maintain the integrity of the performance measurement baselines;

(5) the need and techniques for communication among stakeholders;

(6) the associated project phases for multi-phase projects; and

(7) key management reviews for content, extent, and timing to handle open issues and pending decisions.

3.4.2.3 Project scope statement

Developing a project scope statement is one of the planning processes, aiming to define the works and/or services—what needs to be executed or delivered.

A project scope statement includes:[16]

(1) the requirements and characteristics of the works and/or engineering services to be produced;

(2) the objectives of the project and the works and/or engineering services;

(3) criteria for acceptance of the works and/or services;

(4) project boundaries;

(5) requirements and deliverables;

(6) constraints;

(7) assumptions;

(8) project organization;

(9) defined risks;

(10) schedule milestones;

(11) Work Breakdown Structure (WBS);

(12) cost estimate; and

(13) approval requirements.

The project scope statement is developed from information provided by the employer, such as those contained in invitation to tender, tender documents, contract documents, and those obtained by the contractor's staff, such as the site visit report, etc.

The contractor's project management team shall further refine the project scope statement as the variations and other changes take place during the running of a contract.

3.4.2.4 Three stages of contractor's planning

The project scope statement and the project management plan as a whole are subject to constant updating and revisions as the project progresses. The planning processes are interaltue and shall not, therefore, be taken as a one shot effort.

As a matter of fact the contractors divide planning processes into three major stages in time dimension to serve specific purposes.[25]

The first stage of the planning process is referred to as pre-tender planning, and followed by pre-contract planning and contract planning in that order as is shown in Figure 3.20.

Pre-tender planning focuses on the preparation of an estimate and aims to reach a decision to tender. The ultimate adjudication is required to formulate a tender.

The activities which take place or deliverables are produced during the pre-tender planning process are shown in Figure 3.20.

Pre-contract planning takes place after the award of a contract, immediately prior to the commencement of works on the site. It has become an accepted practice for the contractor to submit to the A/E a construction programme as a deliverable resulting from this pre-contract planning. This may be a presentation based on the contractor's own detailed programme. The purpose of this requirement is that the employer should have some guide as to the intentions of the contractor. On the other hand, the contractor frequently enters key dates for the provision of drawings, instructions and other information which may have impact on his execution and completion of the works. The document later becomes reference material whenever claims arise from delays.

Pre-tender planning:
- Decision to tender
- Pre-tender report
- Site visit report
- Enquiries to subcontractors and suppliers
- Statement of method
- Estimate build-up
- Pre-tender programme
- Preliminaries build-up
- Estimate adjudication
- Analysis of results

Pre-contract planning:
- Pre-contract meeting
- Pre-contract checklist
- Subcontract orders
- Site layout planning
- Requirements schedules
- Master programme
- Commencement arrangements
- Preparation of contract
- Budgets

Contract planning:
- Monthly and six-weekly planning
- Weekly planning
- Daily planning
- Progress reporting
- Updating of progress

Figure 3.20 Contract Planning Processes[25]

Contract planning is the third stage and takes place during the execution of the works by the contractor. It is the contractor's responsibility to complete the contract within the specified time for completion. Contract planning activities establish standards against which progress can be reviewed at regular intervals during the period of time. Figure 3.20 indicates the planning process encompassing the three stages involved.

The detailed procedures involved in pre-contract planning and contract planning will be described in Chapter 5.

3.4.3 Decision to tender

Before a decision can be made as to whether or not to accept the invitation to tender certain factors, which have impact on tendering policy in general, shall be considered:[15]

(1) Current commitments on contracts in progress;

(2) workload in the estimating and surveying departments;

(3) capital available to finance new projects—effect on capital commitment on current projects;

(4) availability of resources in terms of staff, labour, and equipment;

(5) type, location, size, nature and value of the works;

(6) extent or value of own work in relation to that of nominated sub-contractors and suppliers;

(7) competition-number of competitors;

(8) knowledge of A/E, employer, quantity surveyor and/or consultants, nominated subcontractors and suppliers;

(9) the conditions of contract, especially those which shall be carefully considered:

(a) delay damages for the works;

(b) limit of retention money;

(c) defects notification period;

(d) performance security (bond or bank guarantee), and

(e) adjustment for changes in cost and/or legislation.

(10) availability and characteristics of labour in the vicinity of the works. Alternative opportunities, degree of trade union influence and pay policy of competitors;

(11) time available to prepare tender.

The market conditions likely to affect tendering include:

(1) general market conditions in relation to the availability of contract work; number of enquiries received;

(2) availability of finance, current interest rates/bank rate, and

(3) government policy and its effect on the construction industry.

The contractor who is well experienced in certain types of works, such as housing and industrial buildings, would be well advised to maintain competitive tendering for work of a similar type. To consider expanding into building or engineering works of a different nature introduces new risks and demands a different mix of resources. The contractor shall not make any such a decision before he has fully satisfied himself that the risks can be addressed effectively.

The information which should be offered along with an invitation to tender and other factors which should be taken into account by the contractor's management is included as a check list of items as follows.

(1) information required to enable a decision to be made whether or not to tender:
(a) location of the works;
(b) names of employer, A/E and/or quantity surveyor;
(c) other consultants;
(d) description of the works;
(e) commencement and completion dates;
(f) principal trades to be nominated;
(g) proposed contractual arrangement;
(h) details of documents and information to be provided;
(i) deadline for submission of tenders, and
(j) number of tenders invited.
(2) additional information which should be taken into account by management:
(a) financial resources;
(b) market conditions;
(c) constructability (construction problems);
(d) previous experience of similar types of the works;
(e) previous experience with the A/E and/or consultants;
(f) adequacy of tender information;
(g) resources necessary to carry out the works;
(h) workload at hand and that estimated;
(i) risks with weather and season and those imposed by conditions of contract;
(j) time available for preparation of tender.

To sum up, the contractor shall be satisfied that he has adequate time available in the estimating department to allow proper preparation. Overloading the estimating department can lead to errors of judgement and arithmetic. Estimating is an exacting and precise activity and the submission of unrealistic estimates can jeopardize his financial stability.

It is unwise to submit tender below cost attempting to win a contract when pressed by a slump demand for service of construction industry, and hoping to recoup the loss when the demand becomes strong. The contractors shall improve their awareness of risks involved in doing so.

Sub-clause 4.11 of the *Red Book* under the heading of "Sufficiency of the Accepted Contract Amount" reminds the tenderer (the contractor when awarded the contract) that he shall be deemed to: (a) have satisfied himself as to the correctness and sufficiency of the accepted contract amount, and (b) have based the accepted contract amount on the data, interpretations, necessary information, inspections, examinations and satisfaction as to all relevant matters related to the site, as referred in the contract. Unless otherwise stated in the contract, the accepted contract amount covers all his obligations under the contract (including those under provisional sums, if any) and all things necessary for the proper execution and completion of the works and the remedying of any defects.

Words, Phrases and Expressions 3.4

[1] standing list 常备名单
[2] jeopardize vt. 危及,损害(endanger, put in danger)
[3] completion date 竣工日期,完成日期
[4] pre-tender planning 投标(前)规划
[5] pre-contract planning 签约后规划,开工(前)规划
[6] contract planning 合同期间规划,履约规划
[7] estimate build-up 编制估算(书)
[8] pre-tender programme 投标施工计划
[9] estimate adjudication 估算书评价
[10] preliminaries build-up 编制开办费估算(书)
[11] master programme 施工总计划
[12] commencement arrangements 开工准备工作
[13] progress reporting 进展报告
[14] site layout planning 现场布置规划,施工平面设计
[15] pre-tender report 投标报告
[16] site visit report 现场考察报告
[17] decision to tender 投标决策
[18] requirements schedules 要求说明书表格
[19] accepted practice 公认做法,公认惯例,通行惯例
[20] formulate a tender 编制出正式的标书
[21] financial stability 财务稳定性
[22] current workload 目前手头工作量
[23] commitment on contracts in progress 准备为进行中的合同投入的力量

Notes 3.4

疑难词语解释

[1] This system may appear to provide maximum competition.

【译文】 这种办法似乎可以实现最大程度的竞争。

[2] The A/E may have the problem of deciding whether or not to recommend a tender that is not the lowest, and this may conflict with the standing orders of the employer.

【译文】 建筑师/工程师在决定是否(向业主)推荐并非最低价格的标书时会左右为难,若向其推荐,就会违反业主的现行规则。

[3] The system also involves too many contractors in abortive tendering and hence wasted resources.

【译文】 这种办法会使太多承包商的投标活动前功尽弃,因而浪费了资源。

[4] He should accept it only when he is satisfied that he has adequate technical and financial resources available to complete the contract on schedule as required by the employer.

【译文】 只有确信自己的技术和财力足以按照业主要求的计划完成这一合同时,他才能接受这一邀请。

[5] Sub-clause 4.11 of the *Red Book* under the heading of "Sufficiency of the Accepted Contract Amount" reminds the tenderer (the contractor when awarded the contract) that he shall be deemed to: (a) have satisfied himself as to the correctness and sufficiency of the accepted contract amount, and (b) have based the accepted contract amount on the data, interpretations, necessary information, inspections, examinations and satisfaction as to all relevant matters related to the site as referred in the contract. Unless otherwise stated in the contract, the accepted contract amount covers all his obligations under the contract (including those under provisional sums, if any) and all things necessary for the proper execution and completion of the works and the remedying of any defects.

【译文】 红皮书第4.11款以标题"中标合同额的盈利可能性"提醒投标人(一旦中标即为承包商):应当认为承包商:(a)确信中标合同额是正确和足够的,且(b)是根据与合同中提到的所有事项有关的数据、解释、必要的资料、考察、研究,以及对此的确信无疑而确定中标合同额的。除非合同另有说明,中标合同额足以使承包商履行合同为其规定的全部义务(包括可能有的暂列金额为其规定的义务),以及为恰当地施工、竣工并修补任何缺陷所需的全部有关事项。

【解释】 将 Sufficiency of the Accepted Contract Amount 译成汉文时,应从承包商签订合同的目的、投标竞争形势、合同为其规定的义务与责任,以及某些承包商仅为中标而不顾亏损等情况综合考虑。为此,将 sufficiency 译为"盈利可能性"。若译成"中标合同额的充分性",则不能表达"中标合同额"及"充分"的目的是什么。另外,cover 不能译成"包括"。在此句中,cover 的主语是 accepted contract amount,宾语是 obligations 和 all things。在说"甲包括乙"时,甲与乙应属同类。显然,accepted contract amount 和 obligations 或 all things 不属同类。

3.5 Preparation of tender

A contractor shall prepare and submit a tender as required in the tender documents. He usually has to make extensive inquiries and he must be given sufficient time so as to base his tender on adequate knowledge and information and reduce the possible difficulties if he is awarded the contract. It is obvious that the period of time to be allowed depends on the size and complexity of the project. A period of at least four weeks should be allowed for tendering and at least eight weeks is recommended for large projects.

3.5.1 Estimating procedures

3.5.1.1 Examination of information available

As soon as a prospective tenderer has received the tender documents, he will have a senior member examine the preliminaries and general items covered in the bills of quantities to ascertain whether there is any deviation from the standard form of contract or any unusual or unfair conditions. If these deviations or conditions were to be unduly onerous and to expose him to exceptionally high risks, it might be a wise decision not to submit a tender.[15]

Preliminaries refer to the items in a bill of quantities for miscellaneous and preparatory work for which a contractor shall undertake and be paid accordingly, such as ordering materials, mobilizing and deploying his resources, running the site and providing services to the employer's and/or A/E's representative.

The senior member, usually an estimator (may be a professional quantity surveyor in the U.K., commonwealth, Hong Kong, etc.) should extract from the tender documents and other project information available then to him the following details which:

(1) affect the contractor's intended methods of construction;
(2) impose restrictions or limitations of any kind;
(3) affect access to the site;
(4) interrupt the regular sequence of trades or site operations;
(5) affect the time for completion;
(6) require specialist skill, processes or materials;
(7) have a significant effect on the construction programme; or
(8) have major cost implications.

These items have effects on costs to the contractor and shall draw therefore the estimator's attention when preparing estimate.

The estimator shall liaise with the contractor's all other members involved in purchasing, programming and other associated activities and ensure that they should be provided with copies of all the information relevant to the project under the contractor's consideration.

He shall identify all the factors which may have effect on his approach to the pricing of the bills of quantities in question, such as:

(1) standard, correctness and completeness of extracted information;
(2) clarity of the specification and quality requirements set out in the tender documents;
(3) constructability of the works;
(4) extent by which standard details should be used and amount of repetitive work;
(5) amount of information covering ground conditions and sub-structural work; and
(6) problem areas and constraints on construction imposed by the design.

The estimator shall also carefully check if the items of work covered by prime cost sums are clearly stated and can be identified within the information available to him. Many other items need his attention, such as:

(1) if the subcontractors for whom prime cost sums are set aside have been nominated;

(2) if the nominated subcontractors will indemnify the contractor against their design responsibility;

(3) the extent of progress of such design work and coordination with other consultants' drawings;

(4) if attendance has been adequately identified;

(5) if adequate allowance has been made for the contractor's profit and discounts are to be allowed;

(6) if the nominated subcontractors can conform with the contractor's requirements in terms of his tender programme; and

(7) if the contractor's work requirements of that of specialists are clearly defined and measured in the bill of quantities.

3.5.1.2 Method statement and tender programme[15]

The contractor shall prepare a tender programme and a method statement to be included in his tender. If his tender is successful the programme represents his intentions at the time of tender and upon which his pricing of the works was based.

The preparation of method statement now forms an essential requirement of a contractor's procedures. Method statements may be used for (1) tendering; (2) site operations; and (3) stating safety measures.

The method statement for tendering outlines the sequence and methods of construction upon which estimate is based. It should indicate how the major elements of work will be handled and highlight areas where new or alternative methods are being considered and thus provides an aid for the estimator and a basis for him to build up rates and/or prices; enables the contractor's tender team to assess the requirements of the contractor's equipment for inclusion in the tender and the tender to be on practical methods. It also enables the employer to assess alternative proposals at tender evaluation. The method statement for tendering should be supported with details of cost data, gang sizes, contractor's equipment and supervision requirements. It should provide for requirements of temporary works, as well.

The method statement for site operations provides a method statement for the employer's representative and meets the statutory requirements and enables the employer's representative to assess the contractor's methods and sequence of working.

The safety method statement forms part of the safety plan and documents the identified risks and hazards in the construction operations from a safety point of view. The statement shall include the subcontractors' assessment of risks.

Alternative operational method, site organization and work sequences should be evaluated at this stage and decisions made on the intended method of construction. Confirmation of these arrangements needs visit to the A/E and/or to the site.

LABORATORY RENOVATION			METHOD STATEMENT			
OP. NO.	OPERATION	QUANTITY	METHOD	RESOURCES		NOTES
				EQUIPMENT	LABOUR	
1	Concreting to laboratory floors (floors 1~5)	78m³/floor	OPTION1 Place concrete using front discharge dumper trucks. Goods lift to be used as access from ground floor level. Place concrete in 5m wide bays and tamp level. Steel leveling screed finish.	3~dumpers 1~5 m tamp 1-steel screed 1-vibrator RMC supply	3 drivers 5 labourers 1 supervisor	Output per day 26 m³
2	Concreting to laboratory floors (floors 1~5)	78m³/floor	OPTION2 Place concrete using mobile concreting pump located in rear yard area. Pump concrete through window openings at each floor level. Continuous bay pour-one pour each floor.	1-vibrating screed 1-steel screed 1-vibrator RMC supply	6 labourers 1 labourer at pump 1 supervisor	Output per day 50 m³

Figure 3.21 Tender Method Statement In a Tabular Format[31]

The method statement may be presented as a written statement or in tabular form. It is essential to understand the particular purposes of method statements at various stages of the planning process.

Figure 3.21 illustrates the tender method statement presented in a tabular format, which has the advantages in that the requirements of concrete placement to floors of a building. Figure 3.22 and Figure 3.23 gives a sample safety method statement and a hazard assessment statement, respectively.

The programme may be presented as a bar chart or a network diagram. Figure 3.24 illustrates a linked bar chart for the waterway diversion project. Bar lines represent the duration of each operation and the relationship between the operation's initiation (commencement, start) and completion (finish) can be readily observed. Some symbols can be introduced on the bar chart to aid resources control and leveling. The key dates (called milestones) can also be shown in relation to nominated and own subcontractors.

Linked bar charts retain the visual benefits of bar chart with increased emphasis on dependencies. This takes the form of vertical links between the completion of one activity and the start of another. The greater emphasis on coordination and construction sequence allows the technique to be used for more complex projects than the normal bar chart. However, the float concept is generally missing and there is a limit to the amount of linking that is possible.[15]

CONTRACT	CONT. No	Prepared by	Date	Checked by	Date
WATERWAY DIVERSION	C002	David Smith	14 September/Sep	Thomas Lee	16 September/Sep
Operation	**Contractor's equipment**				
Construct sheet piled cofferdam	22 RB crane BSP 900 Pile hammer Komatsu 380 Excavator 2 No. 30T Wagons 150 mm Diesel pump				
Sequence of working	**Supervision and monitoring**				
Construct hardstanding for piling rig in lock mouth using imported quarry waste Erect guide frame and install sheet steel piles Pump out water between hardstanding and piles Fill behind piles for access Remove piling frame Construct top bracing with steel wallings and struts Remove hardstanding and install bottom bracing Erect secure ladder access to bottom of cofferdam	Site engineer Piling foreman Operation to be monitored daily by the site agent Banksman to work with mobile equipment Daily check on crane equipment, load indicators and operation				
	Controls				
	Authorized personnel area only When working over water to remove piling frame, life jackets and safety harness must worn Area to be fenced off at night with chestnut paled fence Warning notice to be displayed either side of cofferdam "Danger deep excavation" and "Danger deep water"				
Emergency procedures	**First aid**		**PPE schedules**		
Send for site first aider and/or call emergency services where necessary Rescuers must not put themselves in danger Follow first aid and drill if appropriate Do not remove evidence Notify site agent	Dinghy to be moored adjacent to cofferdam 2 No. life buoys in wooden locker First aid box in site office 2 way radios		Safety harnesses Hard hats Gloves Welding goggles High visibility vests Life jackets Ear defenders		

Figure 3.22 A Sample Safety Method Statement[31]

3.5.1.3 Site visit

Following the preliminary assessment of the contract to tender for, the contractor will have his staff, e.g. his estimator, visit the site often accompanied by other members of the contractor's team. The team normally extends its visit to embrace the general locality and other matters of significance and produces a site report when their visit ends.

The site report usually includes, but is not limited to, details of the site and its surroundings, access, topographical features, including trees and site clearance work, ground conditions, groundwater level if it can be determined, existing services, including any overhead cables, possible security problems, facilities for disposal of surplus spoil, labour situation, availability of plant and materials, weather conditions and effects of bad weather on site operations, any special problems such as tides, high winds, flooding, noise control and road diversions, any constraints on working on site, such as restrictions on the area available for tower cranes and other equipment, and space for site huts, storage compounds and the like.

CONTRACT	CONTRACT No	Prepared by:			Date of assessment	
WATERWAY DIVERSION	C002	David Smith			12 September 2006	
Operation	Potential Hazards	Risk Assessment			Response Actions	
		H	M	L		
Construct sheet piled cofferdam	Trapping and crushing by moving contractor's equipment		M		Warning device/Banksman	
	Falls from height	H			Guard rails along lock	
	Falls of materials	H			Leave piles proud for edge protection	
	Restricted space for working			L		
	Working over water	H			Safety harness and life jacket to be worn	
Programming for	Training/Certification	Action			PPE Required (specify)	
25 September	All operatives to undergo one-day safety awareness course Operators of contractor's equipment to be certified	Method statement Work permit Assessment			Safety harnesses Hard hats Gloves Welding goggles HV vests Life jackets	

Figure 3.23 A Sample Hazard Assessment Statement[31]

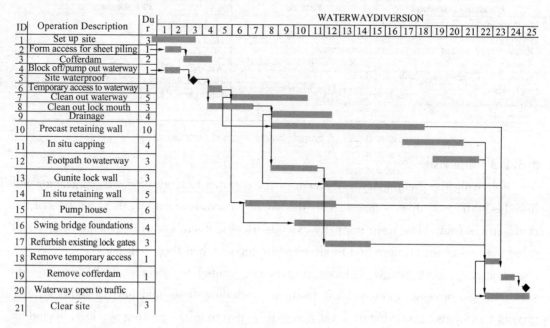

Figure 3.24 A linked Bar Chart for the Waterway Diversion Project[31]

3.5.1.4 Selection of contractor's equipment

It is necessary for the contractor to select the most suitable equipment for the potential contract as early as practically, once the decision on method of construction has been made. This will normally entail discussions between the estimator, contract management and, if possible, the person who will be the contractor's site representative. The choice of contractor's equipment will also be influenced by the availability or otherwise of suitable items of contractor's equipment in his ownership, and a careful study of the site. The agreements reached and the decision made based on the discussions will be recorded and acted upon by the estimator in preparing the estimate.

3.5.2 Costs to the contractor of works

3.5.2.1 Costs

"Cost" of the good or service means differently to the seller and a buyer. In theory "cost" of a good or service refers to the value of the factors of production used by the seller to produce the good or to deliver the service.

The employer and the contractor are often divided over the cost of a building or engineering works.

The *Red Book*, defines, under sub-paragraph 1.1.4.3, "cost" (to the contractor) as all the expenditure reasonably incurred or to be incurred by the contractor, whether on or off the site, including overhead and similar charges (properly allocable thereto), but does not include profit.

In a contractor's perspective, however, the cost of the permanent works as completed is the total value of the preliminaries and general items, labour, materials, goods, equipment, subcontractor's work, provisional sums and prime costs, overheads, and profit.

Overheads can be defined as the costs incurred by the contractor in operating his business which are not directly related to individual items of works. There are two main groups of overheads:[15]
(1) site overheads which include such costs as site supervisory staff, site buildings, temporary works and services; and
(2) head office overheads which cover the costs incurred in operating the business in its entirety and cannot be related directly to an individual contract, and include head office staff and buildings.

Profit is, according to economists, a return to the owners of the contractor's company, not only for their ownership of the business but also for their willingness to take risks (time, energy, etc.) in a market economy. The greater the risk, the greater the anticipated profit needs to be, otherwise the owners as entrepreneurs would move their funds to safer investments offering similar profits[33].

The profits are usually measured in terms of annual net turnover which is total annual turnover less annual costs. It is customary to include an addition of 3 to 5 per cent net turnover.

3.5.2.2 Rate and price

The principles/method of measurement will, by defining what is to be measured, define what is appropriate for valuing each item in the bill of quantities:

(1) A "rate" per unit quantity (such as $4/m^3$ for work measured by volume), which in some languages is referred to as a "unit price", or

(2) A lump sum "price" (such as $4000 for an item of work which is not to be measured).[32]

In calculating rates for insertion in the bill of quantities, careful consideration shall be given to every factor which may influence the cost of the building or engineering works. There can be no substitute for comprehensive company data and feedback from previous work which are in nature similar to the contract being priced. Rates for measured items in the bill can consist of any, or a combination of, the basic elements of labour, materials, equipment, sub-contractors' items, overheads and profit. Each element should be analyzed and estimated separately so that the total cost of each element can be considered by the contractor's management[15].

3.5.2.3 Preparation of estimate

An estimate for a construction project represents the values of a combination of diverse inputs to create the building or engineering works.

Many employers may prepare its own estimates, usually through the engaged "independent" A/E, as a check on contractor's tender price. Significant differences from these estimates may be an indication that the tender documents were not adequate or that the prospective contractor either misunderstood or failed to respond fully to the tender documents. Independent estimates or A/E's estimates are often referred to as *should cost* estimates.

It is essential to understand that the estimate prepared by a contractor as described above is an amount more than the cost to be incurred by the contractor for carrying out the works.

3.5.3 Estimating procedure in China

As described in the *Procedure for Estimating Building Costs of Construction Projects* issued by the Ministry of Construction (MOC) in 2003 and China as Appendix 2 to *Jianbiao* [2003]206, there are two alternatives that can be used, that is *direct work cost based unit rate method* and *all-in unit rate* method. MOC has also defined the components of the cost of building and engineering as shown in Figure 3.25.

3.5.3.1 Direct work cost based unit rate method

An estimator starts estimating with the costs for an item of work of labour, materials, and equipment costs. The cumulative figure for the item is referred to as *direct work cost based unit rate*. *Direct work cost based unit rate* is then multiplied by the bill quantity of the item to produce the direct work costs of the item.

Estimating the site general items then follows. The cost of each site general item is a percentage of the direct work costs. All the percentages and the direct work costs build up the

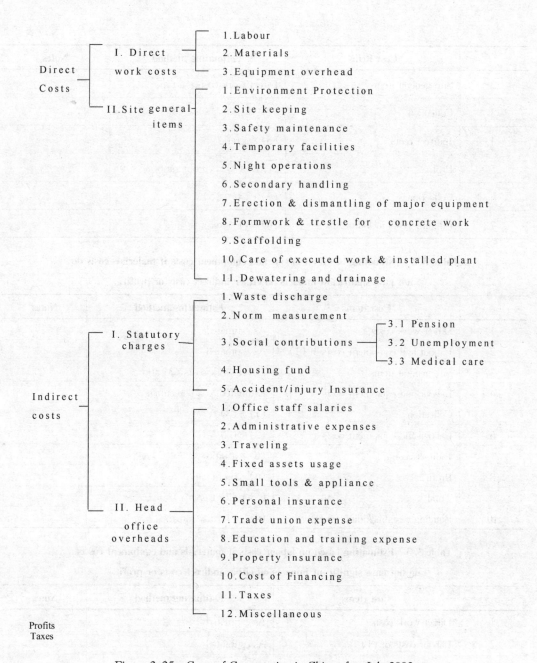

Figure 3.25 Costs of Construction in China after July 2003

direct costs of the works. As far as the indirect costs, profit and taxes is concerned, the steps are tabulated as shown in Table 3.1, 3.2 and 3.3, depending on their relations to labour costs, materials costs and equipment costs.

Table 3.1 Estimating based on direct costs if costs of all labour, materials and equipment have significant impacts on indirect costs and profit.

	Cost items	Estimating method	Notes
1	Direct work costs	As estimated	

续表

	Cost items	Estimating method	Notes
2	Site general items	Direct work costs × a ratio	
3	Collection	(1) + (2)	
4	Indirect costs	(3) × a ratio	
5	Profit	((3) + (4)) × profit ratio	
6	Collection	(3) + (4) + (5)	
7	Building costs including taxes	(6) × (1 + tax rate)	

Table 3.2 Estimating based on labour and equipment costs if materials costs do not have significant impacts on either indirect costs or profit.

	Cost items	Estimating method	Notes
1	Direct work costs	As estimated	
2	Labour and equipment costs of (1)	As estimated	
3	Site general items	Direct work costs × a ratio	
4	Labour and equipment costs of (3)	Estimated as a percentage	
5	Collection	(1) + (3)	
6	Labour and equipment costs	(2) + (4)	
7	Indirect costs	(6) × a ratio	
8	Profit	(6) × profit ratio	
9	Total	(5) + (7) + (8)	
10	Building costs including taxes	(9) × (1 + tax rate)	

Table 3.3 Estimating based on labour costs if materials and equipment costs do not have significant impacts on either indirect costs or profit.

	Cost items	Estimating method	Notes
1	Direct work costs	As estimated	
2	Labour costs of (1)	As estimated	
3	Site general items	Direct work costs × a ratio	
4	Labour costs of (3)	Estimated as a percentage	
5	Collection	(1) + (3)	
6	Labour costs	(2) + (4)	
7	Indirect costs	(6) × a ratio	
8	Profit	(6) × profit ratio	
9	Total	(5) + (7) + (8)	
10	Building costs including taxes	(9) × (1 + tax rate)	

3.5.3.2 All-in unit rate method

An estimator first estimates the *direct work cost based unit rate* for each work item. As far as the indirect costs, profit and taxes for the work item are concerned, the estimating steps are tabulated as shown in Table 3.4, 3.5 and 3.6, depending on their relations to labour costs, materials costs and equipment costs. The sum of *direct work cost based unit rate*, indirect costs, profit and taxes for each work item is the all-in unit rate of the work item. The sum of products of all-in unit rates by bill quantities is the building costs of a construction project.

A ratio of C/C_0 shall be calculated before decision is made on which set of steps, as shown in Table 3.4, 3.5 or 3.6, shall be followed. The C represents the ratio of materials costs to direct work costs, while C_0 the ratio of materials costs to direct work costs as measured in the local standard schedules of unit rates used prior to July 2003.

Table 3.4 Estimating based on direct costs if $C/C_0 > 1$

	Cost items	Estimating method
1	Direct work costs for work item	As estimated as sum of labour, materials and equipment costs
2	Indirect costs for work item	(1) × a ratio
3	Profit	((1) + (2)) × profit ratio
4	Collection	(1) + (2) + (3)
5	Building costs including taxes	(4) × (1 + tax rate)

Table 3.5 Estimating based on labour and equipment costs if $C/C_0 < 1$

	Cost items	Estimating method
1	Direct work costs for work item	As estimated as sum of labour, materials and equipment costs
2	Labour and equipment costs of (1)	As estimated as sum of labour and equipment costs
3	Indirect costs for work item	(2) × a ratio
4	Profit	(2) × profit ratio
5	Collection	(1) + (3) + (4)
6	Building costs including taxes	(5) × (1 + tax rate)

Table 3.6 Estimating based on labour costs if materials and equipment costs do not have significant impacts on either indirect costs or profit

	Cost items	Estimating method
1	Direct work costs for work item	As estimated as sum of labour, materials and equipment costs
2	Labour costs of (1)	As estimated as labour costs
3	Indirect costs for work item	(2) × a ratio
4	Profit	(2) × profit ratio
5	Collection	(1) + (3) + (4)
6	Building costs including taxes	(5) × (1 + tax rate)

3.5.4 Queries and pre-tender meetings

Tenderers should focus on identifying all the doubtful points and clarifying them with the A/E acting on behalf of the employer as soon as possible and before the deadline of submission of tenders. Where queries relate to possible ambiguities, the need for supplementary information or errors in the documents, any verbal replies by the A/E should be recorded and then confirmed in writing by means of addenda notices or amendments issued to all tenderers.[15]

Pre-tender meetings are sometimes arranged to clarify doubtful and uncertain points raised during the tender period. The meetings are best arranged in the form of a group meeting of all tenderers, possibly on site, and all information provided on these occasions should be properly recorded and confirmed in writing to all tenderers. Where this information has contractual significance it needs to be incorporated in the contract documentation.

3.5.5 Amendments

Amendments to the tender documents should be issued only if the A/E is convinced that the tenderers' pricing will be so seriously affected as to distort the balance between their offers. Unless significant changes are necessary, amendments are better delayed and dealt with either as post-tender clarification or as part of the administration of the contract after acceptance of a successful tender.

Tenderers should be notified immediately if it is decided to issue any amendments to the tender documents. Where the amendments are significant, an extension to the tender period may be necessary. Amendments made to the documents are best prepared in the form of addenda letters or notices, which should be numbered consecutively and accompanied by an acknowledgement slip.

3.5.6 Submission/receipt of tenders

Tenders should be sent by registered post, recorded delivery services, or be delivered by hand in a plain sealed enveloped, entitled as directed but not bearing the name of the tenderer. Nowadays, electronic delivery is becoming acceptable. Tenders are confidential and should remain unopened and secure until the designated time for opening. Tenderers have the right to modify their tenders in writing at any stage before the prescribed deadline for submission of tenders.

Tenders received after the prescribed date and time are invalid and should not be considered unless:
(1) a cable or telex stating the tender price is received on time, and
(2) there is clear evidence, such as a postmark, that the complete tender documents have been dispatched within a sufficient margin of time to presume its due arrival.

3.5.7 Cost of tendering

The cost of tendering is an appreciable part of a contractor's expenditure. Whether or not the tender is successful, this cost must be included in tender prices generally, and thus eventually must be borne by employers generally. Contractors sometimes tender when they do not wish to undertake the work in the belief that their failure to do so may prejudice future opportunities. When appropriate, the A/E should invite such contractors to withdraw from the selected list, and make it clear that their failure to tender will not prejudice future invitations.

3.5.8 Time for completion

Varying opinions are held about the advisability of the time for completion being determined by the employer, as in the majority of circumstances it can be argued that contractors are better able to decide a reasonable completion time. Furthermore, time and cost are interrelated. One approach is to insert a completion date in the contract documents, but to inform contractors that they may submit an alternative tender based on their own selected contract period. However, the comparison by the A/E of tenders based on different contract periods is very difficult.

3.5.9 Tender prices

3.5.9.1 Difference between tenders

The tender prices from even equally competent contractors may differ significantly, not so much because routine operations are differently priced, but because of different tendering strategies, construction methods and techniques, and because of differing values attached by the contractors to risks and the methods of dealing therewith. Profit and/or general overhead margins may vary from a contractor to another depending on their tendering policies at the time.

3.5.9.2 Distinction between estimates and tenders[15]

Estimating is different from tendering. *Estimating* is defined as the technical process of predicting costs of construction, while *tendering* is a separate and commercial function based upon the net cost estimate, and culminates in an offer to carry out defined work under prescribed conditions for a stated sum of money.

An estimate can be a reasonably accurate calculation and assessment of the probable cost of carrying out defined work under known conditions. If the estimate is not reasonably accurate it will be of little value and could subsequently result in major problems for the contractor. However, the degree of accuracy obtainable will be affected by the particular circumstance of each project.

3.5.9.3 Qualified tenders

To enable the A/E to compare tenders on a common basis, it is important that tenders are submitted without qualification or alteration to any of the tender documents. It is often stipulated in the instructions to tenderers that no qualifications will be permitted, but if it is not it should be made clear how qualified tenders will be dealt with in tender evaluation (referred to as adjudication in the United Kingdom).

3.5.9.4 Alternative tenders

The instructions to tenderers should state whether offers based on alternative designs will be considered, and if so the procedure to be followed. Sometimes employers require any alternative proposal to be submitted in confidence before the end of the tender period so that the A/E can give a preliminary view on whether or not it is likely to be acceptable.

Words, Phrases and Expressions 3.5

[1] method statement 施工方法说明书(相当我国过去的"施工方案")

[2] on a…basis 以……为准则，以……方式（an underlying condition or state of affairs）

[3] indemnify vt. 保障……不承担义务、责任或不蒙受损失，是 indemnity 的动词形式，常与 from, against 连用（make sb., oneself safe from, against sth.）。

[4] attendance n.（承包商对分包商的）照看与管理（act of attending (upon sb.)）

[5] tender programme 投标时提交的施工计划

[6] alternative operational method 可选用的作业方法

[7] major elements of work 主要的分项工程

[8] gang sizes 班组人数，班组大小

[9] independent estimates (A/E's estimates) 独立估算，标底

[10] *should cost* estimates 合理费用（成本）估算（An estimate of the cost of a product or service used to provide an assessment of the reasonableness of a prospective contractor's proposed cost. 对产品或服务成本进行的一种估算，用途是估计预定承包商报价中提出的成本是否合理。）

[11] pre-tender meetings (投)标前会议

[12] submission/receipt of tenders 递交与接收标书

[13] cost of tendering 投标费用

[14] time for completion 竣工时间，完成时间

[15] qualified tenders 附加条件的标书，有保留的标书

[16] direct work cost based unit rate method 工料单价法

[17] all-in unit rate method 综合单价法

Notes 3.5

疑难词语解释

[1] Enquiries made by tenderers to either the employer or the A/E should be dealt with on a strictly formal basis.

【译文】 投标人向业主或建筑师/工程师提出的询问都应严肃地对待。

［2］ Whether or not the nominated subcontractors will indemnify the contractor against their design responsibility.

【译文】 指定分包商是否愿意保障承包商不因分包商的设计责任而承担责任或蒙受损失。

3.6 Selecting a contractor

3.6.1 Tender opening

Tenders should remain unopened and made secure until the appointed time for opening. At least two persons representing the employer should be present when the tenders are opened. They should record the totals of the bills of quantities and pass them to the A/E for assessment. The A/E should then report his findings to the employer and make appropriate recommendations.

3.6.2 Covering letters

It is common for a tender to be accompanied by a covering letter, the contents of which may be important.

For most part the covering letter forms no part of the contract. It was not referred to in the tender itself. A/E confronted with such a situation when examining tenders must check whether such a covering letter is intended to form part of the contract and, if so, at least ensure that it is specifically referred to in the Letter of Acceptance.

In certain circumstances, a covering letter can be considered as part of the tender, in which case both parties should be aware that such matters as proposed methods of construction or proposed subcontractors who may be described in the letter become legally binding.

3.6.3 Examination of tenders

The tender rates are those entered by the tenderers in the bill of quantities. The A/E should check all the tenders for arithmetical errors. For the purposes of comparing tenders any error in the extension should be corrected. Errors in the summation of extensions should also be corrected to give the correct total of the priced bill of quantities. In order to preserve parity of tendering, correction of tender rates after submission of tenders should not normally be permitted. In these circumstances a contractor may be given the option of withdrawing his tender.

The A/E should carefully examine the tender rates, and if he finds any of them which is excessively high or low, he should carefully consider the consequences on the outcome of the contract, and advise the employer accordingly in his report. If apparent errors occur due to a misunderstanding of the tender documents and these are common to several tenders, it may be necessary to invite new tenders.

Pricing of a tender in a way which will result in high payments to the contractor early in the contract period-sometimes called front loading-should be considered carefully, and the financial consequences recognized when tenders are compared.

3.6.4 Notification to tendering results

All tenderers should be notified of the tendering results, including the amount of the successful tender, as soon as possible because their future tendering policy may be influenced by the results. This notification frequently consists of a list of tenderers and a separate list of tender prices thus enabling the tenderers to note their ranking without specific reference to the others.

When it becomes clear during examination of the tenders that certain tenders will not be accepted, the tenderers should be advised as soon as possible. Figure 3.26 is a sample notification of unsuccessful tender. The remaining tenders will be subject to closer examination but as soon as possible after the employer has accepted a tender, the unsuccessful tenderers should be notified.

Our Ref:

27th August 2005

Contractor's address

Attn: Mr. Zhang Xiaobing

Dear Sirs:

Main Contract for Proposed Shopping Centre on 45 Wudaokou, Haidian District, Beijing

Further to my letter of 21 July 2005, I regret to advise that your tender has not been accepted. Yours faithfully,

Li Daguan
Authorized person

Figure 3.26 A Notification of Unsuccessful Tender

3.6.5 Post-tender meetings and negotiations

Following examination of the tenders, it may be desirable for the A/E and employer to meet. Such meetings should serve to clarify any points of doubt or difficulty such as resources, methods of construction, organization, qualifications and alternative designs. The A/E should not meet tenderers without the employer's prior agreement.

3.6.6 Lowest evaluated tender

In evaluating the tenders, the employer, A/E or a tender board will adjust the tender price as follows:
(a) making any correction for errors;
(b) excluding provisional sums and the provision, if any, for contingencies in the summary

bill of quantities, but including daywork, where priced competitively;

(c) converting the amount resulting from applying (a) to (b) above and (f) below, if relevant, to a single currency;

(d) making an appropriate adjustment on sound technical and/or financial grounds for any other quantifiable acceptable variations, deviations or alternative offers;

(e) making an allowance for varying times for completion offered by tenderers, if permitted in the tender documents; and

(f) applying any discounts offered by the tenderer for the award of more than one contract, if tendering for this contract is being done concurrently with other contracts.

An adjusted tender price is referred to as e*valuated tender amount* or *evaluated tender price*. The employer, A/E or a tender board shall determine for each tender the evaluated tender amount.

The A/E should not hesitate to recommend a tender other than the lowest evaluated tender if he concludes that it is in the best interest of the employer to do so. In making such a recommendation a clear explanation is essential so that the employer fully understands the consequences of his rejecting the A/E's recommendation. In open tendering it may be necessary to reject a tender, either because of inadequate experience or because of doubtful financial standing of the tenderer. In selective tendering it may be difficult to justify the rejection of the lowest evaluated tender, especially in the public sector where existing rules or standing orders may specify that the lowest evaluated tender must be accepted.

The employer usually reserves the right to accept or reject any variation, deviation, or alternative offer. Variations, deviations, alternative offers, and other factors that are in excess of the requirements of the tender documents shall not be taken into account in tender evaluation.

3.6.7 Tender report

The A/E's advice to the employer should take the form of a report (is referred to as *tender adjudication report* in the UK) setting out how he has scrutinized the tenders and giving his conclusions and recommendations regarding which tender to accept. The scope and detail of this report will vary according to the circumstances, but frequently the following are included:

(a) a tabular statement of the salient features of the tenders received, e.g. tenderer's name, total of priced bill of quantities before correction of arithmetical errors, validity and qualifications;

(b) arithmetical errors discovered and the effects of corrections on the bill price;

(c) details of discussions held with any of the tenderers;

(d) a concise summary of the analysis of each tender or, say, the three or four lowest evaluated tenders; reasons for considering any tender invalid, the consequences of any qualifications and discussion of any methods of construction proposed;

(e) a tabular comparison of various sections of the bill of quantities and main rates with comments on unusual rates;

(f) a comparison of tenders with the A/E's estimate;

(g) recommendations of the most acceptable tender;

(h) recommendations for dealing with errors, qualifications and any points arising from discussion with the recommended tenderer, whether by amending the tender or by a counter-offer by the employer;

(i) a financial statement indicating likely programming of payments to be made by the employer.

Words, Phrases and Expressions 3.6

[1] selecting a contractor 选定承包商,定标

[2] tender opening 开标

[3] cover(ing) letter 首封函

[4] extensions n. [会计用语] (从另一栏) 转来的金额, 算出的金额 (如发票上所示单价与数量相乘而得的总金额) (a part constituting addition)

[5] parity n. 平等 (equality, as in amount, status, or character)

[6] front loading 前高后低报价, 不平衡报价 (A "front-loaded" or "unbalanced" tender is one in which some rates or prices are relatively high and others low in relation to the A/E's estimate of the real cost of work to be performed. The imbalance may be due to a genuine error on the part of the tenderer or a misconception of the risks involved. But more often it is due to the tenderer's attempt to increase cash flow over his true needs by inflating the rates or price of those items of work occurring early in the programme, or of those items of work that he feels will be used more than scheduled. Any seriously unbalanced pricing should be "clarified" by the employer with a view to its justification by the tenderer and also to determine whether the successful tenderer should provide additional performance security to cover the employer's exposure in the event of subsequent default. "前高后低" 或 "不平衡" 报价乃是有些待做工作的价格高于建筑师/工程师对其真正费用的估计, 而另外一些则低于建筑师/工程师的估计者。此种不平衡可能是由于投标人方面偶然计算错误或未正确理解其中之风险所致。但更多的情况是, 投标人有意这样做。投标人企图通过提高先开工或他觉得实际数量有可能超过计划的工作细目价格以增加他实际需要的现金流。对于任何严重不平衡的报价, 业主应当让投标人说明其中理由, 给予澄清。业主同时还应当决定是否让中标的投标人增加履约保证金, 用于日后由于对方违约而致使自己蒙受风险时的开销。)

[7] lowest evaluated tender 最低评估价标书, 最低评标价

[8] tender (adjudication) report 评标报告

[9] financial statement 财务报表

Chapter 4　Contract Administration

4.1　Contract formation

4.1.1　Forming a contract

4.1.1.1　Forming a contract

Since contracts are so important in defining the rules by which the construction industry operates, it should be obvious that when two parties enter into a contractual relationship, each would know and acknowledge that fact. However, this is not always the case. When one of the parties denies that a contract exists, it becomes important to understand when, and how, legally binding contracts are formed. A contract is formed when one party unconditionally accepts an offer (tender, proposals, quotations) made by the second party.

The procurement agent or contract administrator may seek support early from specialists in the disciplines of contracting, purchasing, and law prior to the formation of a contract.

4.1.1.2　Essential elements of a contract[13][17][20]

Three elements for contract formation are necessary: an offer, an acceptance and consideration.

☐Offer

It can be defined as a manifestation of interest or willingness to enter into a bargain made in such a way that the receiving party realizes that furnishing unconditional acceptance will seal the bargain. The word "unconditionally" implies that full and complete agreement has been established between the parties. A manifestation of willingness with reservation of any kind is not an offer. For example, a house painter who declares, "I'll paint your house for a price of $3,000 during the third week of September, provided my other work will let me," or words to that effect, has not made a binding legal offer because the manifestation of willingness is qualified or "hedged".

☐Acceptance

The acceptance is the second element that must exist to form a contract. For the acceptance to have any relevance and legal meaning, it must be an acceptance of whatever was offered. A form of acceptance accompanied by ifs and buts or provided that changes the offer in any significant respect is not an acceptance at all but constitutes a counteroffer, which in its turn requires unconditional acceptance by the first party.

Unconditional acceptance must be communicated to the party making the offer-silence does not mean consent. It shall be considered having been communicated (and acceptance is considered complete) at the moment it is posted if it is sent by post; it is not prejudiced by loss or delay in the post. Proof of posting is therefore important. To be effective, an acceptance must be made within any limit of time stated in the offer, or, if none is stated, within a reasonable time.

An exchange of offers and counteroffers between two parties constitutes a negotiation. In a negotiation, only the final offer and acceptance matter in respect to contract formation. A contract between two parties cannot be legally binding until and unless there is meeting of the minds-that is, the mutual agreement is not made under duress-at the time the contract is formed. Both parties must understand and accept that they have mutually agreed to be bound by the same set of terms and conditions or, in other words, by the final offer and acceptance. The trouble starts when the parties later discover that they did not have a common understanding of the agreement. Such is the genesis of many building or civil engineering contract disputes.

☐ Consideration

The third element necessary for contract formation is the consideration. In building and civil engineering construction, the consideration may be a sum of money, but not always. It can just as well be some other thing, such as the discharge of an obligation that has a value. The value may not be great. The main point is that consideration for both parties to the contract must always be present in one form or another in order for a contract to be formed. One way to think of consideration is that each party must have a rationale for entering into the contract and an expectation of receiving something of value for performing the contract satisfactorily. In a building or civil engineering contract, the employer's consideration is getting the works executed, completed and defects therein remedied and the contractor's consideration is receiving the contract price.

4.1.2 Other elements essential to a valid contract[13][17][20]

In addition to the agreement evidenced by offer and acceptance there are other essential elements of a contract. Some affect every contract, but others are not likely to be met in the normal run of building or engineering contracts. Some of the requirements cause a contract to be void automatically if they are not fulfilled. Others result in a contract that can be voidable (or partly voidable) as of right by one of the parties-usually the one who has been aggrieved or put at a disadvantage by the non-fulfillment. The main requirements are:

(a) Intent by the parties to create a legal relationship between them. The intent is to be legally bound by the contents of a contract and must be clearly apparent. There is a very strong presumption in construction and other commercial contracts that such intent does exist. This prerequisite is therefore of greater importance in other branches of the law.

(b) A genuine consent of the parties. There must be an agreement between the parties to the contract based on a definite offer by one of the parties and an unqualified acceptance of the offer by the other party. A conditional acceptance or a counter-offer may supersede the original offer and cancel its effect. Once an offer is accepted, a binding contract is formed, subject to the remaining prerequisites, even if a provision is included in the contract for the subsequent execution of formal documentation, as provided in clause 9 of the *Red Book of Fourth Edition* and sub-clause 1.6 of the *Red Book*. When seeking to determine what the parties consented to do by their contract, one should not look into the parties' minds but should ascertain their intentions by the outward expression as conveyed by the written or spoken words of that contract. It is also worth noting that where a person signs a written or a printed standard form of contract, he is bound by it, whether or not he has read it, understood it or accepted it.

(c) Legal capacity of the parties to enter into a contract. There must be the capacity to contract. Any sane adult person can enter into a contract and such capacity is referred to as that of a "natural person". Corporations, on the other hand, have a judicial capacity referred to as that of an "artificial person" and created by law through a separate identity distinct from their members.

(d) Legality of the objective of the contract. The objectives of a contract must be lawful, otherwise it is void.

(e) Limitation periods. The cause of action accrues on the date of the breach of contract and it is independent of the date of any damage which may have occurred. In construction contracts, however, the date of the breach may or may not be that date when the act causing the breach was committed, since it could be extendible to the date of practical completion of the works. This extension depends on the nature of the breach and on the circumstances of the case. By comparison, under the law of tort, the cause of action accrues only when damage is suffered, so that the cause of action may not arise until long after the relevant work was carried out. The periods of time which limit an action under the law of tort vary from one country to another.

(g) Performance of a contract. When the parties to a contract perform the obligations undertaken by its content, the contract is then discharged through performance. In a construction contract, performance of the contract means, on the one hand, completion of the work as well as any matter relating to the obligations of the contractor for maintenance of defects and, on the other hand, payment by the employer. The employer, however, does in general retain his rights to sue for "breach of contract" should a latent defect be discovered within the period of limitation specified under the relevant statute, unless the contract provides otherwise.

4.1.3 Liability in the construction process[13]

The participants in the construction industry are exposed to four broad classes of liability in one or more separate ways.

4.1.3.1 Contract liability

The most prominent and obvious way a participant in the construction process becomes exposed to potential liability is by becoming a party to a legal contract. This first broad class of liability is called contract liability and results when a party to the contract breaches the contract by failing to conform to one or more of its provisions.

4.1.3.2 Tort liability

The second broad class of liability flowing to persons engaged in construction is tort liability based on tort law. The general tort concept is part of common law. Tort liability does not depend on the existence of a contract.

4.1.3.3 Statutory liability

The third broad class of liability is that imposed by law or statute and is called statutory liability. This class of liability flows directly from the provisions of enacted laws or statutes that apply in specific localities as well as from federal laws that apply throughout the United States. As is true with contract and tort liabilities, statutory liabilities may be either express or implied.

4.1.3.4 Strict liability

A second liability descriptor often used that can apply to all three broad classes of liability is strict liability, which means that it is not necessary to prove fault or negligence to establish that a person or entity is liable for some act or failure to act. The act or failure to act itself is all that is necessary to establish the liability.

Strict liability is usually associated with tort liability situations, but it can apply to other classes of liability as well.

4.1.4 Applicable law of a contract[13]

In a contract for building or engineering works that involves the employer from a country and the contractor from another, the law under which the parties' rights and obligations are determined may become a problem. It could be the law of the country where the contract is entered into, where the works is constructed, or where either of the parties to the contract comes from. It could also be the law of the jurisdiction where a significant portion of the works is manufactured, or that where the works is financed or simply the law which the parties consider well-suited to govern the particular contractual relationship.

It has been accepted in the world that subject to few limitations, the parties are free to choose on their own the law applicable to their contract. This freedom to choose is referred to as the principle of autonomy of the parties. This principle has also been adopted in international conventions.

The law which governs a contract between certain parties and by which questions as to the validity, application and interpretation of its terms are addressed is referred to as the "applicable law of the contract". In some jurisdictions, the terms "proper law of the contract" or "governing law of the contract" are used, instead.

The statement concerning the applicable law of the contract in the third (fourth) edition of the *Red Book* reads:

"*Language/s and Law*

5.1 *There is stated in Part II of these Conditions*:

……

(*b*) *the country or state the law of which is to (shall) apply to the Contract and according to which the Contract is to (shall) be construed.*"

The counterpart of the *Red Book* published in 1999 is like this:

"1.4 *Law and Language*

The Contract shall be governed by the law of the country or other jurisdiction stated in the Appendix to Tender."

Sometimes, the applicable law of certain specific matters arising out of the contract could be different from the applicable law of the contract as a whole.

If the parties to a contract are not certain what law is applicable to the contract, the principles of a branch of law referred to as "private international law" could be followed in selecting one. The law is sometimes called "the conflict of laws". The principles attempt to answer as to what law is the most appropriate and which forum is appropriate to determine a particular issue in an international or inter-jurisdictional context. This branch of the law forms part of the legal system of every jurisdiction and therefore, there are as many systems of conflict of laws as there are jurisdictions.

Three alternatives are available as to the determination of the applicable law of a contract:

(a) where there is an express choice of the applicable law;

(b) where there is an inferred choice of the applicable law; and

(c) where there is no choice of the applicable law.

When entering into a contract for building or engineering work, the parties may expressly choose the law which they wish to apply to their contract. This choice may be expressed by a simple statement naming the country to which the chosen law belongs.

It is wise for the parties to exercise their discretion carefully and to choose an appropriate system of law as the applicable law of the contract. Of course, what may be "appropriate" for one party may not be so for another. Because of the "consumer sovereignty" nature of the global construction market, the choice of the applicable law of the contract is generally made by the employer.

Words, Phrases and Expressions 4.1

[1]　contract formation 合同订立

[2]　enter into 使自己成为一方当事人(make oneself a party to or in; *enter into an*

agreement)

[3] enter into a contract 缔结合同（注意与sign a contract 意义不同，sign a contract 是"在合同上签字"，或"签署合同"）

[4] offer（tender, proposals, quotation, etc.）在合同法中称做"要（yāo）约"，但在不同行业的合同中，名称各异，如"标价"、"建议"、"报价"、"报盘"、"发价"等。

[5] accept *n*. 在合同法中称做"承诺"，具体的行业都有自己的称呼。

[6] bargain *n*. 交易，买卖（agreement to buy or sell *sth*., exchange *sth*., do *sth*. made after discussion; *sth*. obtained as the result of such an agreement）

[7] enter into a bargain 做买卖，参加交易

[8] unconditional acceptance 无条件承诺，无条件接受要约（标价、建议、报价、报盘、发价等）

[9] seal *vt*. 决定，解决（decide or settle irrevocably）

[10] seal the bargain 做成买卖，成交

[11] hedge *vt*. 设防，两面下注避免损失（protect against a possible loss by counterbalancing(bets, etc.)）

[12] effect *n*. 大意（main idea or meaning）

[13] words to that effect 表示上述意思的话，表示那样意思的话 words to this effect 表示这个意思的话

[14] counteroffer *n*. 反要约，还价，发还价（a return offer made by one who has rejected an offer）

[15] duress *n*. 胁迫（compulsion by threat, forcible restraint or restriction, unlawful restraint）

[16] genesis *n*. 开始，起源（beginning, starting point）

[17] run *vt*. 组织，管理，使活动（organize, manage, cause to be in operation）

[18] aggrieve *vt*. （通常用被动式）使苦恼，使悲伤（usually passive) grieve

[19] legal capacity *n*. 法律身份

[20] object(ive) of a contract *n*. 合同标的

[21] tort *n*. 侵权（行为），民事过失（a civil wrong, not including a breach of a contract, for which the injured party is entitled to compensation）

[22] contract liability 合同责任

[23] tort liability 侵权行为责任

[24] statutory liability 法定责任

[25] strict liability 后果责任，严格赔偿责任，严格责任（liability imposed without regard to fault）

[26] applicable law of a contract 合同适用法

[27] govern *vt*. 支配，影响；指导（determine, influence; direct）

[28] principle of autonomy of parties 当事人自决原则

[29] proper law of the contract 合同适当法

[30] governing law of the contract 合同支配法
[31] private international law 国际私法
[32] conflict of laws 冲突法，国际私法

Notes 4.1

疑难词语解释

[1] Some affect every contract, but others are not likely to be met in the normal run of building or engineering contracts.

【译文】 某些要素影响到每一项合同，而另外一些则在正常履行建筑或工程合同时不可能遇到。

[2] A form of acceptance accompanied by ifs and buts or provided that that changes the offer in any significant respect is not an acceptance at all but constitutes a counteroffer.

【译文】 带有"如果和但是"或者"条件是……"的承诺，只要在某一重要方面改变了要约，就不能算是承诺，而是一种反要约或还价。

【解释】 这里 ifs 和 buts 是名词，provided that 本来是连词，意思是"以……为条件"，但是这里也当做名词用。provided that 后面的 that 引起一个修饰 ifs and buts or provided that 的一个定语从句。

4.2 Pre-contract arrangements

4.2.1 Letter of Intent

When it is not immediately possible to issue a formal letter of acceptance to the successful tenderer, it may be useful for the employer or A/E to issue a letter notifying of the employer's intention to enter into a contract with him based on his tender. Sometimes the letter of intent will ask the successful tenderer to undertake certain preliminary work. Such a letter should be precise and contain at least:

(a) a statement of intent to accept the tender at a future date;
(b) instructions to proceed (or not to proceed) with ordering materials, letting subcontracts and so on;
(c) a statement of what costs, if legitimately incurred, will be reimbursed to the contractor if eventually no contract is made;
(d) a limit to financial liability before formal acceptance;
(e) a statement that formal acceptance will render provisions of the letter of intent void;
(f) a request for acknowledgement of receipt and agreement to the conditions of the letter of intent.

The successful tenderer should treat the letters of intent with great caution since they rarely contain any express undertaking by the employer to pay for any work done in anticipation of the contract. If the contract does not proceed, the tenderer may find it difficult to obtain payment. At worst, the tenderer could be involved in expensive court proceedings. The letter as shown in Figure 4.1 is a suggested acknowledgment which may help to safeguard tenderer's interests.

Contractor's address
12 October 2006

A/E's address

Dear Sir,

Re: Employer's intention to enter into a contract with us for Good Luck Plaza

Thank you for your letter of 6 October 2006 informing us of the employer's intention to enter into a contract with us for this project on the basis of our tender of 6 October 2006.

You request us to start the access to the site in anticipation of the contract. However, before undertaking this preparatory work we would like to have an express undertaking from the employer to reimburse the cost of any work so undertaken, together with a reasonable allowance for overheads and profit, should the contract not proceed for any reason.

On receipt of this undertaking, we shall proceed immediately in accordance with your request.

Yours sincerely,

Zhang Dacheng

Project manager

Figure 4.1　A Letter of Intent[23]

4.2.2　Letter of Acceptance

The employer should issue a letter of acceptance as finally modified and agreed between the employer and the selected contractor. This should inform the contractor that all future correspondence regarding the administration of the contract, other than any formal requirement to the contrary in the contract, should be addressed to the A/E. Alternatively, the employer may send to the selected contractor a counter-offer. This is an offer comprising the contractor's original tender but incorporating any modifications or amendments which the employer wishes to make.

The letter of acceptance or counter-offer may be issued by the A/E, on the written authority of the employer, provided that the existence of such authority is made known to the contractor.

A legally binding contract is established when a tender is accepted by the employer or the employer's counter-offer is accepted by the contractor. Formal acceptance in writing is not legally essential, but it is prudent and commercially preferable so as to prove the scope and existence of the contract. A sample letter of acceptance is given in Figure 4.2.

Our Ref: 05083/SPC/01
Land View Construction Co., Ltd.
1037 Kimball Bridge Road
Roswell, GA 30l123

BY FAX & POST
27 March 2005

Attn: Mr. John Sims

Dear Sirs,

<u>Letter of Acceptance-Main Contract for Proposed Multi-Service Centre on LOT 1075 North Pointe Road, Roswell</u>

We refer to your submission of tender for the captioned works on 23rd September 2004 and the following post-tender correspondences:

1. Letter dated 25th September 2004 from SRSS to Land View Construction Co., Ltd.
2. Letter dated 27th September 2004 from Land View Construction Co., Ltd. to SRSS.
3. Letter dated 5th October 2004 from Land View Construction Co., Ltd. to Design 2.
4. Fax transmittal dated 25th November 2004 from Design 2 to Land View Construction Co., Ltd.
5. Letter dated 21st November 2004 from Land View Construction Co., Ltd. to Design 2.
6. Letter dated 22nd January 2005 from Land View Construction Co., Ltd. to Design 2.

On behalf of the Nursing House Association of Roswell, we are pleased to advise that your tender in the amount of $3,267,727.00 (including a contingency sum of $100,000.00) has been accepted. The contract period is 320 calendar days from the Date for possession which will not be later than 30th April 2005.

Unless and until a formal agreement is prepared and executed, your submitted tender together with this letter shall constitute a binding contract between you and the Employer.

Contract documents are being prepared and you will be notified when they are ready for execution. All post-tender correspondence above-mentioned shall form part of the contract documents.

Please acknowledge your receipt of this letter of Acceptance by countersigning and returning the duplicate copy of this letter.

Yours faithfully,

John C. Bennett
Authorized person

JB/as
Cc: NHA-Miss Agatha Carter
     SRSS-Mr. Harry Paley

Figure 4.2   A Letter of Acceptance

4.2.3 Agreement

The form of agreement, if the employer requires, should be drawn up and executed by the employer or the A/E and contractor at this stage. If, in the rush to get work started, the agreement is left until later, difficulties may arise over the precise wording, particularly if the tender has been modified by agreement between the parties during the period between its submission and the issuing of the letter of acceptance.

The contractor should check the contract documentation before he signs them, in particular, check the deletions and entries in the appendix. The information to be stated there should have been set out in the bills of quantities on which he tender.

The contractor should also check that any amendments to the printed conditions have been made as notified to him at the time of tendering, and that no amendments have been of which he has not been notified. The contractor and the employer shall initial any deletions and amendments.

4.2.4 Contract price

Majority of standard forms of contract define "contract price" as the sum stated in the Letter of Acceptance as payable to the contractor in consideration of his execution, completion of the works and remedying of defects therein in accordance with the provisions of the contract.

Contract price may be adjusted by agreement between the employer and the contractor in order to reflect any changes that may occur and affect significantly the cost of works in accordance with the contract.

In practice, the actual quantities of work to be performed may vary from the quantities estimated by the A/E and stated in the bill of quantities. The contract price is the total sum of money due to the contractor for fulfilling the contract. It is the amount obtained by applying the tender rates to the quantities of work actually performed, together with preliminary items, amounts for variations, contract price fluctuations, claims and all other amounts to which the contractor is properly entitled under the provisions of the contract.

The *Red Book* defines contract price, through sub-paragraph 1.1.4.2 and sub-clause 3.5, 12.1, 12.2, 12.3 and 14.1, as the price, inclusive of adjustments in accordance with the contract, that the engineer and the contractor agree or the engineer determines, after due consultation with each party in an endeavour to reach agreement, by evaluating each item of work, applying the appropriate rate or price for the item to the measurement agreed or determined in accordance with the contract.

Words, Phrases and Expressions 4.2

[1] letter of intent 意向书(An obligation instrument that can be used to protect price and

availability of long-lead items and for other purposes as specifically stated. 可用来保护长期物品价格和货源，以及其他具体指定用途而承担义务的一种文件。)

[2] preliminary work 准备工作

[3] statement of intent 意向说明书

[4] instructions to proceed with ordering materials, letting subcontracts 着手订购材料，发出分包合同的指示

[5] render vt. 致使(cause to be (in some condition))

[6] undertaking 承诺

[7] in anticipation of 在……之前

[8] proceedings n. 诉讼(legal actions against sb.)

Notes 4.2

疑难词语解释

[1] A legally binding contract is established when a tender is accepted by the employer or the employer's counter-offer is accepted by the contractor. Formal acceptance in writing is not legally essential, but it is prudent and commercially preferable so as to prove the scope and existence of the contract.

【译文】 当业主接受了标书，或承包商接受了业主的还价，就订立了一份有法律约束力的合同。正式书面协议书的法律效力不大。但是，为了谨慎起见，最好还是要这份协议书，以便见证合同的范围与存在。

[2] The form of agreement, if the employer requires, should be drawn up and executed by the employer or the A/E and contractor at this stage.

【译文】 如果业主要求签署协议书，则协议书格式应由业主同承包商或建筑师/工程师同承包商起草并签字盖章，使之生效。

[3] The contractor and the employer shall initial any deletions and amendments.

【译文】 承包商与业主应在所有删除和修改之处签字。

[4] Formal acceptance will render provisions of the letter of intent void.

【译文】 正式接受函会使意向书某些条文失效。

[5] Express undertaking by the employer to pay for any work done in anticipation of the contract.

【译文】 业主明确承诺为所有在签约前完成的工作支付款项。

4.3 Contract administration

4.3.1 Contract as an agreement

A contract can be defined as an agreement between two or more parties to do or not to do something and it, as a result of the intention, creates a legal relationship which is enforceable by law.

The agreement can be simple or complex, and can reflect the simplicity or complexity of its objectives, i.e. the building or engineering works in the case of a construction contract. Contracts includes terms and conditions, and can include other items such as the seller's proposal or marketing literature, and any other documentation which the buyer is relying upon to establish what the seller is to perform or provide.

Various statements, promises and stipulations are grouped together in a contract under the "terms". It is the terms of a contract that specify the extent of each party's duties, obligations and rights.

The legally binding nature of a contract usually means that it will be subjected to a more extensive review and approval process and makes the parties be acutely aware of the legal implications of actions taken when administering a contract.

The review and approval aims primarily to ensure that the wording describes products, services, or results which will satisfy the buyer's identified needs. In the case of major public projects the review process may involve the public in reviewing the agreement.

A contract is subject to remedy in the courts and can be amended any time prior to contract closure by agreements between the parties. Such amendments may not always be equally beneficial to both the seller and the buyer.

Any standard form of contract shall be tailored to suit the specific needs of a project.

Careful preparation and wording of the conditions of the contract helps identify some risks and then avoid or mitigate them.

The buyer-seller relationship may exist at many levels on any one project and between any two organizations, e.g. the employer (buyer) and the A/E (seller), the employer (buyer) and the contractor (seller), the contractor (buyer) and subcontractors (seller), the contractor (buyer) and suppliers (seller), etc..

This chapter assumes that a formal contractual relationship is created and exists between the buyer (employer or contractor) and the seller (contractor or subcontractor or supplier).

4.3.2　Contract administration[16]

Entering into a contract for works or services is one method of allocating the responsibility for managing or assuming potential risks.

Complex works or services may involve multiple contracts or subcontracts simultaneously or in sequence and need an active management. In such cases, each contract life cycle can end during any phase of the project life cycle.

The purpose of administering the contract by both parties is similar, that is to ensure that his own legal rights are protected and obligations of the other party are met. On larger projects where multiple contracts exist it is essential to manage interfaces among the providers of various goods or services.

Contract administration commences once the employer has issued the tender documents and the contractor has decided to tender for the contract. It ends in most cases when the contract is discharged in a manner as provided in the contract. Therefore, the root causes of success or failure of contract administration can be traced back to its commencement.

4.3.3 A/E as contract administrator

The employer usually delegates contract administration onto the A/E or an independent consulting firm.

The A/E reviews and documents how well a seller (contractor, subcontractors and suppliers) is performing or has performed based on the contract and established corrective actions. Also, the performance is documented as a basis for future relationships with the seller. Seller performance evaluation by the buyer is primarily carried out to confirm the competency or lack of competency of the seller, relative to performing similar work on the project or other projects. Similar evaluations are also carried out when it is necessary to confirm that a seller is not meeting the seller's contractual obligations, and when the buyer contemplates corrective actions. Contact administration includes managing any early termination of the contracted work (for cause, convenience, or default) in accordance with the termination clause of the contract.

4.3.4 Financial management[15][16]

Contract administration also has a financial management component that involves monitoring of payments to the seller (contractor, subcontractors and suppliers). This ensures that payment terms defined within the contract are met and that seller compensation is linked to seller progress, as defined in the contract.

4.3.4.1 Financial management by the A/E

The resident engineer (the A/E's site representative) should at an early stage agree ground levels with the contractor and suitable arrangements for dealing with daywork sheets and claims for additional costs. An accurate record of all drawings, both original and revised, should be maintained and all instructed variation when valued and filed. Continuous records should be kept of all important matters such as labour employed, equipment in use, weather conditions and causes of delay all of which could subsequently be used to verify the contractor's claims.

During the currency of a contract a competent staff member should effectively keep cost and expense under control by constantly checking cost items which have been incurred and advising the resident engineer on the situation in adequate time for any necessary corrective action without adverse effects on the project.

The employer should be informed of his financial commitments and when he will be required to make payments. The resident engineer and his staff must effectively keep in check expenditure on variations, provisional sums, etc., all set against quotations, quality control, time for completion and claims.

4.3.4.2 Financial management by the contractor

If a particular contract is profitable depends basically on knowing the value of work executed at any specific date, as opposed to the cost actually incurred in earning that value. The difference between the two figures will be the amount available to allocate to the contractor's off-site overheads, to fund its working capital and make a profit. In an adverse situation the difference may show that off-site overheads are not being covered and that there will be no profit. In the worst situation, the actual cost incurred on the site may exceed the value of work.

(1) Cost control

It is the profit or loss that is the primary concern of the contractor's site quantity surveyor or other competent staff member. He works closely in this capacity with the contractor's representative, who is monitoring performance, comparing it against pre-determined objectives and taking remedial action where necessary. The factors such as the operatives, materials, machines and subcontractors shall be given sufficient attention. Equally important are the non-tangible items such as progress and productivity, cost, quality, safety, information, methods and the performance of subordinate management staff.

(2) Data as input to cost control

The main sources of data available to the contractor's quantity surveyor are the bill of quantities, estimates of cost, method statement and the master programme. During construction these sources of data will be supplemented by interim valuations; up-to-date accounts of labour, equipment, materials and subcontracted work; salaries and all other site costs; and finally the programme of work executed compared with the assumptions upon which the tender was based.

(3) Earned value technique (EVT)[16]

This technique compares the cumulative value of the budgeted cost of work executed (earned) at the original allocated budget amount to both the budgeted cost of work scheduled (planned) and to the actual cost of work executed (actual). This technique is especially useful for cost control, resource management, and production.

An important part of cost control is to determine the cause of a variance, the magnitude of the variance, and to decide if the variance requires corrective action. The earned value technique uses the cost baseline contained in the project management plan to assess project progress and the magnitude of any variations that occur.

EVT involves developing three key values for each item of work:

□Planned value (PV) is the budgeted cost for the work to be executed up to a given point in time.

☐ Earned value (EV) is the budgeted amount for the work executed during a given time period.

☐ Actual cost (AC) is the total cost incurred in completing the work during a given time period. This AC must correspond in definition and coverage to whatever was budgeted for the PV and the EV (e.g., direct hours only, direct costs only, or all costs including indirect costs).

The PV, EV and AC values are used in combination to provide performance measures of whether or not work is being accomplished as planned at any given point in time. The most commonly used measures are cost variance (CV) and schedule variance (SV). The amount of variance of the CV and SV values tend to decrease as the project reaches completion due to the compensating effect of more work being accomplished. Predetermined acceptable variance values that will decrease over time as the project progresses towards completion can be established in the cost management plan.

Cost variance (CV) = $EV - AC$. CV at the end of the project will be the difference between the budget at completion (BAC) and the actual amount spent.

Schedule variance (SV) = $EV - PV$. SV will ultimately equal zero when the project is completed because all of the planned values will have been earned.

Figure 4.3 uses S-curves to display cumulative EV data for a project that is over budget and behind the programme. An example can illustrate the use of EVT in measuring the performance.

Assume a contractor wins a contract for earthwork. The volume which he is supposed to excavate is $10,000$ m^3. The rate for the excavation work entered in the bill of quantities by the contractor and accepted by the employer is ¥45.00/m^3 and the extended amount shall be ¥45.00/m$^3 \times 10,000$ m^3 = ¥$450,000$. The Time for Completion is 25 days. It is easy to know that the contractor shall excavate an average volume of 400 m^3 a day.

The contractor's quantity surveyor made a measurement in the morning of the seventh day of commencement and the measurement is $2,000$ m^3, and checked the cost and expense which the contractor has incurred and it has been found at ¥$120,000$.

The quantity surveyor made a few simple calculations and figured out that:

$$EV = ¥45.00/\text{m}^3 \times 2,000 \text{ m}^3 = ¥90,000$$

$$PV = ¥45.00/\text{m}^3 \times 400 \text{ m}^3/\text{day} \times 6 \text{ days} = ¥108,000$$

As far as AC, the total cost incurred in excavating the $2,000$ m^3, is concerned it is:

$$AC = ¥120,000$$

Thus it is easy to know, based on the three amounts, that:

$$CV = EV - AC = ¥90,000 - ¥120,000 = -¥30,000 < 0,$$

which indicates that the contractor has spent more than he should have spent.

$$SV = EV - PV = ¥90,000 - ¥108,000 = -¥18,000 < 0,$$

which indicates that the progress is behind the programme. The delay can be determined by means of a calculation as follows:

The volume needs to be excavated so that it can be valued at ¥18,000 amounts:

$$¥18,000/(¥45.00/m^3) = 400 \text{ m}^3,$$

Then, the duration needed to excavate a volume of 400 m³ shall be:

$$400 \text{ m}^3/(400 \text{ m}^3/\text{day}) = 1 \text{ day},$$

which reveals that the progress is 1 day behind the programme.

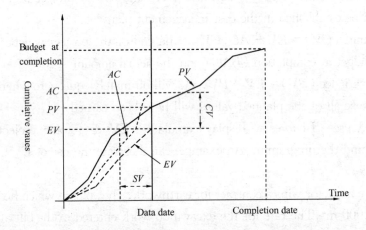

Figure 4.3 Illustrative Graphic Performance Report

The earned value technique in its various forms is a commonly used method of performance measurement. It integrates project scope, cost (or resource) and schedule measures to help the project management team assess project performance.

4.3.4.3 Financial reporting

Reliable and regular financial reporting is indispensable in cost control. The reports should be produced ideally on a monthly basis. A basic financial report of a contract should contain:

(1) initial tender figures and expected profit;
(2) forecast figures at completion for value and profit;
(3) current payment application by the contractor;
(4) current certified value;
(5) adjustments to the certified valuation;
(6) costs to date and the accounting period in question; and
(7) cash received to date, retention deducted and certified sums unpaid.

A typical financial report summary is shown in Figure 4.4.

Typical Financial Report Summary[33]

(A) General Information
(a) Contract Name: New Factory
(b) Contract Reference: 691
(c) Contract Address: West Street, Shatin
(d) Client: Rapid Development Ltd.
(e) Contractor: Better Build Contractors Ltd.
(f) Architect: Design Partnership, South Street, Wanchai

	HK $
(B) Bills of Quantities Breakdown	
Bill No.1: Preliminaries (excluding dayworks and contingencies)	812,461
Dayworks	85,650
Contingencies	100,000
Bill No.2: Preambles	
Bill No.3~15: Measured work	5,632,682
Bill No.16: Provisional sums	361,000
Bill No.17: *PC* sums (including profit and attendance)	<u>1,691,856</u>
CONTRACT SUM	8,683,649
(C) Final Account Breakdown	HK $
(a) Remeasurement Account	(-1,237,540
	(+1,418,962
(b) Variation Account	(-582,946
	(+672,891
(c) Provisional Sums	380,691
(d) Dayworks	31,968
(e) Nominated Subcontractor's & Supplier's Final Accounts (including profit and attendance)	1,721,684
(f) Fluctuations	550,281
(g) Claims (agreed)	45,291

Figure 4.4   A financial report summary

Words, Phrases and Expressions 4.3

[1]   quantity surveyor 工料估算师 (Quantity surveying is a traditional profession in the United Kingdom and a number of commonwealth countries or jurisdictions. In the past quantity surveyor is a person who was concerned with the measurement and valuation of construction work, but currently the practice of quantity surveying extends well beyond these basic activities. Modern quantity surveying is concerned with the financial and economic aspects of construction. Its activities extend from the provision, for the prospective building owner, of reports on feasibility, through early

stage estimates and cost-planning of the evolving design, into the preparation of bills of quantities, the obtaining and evaluation of tenders, the negotiation of contractual arrangements and the complete administration of the financial aspects of construction contracts. Within construction firms quantity surveyors provide the commercial counterparts to the above and sometimes undertake project planning and production management. 工料估算是英国和一些英联邦国家或地区年头久远的职业。过去，工料估算师就是计量建筑工程，估算其价值。如今，这一行业已远非限于上述基本活动。现代的工料估算受理的是建筑业的财务与经济方面。活动范围已从为上门的建筑物业主编写可行性报告、早期阶段费用估算、设计各阶段的费用规划，扩大到编制工程量清单，招标与评标，商定合同安排，完成施工合同财务方面的管理。建筑公司的工料估算师则与建筑物业主雇用的工料估算师打交道，地位对等，有时承担项目规划和生产管理的任务。)

4.4 Contract documents and their priority

4.4.1 Contract

The *Red Book* recognizes that the conditions of contract will be accompanied by a contract agreement, drawings, specification, bill of quantities, tender and appendix, and a written acceptance of the tender.

4.4.2 Priority of contract documents

The several documents forming the contract are to be taken as mutually explanatory of one another, but in case of ambiguities or discrepancies the same shall be explained and adjusted by the A/E who shall thereupon issue to the contractor instructions thereon.
Sub-clause 1.5 of the *Red Book* sets out that for the purposes of interpretation, the priority of the documents shall be in accordance with the following sequence:
(a) the contract agreement (if any),
(b) the letter of acceptance,
(c) the letter of tender,
(d) the particular conditions,
(e) the general conditions,
(f) the specification,
(g) the drawings, and
(h) the schedules and any other documents forming part of the contract.
If the contractor finds any ambiguity or discrepancy in the documents, he shall write to the A/E for necessary clarification or instruction. Figure 4.5 is a letter from the contractor to the architect for this purpose.

Contractor's Address
23 October 2006

A/E's address

Dear Mr. Zhao,

Re: Discrepancies in the contract documents

On examination of the Contract Drawings and the Bills of Quantities we find the following discrepancies:

1. Item C, page 12, Bill No.3 gives the roof insulation boards as aerated concrate boards whereas Drawing No.060812 shows laminated plastic boards.

2. Drawing No.970815 shows a recess in the east wall of Room 43 which is not shown on Drawing No.060811.

Will you please issue your instructions with regard to these discrepancies as soon as possible as it will be impossible for us to order the necessary materials until these problems are resolved.

Yours sincerely,

Liu Yugui
Project manager

Figure 4.5 Letter where there are discrepancies or divergences in the contract documents

Sub-clause 1.8 of the *Red Book* requires that if a party becomes aware of an error or defect in a document which was prepared for use in executing the works, the party shall promptly give notice to the other party of such error or defect.

When the contractor comes across any errors in a document, it is strongly advised for him to do so, particularly where the error is against the contractor. In this case it is suggested to write to the A/E a letter like the one shown in Figure 4.6.

Architects/engineers are not infallible and errors will sometimes occur in the documents which they provide, such as the setting-out drawings. If the contractor finds inaccuracies he should draw their attention to the problems at once. To this end, letter shown in Figure 4.6 can be used or adapted. If such drawings are not clear or need amplifying so as to enable the contractor properly to set out, the A/E is bound to issue the necessary further drawings or details.

Contractor's address
30 October 2006

A/E's address

Dear Mr. Huang,

Re: Questionable quantities

On examination of the Bills of Quantities it is apparent to us that Item D on page 21 of the Bill No. 5 relating to roof insulation is questionable in that it seems to give only two thirds the total area of roof insulation actually required.

In accordance with Sub-Clause 7.1 of the Instructions to Tenderer, this error/departure will require to be corrected and the correction treated as a variation under Sub-Clause 13.1 of the General Conditions of Contract.

Please confirm that this will be done and that the amount of the correction will in due course be included in certificates when the work is carried out.

Yours sincerely,
Liu Yugui
Project manager

Figure 4.6   Letter to Architect Regarding Errors in Bills of Quantities

Contractor's address
3 November 2006

A/E's address

Dear Mr. Luo,

Re: Setting out dimensions

With reference to your Site Plan showing dimensions for setting out the Works, we have to inform you that following our own site survey, we have found the following errors and omissions on your drawings:

1. Dimension between the center lines A and P is shown to be 176 m in Site Plan, while the site can not accommodate the size. For your consideration, our re-examination of Drawing No. A060801, Ground Floor Plan, indicates that the same dimension is 116 m.

2. Drawing No. S060815 shows a staircase between the center lines 4 and 5 which is not shown on Site Plan.

Will you please revise the drawing to show correct dimensions and issue it to us so that we may proceed with the setting-out of the Works. Since the setting-out is scheduled to commence on Monday next it is essential that we receive your revised drawing no later than 8:00 am of that day.

Yours sincerely,

Liu Yugui
Project manager

Figure 4.7 Letter pointing out inaccuracies in setting out dimensions

Words, Phrases and Expressions 4.4

[1]set out *vt*. 展示(show, put on display) *set out the Works* 放线（将工程在地面上展示出来）

4.5 Contract administration during construction

4.5.1 Commencement date

Shortly after the contractor receives the letter of acceptance (within 42 days), the A/E must notify the contractor (give him not less than 7 days' notice as provided under sub-clause 8.1 of the *Red Book*) in writing of the date for commencement of the works in order to initiate the execution of the works and to declare the date from which the time for completion will run. Thereafter the contractor is responsible for proceeding with due diligence and expedition and without delay and completing the works in accordance with the contract.

4.5.2 Statutory notifications

The contractor must give all statutory notifications of his new activity to the appropriate bodies, such as the health bureau, fire department, water authority, municipal administration bureau, electricity board, environment protection bureau, etc, but he may assume that all planning applications relative to the works and specified temporary works have been obtained by or on behalf of the employer.

Sub-clause 1.13 of the *Red book* requires the contractor to give all notices, pay all taxes, duties and fees, and obtain all permits, licenses and approvals, as required by the laws (means, in the context of the *Red Book*, all national (or state) legislation, statutes, ordinances and other laws, and regulations and by-laws of any legally constituted public authority) in relation to the execution and completion of the works and the remedying of any defects.

4.5.3 Compliance with laws

All work must be undertaken in accordance with the laws in force in the country where the work is performed. In addition the contract documents may refer to specific acts or regulations of special significance with which the contractor must comply. It is the contractor's responsibility to assess the significance of these and other legal obligations and to make appropriate allowances in his tender.

Sub-clause 1.13 of the *Red Book* obliges the contractor to comply with, and give all notices required by or under laws as therein defined. If compliance involves a change from what is set out in the contract documents this is to be treated as a variation.

As with discrepancies in the contract documents the contractor is under an implied obligation to find divergences between what the contract documents require and any relevant statutory requirements. If they are found the contractor must notify the A/E immediately by a letter as shown in Figure 4.8.

Contractor's address
3 November 2004

A/E's address

Dear Sir,

Re: Divergence between statutory requirements and contract documents

We must draw your attention to a divergence between the Contract Documents and statutory requirements as follows:

1. Drawing No. A040821 shows two windows in the west flank wall which open to the neighboring property, while the Fire Protection Code prohibits such windows.

2. Specifications require that the fire doors be made of asbestos fiberboard, while it is mandatory, according to the Health Regulations, that they be made of glass fiberboard.

Please let us have your instructions by 7 November 2004 so that we can proceed with the work, without disruption or delay to the progress of the Works.

Yours sincerely,

Liu Yugui
Project manager

Figure 4.8 Letter Indicating Divergence between Statutory Requirements and Contract Documents

Contractor's address
3 November 2004

A/E's address

Dear Sir,

Re: Emergency work to comply with statutory requirements

We have received from the District Council a Dangerous Structure Notice under s.62 of the Urban

Development Code relating to the flank wall of the workshop. This notice requires us to take immediate action to ensure stability of the wall. We are now proceeding to do this by the provision of raking shores as advised by the District Surveyor.

The additional work constitutes compliance with laws under Sub-Clause 1.13 of the General Conditions of Contract and a variation under Sub-Clause 13.1. Therefore, under Clause 12.3, we shall require the work which we are carrying out to be valued as if it were a variation instruction issued under Clause 13.1.

We shall be glad to receive your immediate confirmation and any further instructions which you may consider necessary to ensure the permanent stability of the wall. We must also give you notice that the shores which we are compelled to erect will seriously interfere with our work while they remain in position and please, therefore, treat this compliance as a cause of delay giving an entitlement to extension of time under Sub-clause 8.5 and this letter as a notice under Sub-Clause 8.4.

Yours sincerely,

Liu Yugui
Project manager

Figure 4.9 Emergency Work to Comply with Wtatutory Requirements

Sometimes the contractor will have to carry out emergency work in order to comply immediately with a statutory notice. In that event the contractor should confine himself to carrying out only essential work and must inform the A/E forthwith. If the contractor does anything more than strictly necessary to deal with the emergency he will not be entitled to payment for the extra work, nor will he be so entitled if he is responsible for the emergency. Figure 4.9 gives an example letter.

4.5.4 Finance

The employer sets his financial arrangements in motion in line with the cash flow projection of the contract provided by the A/E, who will keep him informed of the financial requirements of the contract as it proceeds.

Sub-clause 2.4 under the *Red Book* requires the employer to submit, within 28 days after receiving any request from the contractor, reasonable evidence that financial arrangements have been made and are being maintained which will enable the employer to pay the contractor the contract price (as estimated at that time) in accordance with the contract. If the employer intends to make any material change to his financial arrangements, the employer shall give notice to the contractor with detailed particulars.

4.5.5 Programme and methods of construction

The contractor's main time obligations are to complete by a specified date, to progress the works to a specified standard (due diligence or due expedition) and to manage progress by taking steps to mitigate delays. The existence and extent of these obligations is determined by the terms of the contract.

On the other hand, construction is a constantly changing and developing process. Thanks to the developments in communications and in computers the constantly updating software for project management has made it much easier to produce and keep up to date the programmes for construction.

At least three uses have been identified of the programme are[27]:
(1) to monitor progress by comparing actual duration against planned duration;
(2) a management tool, to decide the priority to be given to activities to achieve the target completion dates.
(3) to establish the contractor's entitlement to extension of time.

The exact use of the programme in contract administration and whether or not it creates obligations and rights depends upon the terms of the contract.

Both the employer and contractor will monitor the progress of the works in order to anticipate whether or not the date is likely to be achieved.

Sub-clause 8.6 in the *Red Book* provides that the engineer has the authority to instruct the contractor to take measures to meet the completion date if actual progress is too slow to achieve required completion or if progress is behind programme.

The A/E has also the authority to instruct revision to the programme if progress does not match the programme.

In addition to the primary obligation to complete, the contractor usually has an additional primary obligation to progress the works. Failure to comply may form grounds for termination. The *Red Book* requires the contractor to proceed with due expedition and without delay under sub-clause 8.1. Sub-clause 8.3 requires the contractor to proceed in accordance with the programme. The employer's right to terminate the contract may arise from the contractor's failure to proceed in accordance with the programme without reasonable excuse as stated in sub-clause 15.2 (c).

If a programme is to be used to monitor compliance with two primary obligations, then the contract must provide for the contractor to submit a detailed first programme which is realistic representation of his plans and provide for regular revisions to record actual progress and revisions whenever progress is not in accordance with the programme.

Three types of programme are commonly used. The progress programme helps identify problems of progress; the prediction programme facilitates the prediction of future progress; and the management programme allows assessment of the effect on future progress of different corrective actions.

Progress programme in common bar chart form is used to monitor compliance with the obligation to complete by the specified date. Progress is monitored by comparing actual progress of each activity with the estimated durations and timings on the bar charts.

Prediction programme provides a means of updating the prediction of future progress by providing a model of the tiome dimension of the project. It does this by linking individual activities by construction logic links. The effect of delay or acceleration of one activity on the obligation to complete by a specified date can now be estimated. A change in method of working can be taken into account by changing the model of logic links. The obligation to progress the work can also be effectively monitored. The linked bar chart is the most common prediction programme. If actual progress differs from planned, then the programme needs to be revised to show actual progress of activities, revised to show changes in the construction logic and the incidence of actual events.

The management programme is used for analysis and implementation of corrective action. It is the only programme that allows all three obligations to be effectively managed. The most common form of programme is the critical to completion to be identified. Those activities that are not critical will be identified by the extent of "float". In practice, float is an essential requirement of a management programme. Resources are of course limited on construction projects. Giving priority to critical or near critical activities to achieve planned completion, is only possible if lack of action on other activities can be accommodated by available float. To be effective, the management programme needs to be based on up-to-date progress, represent an accurate prediction of future events as well as accurately model the time characteristic of project by logic links. It is an important aspect of a management progrmme that it should identify the activities by other and that the contract obligations of others is measured by reference to actual progress and the update main programme. This is difficult to achieve in practice without a consistent project approach to the use of the management progamme.

A distinction needs to be made between the programme required by the contract and the programme required to manage the contract.

Sub-clause 8.3 of the *Red Book* requires the contractor to submit a detailed time programme to the engineer within 28 days after receiving the notice of commencement and a revised programme whenever the previous programme is inconsistent with actual progress or with the contractor's obligations. Each programme shall include:

(a) the order in which the contractor intends to carry out the works, including the anticipated timing of each stage of design (if any), contractor's documents, procurement, manufacture of plant, delivery to site, construction, erection and testing,

(b) each of these stages for work by each nominated subcontractor,

(c) the sequence and timing of inspections and tests specified in the contract, and

(d) a supporting report which includes:

(i) a general description of the methods which the contractor intends to adopt, and of the major stages, in the execution of the works, and

(ii) details showing the contractor's reasonable estimate of the number of each class of contractor's personnel and of each type of contractor's equipment, required on the site for each major stage.

Obviously, the sub-clause does not specify the form that the programme should take. No obligation is specified to represent the stated method of working on the programme in the form of logic links between activities, to show the construction logic. A programme in bar chart form, together with the supporting report, suffices to represent the obligation under sub-clause 8.3. A programme in linked bar chart form may be used for the purpose of prediction and to fulfill the role required of the programme under sub-clause 8.3, 8.6 and 4.21(h).

The programme, the methods, and the details to be provided by the contractor are subject to the A/E's approval. Unacceptable proposals must be amended as necessary and approval obtained at the earliest possible date if work is not to be delayed.

Letters as shown in Figures 4.10 and 4.11 are suitable for use when submitting the programme or any revision of it to the engineer.

Contractor's address
23 October 2005

A/E's address

Dear Mr. Hu,

Re: The master programme

In accordance with Sub-Clause 8.3 of the General Conditions of Contract we enclose two copies of our master programme for the execution of the Works. You will note that we have presented this in both bar chart and network form. Please let us have your approval of this programme as laid down in the conditions of contract.

Yours sincerely,

Liu Yugui
Project manager

Figure 4.10　Letter Regarding Programme

Contractor's address
29 October 2005

A/E's address

Dear Mr. Fang,

Re: Revision of master programme

In accordance with Sub-Clause 8.3 of the General Conditions of Contract and your extension of time granted on 27 October 2005 we now enclose two copies of our amended and revised master programme relating to the new Completion Date.

Please acknowledge receipt.

Yours sincerely,

Liu Yugui
Project manager

Figure 4.11 Letter Regarding Revision of Master Programme

Words, Phrases and Expressions 4.5

[1] statutory notifications 向水、电、消防等部门发出的通知
[2] contractor's documents 承包商文件 ("Contractor's Documents" means the calculations, computer programs and other software, drawings, manuals, models and other documents of a technical nature (if any) supplied by the Contractor under the Contract. "承包商文件"指由承包商根据合同提交的所有计算书、计算机程序和其他软件、图纸、手册、模型和可能有的其他技术性文件。)
[3] time obligations 在时间方面的义务
[4] form grounds for termination 构成终结合同的理由
[5] progress programme 显示工程实际进展的计划，实际进展计划
[6] prediction programme 预测计划
[7] management programme 管理工程进展的计划，进展管理计划
[8] linked bar chart 连接横道图

4.6 Division of responsibility

On most contracts two independent but related management systems, including contract administration, exist at the site: that of the A/E and that of the contractor.

4.6.1 A/E's responsibilities

4.6.1.1 A/E

Where the employer and the contractor have entered into a contract for the execution, completion of the works and the remedying defects therein in the conventional manner, the A/E must be nominated in the contract documents as 'the A/E' and he exercises the powers reserved to him in that capacity for the administration and timely completion of the contract. In carrying out certain functions he is also the agent of the employer, thereby acting in a dual capacity. In administering the contract the A/E's decisions must be scrupulously fair and impartial as between employer and contractor, and must, to the extent that the contract documents provide, be based on the terms and conditions specified therein. Any restrictions imposed by the employer on the A/E's authority to exercise his unfettered judgment or the full powers vested in him by the contract documents are undesirable and not in the employer's interest. Nevertheless, if any such restrictions exist they must be notified to the tenderers, otherwise the employer may be entering into the contract under false premises. The position of the A/E in relation to the employer's auditor should be made known.

The A/E will need to clarify the terms of his appointment in writing to avoid any future problems. Most architects/engineers use a standard appointment form to simplify the procedure and reduce the risk of omissions. The letter, as shown in Figure 4.12, illustrates one possible approach to an appointment enquiry.

Land View Development Co., Ltd   27 April 2003

Dear Sirs,

Re: Buford Industrial Park

We thank you for your letter of 20 April. We are pleased to hear that you are considering appointing us as the engineer for the above project. You are most welcome to visit us to discuss the work or, alternatively, you might find it more convenient to meet our representative at the proposed site. We can supply you with details of some of the projects undertaken by this office so that you can visit them and see the types of work for which we have been responsible.

A copy of our engineer's appointment form is enclosed for your information.

The precise terms of our appointment can be agreed between us if you decide that you wish us to undertake the commission.

Yours faithfully
Infrastructure Consultancy, Ltd
SAC Architects

Figure 4.12 An Appointment Enquiry Letter

As set out under sub-clause 3.1 of the *Red Book*, the engineer shall carry out the duties assigned to him in the contract, but have no authority to amend the contract.

The engineer may exercise the authority attributable to him as specified in or necessarily to be implied from the contract. On certain matters, the engineer is required to obtain the approval of the employer before exercising a specified authority. However, whenever the engineer exercises a specified authority for which the employer's approval is required, then (for the purposes of the contract) the employer shall be deemed to have given approval.

Whenever carrying out duties or exercising authority, specified in or implied by the contract, the engineer shall be deemed to act for the employer. The engineer has no authority to relieve either party of any duties, obligations or responsibilities under the contract; and any approval, check, certificate, consent, examination, inspection, instruction, notice, proposal, request, test, or similar act by the engineer (including absence of disapproval) shall not relieve the contractor from any responsibility he has under the contract, including responsibility for errors, omissions, discrepancies and non-compliances.

4.6.1.2 Specialist firms

The A/E may recommend to the employer specialist firms to provide skills and advice in connection with certain aspects of the contract, such as inspection and testing of the works, and taking samples of materials and testing on the site and/or off the site, as well as for the inspection and testing of fabricated work. Specialist firms may also become nominated subcontractors, if the contract so provides, for both the detailed design, manufacture and installation of specialist work and/or plant items.

The A/E may also recommend the appointment of a quantity surveyor to measure and value the work where he does not have the necessary in-house expertise. The letter as shown in Figure 4.13 provides one method that can be adopted by an A/E when approaching a quantity surveyor.

Smith and Bailey 2 July 2003
Quantity Surveyors
1072 Glen Abbey Street

Dear Sirs,

Re: John Creek-Buford Industrial Park Phase 2

Our clients, John Creek Development Group, have requested that we appoint you to carry out the quantity surveying duties for the roads and services on phase 2 of the industrial park development at John Creek. The phase consists of the provision of roads, water mains, sewers and ducts for other services. We anticipate that the works will commence in December 2003 and the contract period will be approximately 9 months.

Will you please let us have your acceptance of the appointment and inform us of the basis upon which you wish to calculate your fees.

Yours sincerely

Figure 4.13 A Letter for Approaching a Quantity Surveyor

4.6.1.3 Consultation with other bodies

The A/E will often have to consult with other bodies such as local authorities, statutory undertakings and local consultative groups. Some of the bodies have statutory powers and the A/E needs to seek approval for certain aspects of the works, while other organisations are set up to safeguard local interests and a consultative process is likely to ensue. The letter in Figure 4.14 illustrates possible approaches to a planning authority.

Director of Urban Planning Bureau 16 July 2004
Bureau's address

Dear Sir,

Re: Buford Industrial Park

We refer to discussions held in your Bureau with Mr. Zhang and Ms. Sun on 23 June 2004.

In accordance with the agreement reached at the above meeting, we have revised the proposals and enclose three sets of drawings, numbers GS 305/1N to 8N. These are in substitution for the original drawings, numbers GS 305/1Nr to 8Nr, and are submitted for consideration by your committee.

We thank you for your cooperation in this matter and look forward to receiving planning approval on 30 July 2004.

If there are any further queries, we would appreciate a telephone call in order that they can be resolved quickly.

Yours faithfully
SAC Architects

Figure 4.14 A Letter Seeking for Speedy Approval

4.6.1.4 Delegation of authority

The conditions of contract may distinguish between those matters that the A/E must settle and those that his site representative (the resident engineer) may resolve. The extent to which the A/E can delegate authority to the A/E's representative is laid down in the contract. Within the limits so prescribed the A/E may at his own discretion delegate authority to the A/E's representative. He should be influenced in this matter not only by the character of the work, but also by the capability and experience of the staff he employs on the site. The A/E should inform the contractor in writing of any delegation he makes of his powers to the A/E's representative. The contractor should be given the right to appeal to the A/E against decisions of the A/E's representative. It is good practice for the A/E to notify the contractor of the names and positions of responsible persons he will have to deal with during the execution of the contract.

4.6.1.5 A/E's representative.

The function of the A/E's representative is to watch and supervise on a daily basis the execution, completion of the works and the remedying of the defects therein. Depending on the size of the project, the A/E's representative may have assistant staff under him. Such staff, particularly the clerks of works and inspectors, should be selected with regard to their practical experience of the type of work to be supervised. While his assistants deal mainly with detail, the A/E's representative must plan ahead and discuss future parts of the works with the contractor's representative (project manager) to ensure that the phasing of the works is properly planned to suit the approved programme. This close collaboration of the A/E's representative and contractor's representative also facilitates consideration of changes proposed by the contractor, and the subsequent submission of such proposals to the A/E for his approval. The principal duties of the A/E's representative are:

(a) to organize his work to suit the approved programme;

(b) to co-operate closely with the contractor on matters of safety;

(c) to supervise the works to check that they are executed to correct line and level and that the materials and workmanship comply with the specification;

(d) to examine the methods proposed by the contractor for the execution of the works, the primary object being to ensure the safe and satisfactory execution of the permanent work;

(e) to execute and/or supervise tests carried out on the site, and to inspect materials and manufacture at source where this is not done by the A/E's head office staff;

(f) to keep a diary constituting a detailed history of the work done and of all happenings at the site, and to submit periodic progress reports to the A/E;

(g) to measure in agreement with the contractor's staff the quantities of work executed, and to check day work and other accounts so that the interim and final payments due to the contractor may be certified by the A/E;

(h) in the case of any work for which the contractor may claim payment as additional work, to agree with the contractor and record all relevant circumstances so as to ensure that agreement exists on matters of fact before any question of principle has to be decided by the A/E;

(i) to record the progress of the work in comparison with the programme; and

(j) to record on drawings the actual level and nature of all foundations, the character of the strata encountered in excavation and full details of any deviations from the drawings which may have been made during the execution of the works, i.e. to produce all record drawings.

4.6.1.6 Progress reports

The A/E should maintain a check on progress through regular reports submitted to him by the A/E's representative. These reports must be in a form that will give the A/E a clear and concise picture of the progress made and the extent to which this is ahead of or behind the contractor's programme; they should be accompanied by such progress charts or diagrams as may be necessary for this purpose. If progress is behind programme on any items of work the report should state the reasons for the delay and the steps that the contractor is taking to remedy matters. The A/E's representative may also be required to submit financial reports.

4.6.1.7 Instructions to contractor

All instructions from the A/E's representative to the contractor's representative should be given in writing, either directly or as confirmation of verbal instructions. It is good practice to avoid giving instructions to other individuals within the contractor's organization.

4.6.1.8 Inspectors' duties

The A/E's representative shall employ experienced inspectors to undertake general supervision of the contractor's work, and junior inspectors as detail checkers on the mixing of concrete and any such work requiring constant supervision. The duties of inspectors are of great importance and demand wide experience, practical knowledge, integrity and tact in dealing with the foremen and workmen employed by the contractor. Selection of suitable staff is therefore a matter of great importance not only to secure satisfactory work but also to ensure smooth working of the contract. Inspectors must be above suspicion.

4.6.2 Contractor's responsibilities

4.6.2.1 Performance of the contract

The contractor is responsible for design (to the extent specified in the contract), execution, completion of the works and remedying defects therein in accordance with the requirements of the contract documents. There should be a strong bond of common interest between the A/E and the contractor because both should wish to see good construction materialize and want a successful outcome to crown their labours. However, neither must forget that the independence of the other should be respected; each is entitled to his own freedom of thought, outlook and need for privacy.

4.6.2.2 Freedom of the contractor

It is in the best interest of all parties that the contractor should be as free as possible under the terms of the contract to execute the Works in the way he wishes. His preference for a particular design of and method of carrying out the temporary works and his idea of the order of sequence of construction may differ from those of others. However, he is better content, and therefore works better, when he uses his own ideas. Frequently there may be circumstances that necessitate restrictions. Part of the work may be wanted first for reasons outside engineering. In such cases the A/E usually specifies the necessary requirements, but it may still be possible and desirable to allow the contractor to submit alternative proposals for meeting them.

4.6.2.3 Subcontracts

Building and civil engineering contracts normally provide that the contractor shall not place any subcontracts without the approval of the A/E, and that the contractor shall remain liable for all the acts and defaults of subcontractors. There is a need to distinguish between non-nominated or domestic subcontracts, where the contractor selects the subcontractor for his own reasons, and nominated subcontracts, where the subcontractor is selected by others.

4.6.2.4 Nominated subcontracts

Where the nominated subcontract system is adopted, the A/E obtains a tender from a selected firm for the supply of plant or materials or for the execution of special work, and includes in the bill of quantities a prime cost sum or provisional sum for the work in question. After the main contract has been placed the contractor is instructed to accept the tender of the nominated subcontractor. As the contractor then becomes liable for any default or financial failure of such a subcontractor, he should have the right to decline the employment of any nominated subcontractor against whom he has just cause for objection. It is therefore desirable that the contractor should be called into the discussions on the employment of nominated subcontractors at an early stage.

The original tender of the nominated subcontractor should be based on the obligations and liabilities which the contractor has towards the employer under the terms of the main contract, otherwise the contractor may be involved in financial and other liabilities which he could not have foreseen when tendering. The contract usually provides that the contractor may add a specified or tender percentage to cover his responsibilities and services in connection with nominated subcontracts.

4.6.2.5 Design of the temporary works

The contractor should submit drawings and design calculations for important temporary works to the A/E, and the latter should scrutinize them with care. This check in no way relieves the contractor of his responsibility for the adequacy of the design and construction of the temporary works. However, it does enable the A/E more effectively to discharge his overall professional responsibility to the employer for the satisfactory execution, completion of the works and remedying of the defects therein.

Such independent scrutiny provides valuable extra insurance against mistakes in the design of the temporary works. The contractor should welcome it, and the A/E should insist that it is done, in the interests of the safety of personnel and the works. In many parts of the world the operation of this practice has been made more complex by recent health and safety legislation.

4.6.2.6 Quality standards

More human lives depend on the safety of buildings and civil engineering structures than on the product of most other industries and quality standards must be jealously guarded throughout construction. At the planning stage the contractor must make sure that the construction methods and plant employed by him can produce work of a quality not lower than the standard specified. On the site proper supervision and control are needed to ensure that output and quality do not come into conflict. To this end many of the larger contractors have set up extensive research and development facilities and quality assurance schemes which help greatly to combine increased output with high standards of quality.

4.6.2.7 Materials and workmanship

The contract documents specify in detail the quality of materials and workmanship required and the tests to be made regularly to ensure that the finished work complies with the specification. The *Red Book* makes provision in general terms for the execution of and payment for tests of materials and workmanship required by the engineer.

4.6.2.8 Inspection by A/E

It is the contractor's duty to ensure that every facility is given to the A/E to enable him to inspect materials and manufacture at all stages both on and off the site.

4.6.2.9 Setting out

The contract documents must define clearly the responsibility for setting out the works. The *Red Book* vests it entirely with the contractor, regardless of any check the A/E's representative may make. The complexity and accuracy of the survey methods used in setting out depend on the site conditions and the type of work under construction. For example, survey methods giving a high degree of precision must be employed in the setting out of tunnels or for determining the span distances in bridge construction, while simpler methods will usually suffice for locating roads in a new housing scheme.

4.6.2.10 Progress reporting

Sub-clause 4.21 of the *Red Book* requires the contractor to prepare monthly progress reports and submit to the Engineer in six copies. The first report shall cover the period up to the end of the first calendar month following the commencement date. Reports shall be submitted monthly thereafter, each within 7 days after the last day of the period to which it relates.

Reporting shall continue until the contractor has completed all work which is known to be outstanding at the completion date stated in the taking-over certificate for the works. Each report shall include:

(a) charts and detailed descriptions of progress, including each stage of design (if any), contractor's documents, procurement, manufacture, delivery to the site, construction, erection and testing; and including these stages for work by each nominated subcontractor;

(b) photographs showing the status of manufacture and of progress on the site;

(c) for the manufacture of each main item of plant and materials, the name of the manufacturer, manufacture location, percentage progress, and the actual or expected dates of:

(i) commencement of manufacture,

(ii) contractor's inspections,

(iii) tests, and

(iv) shipment and arrival at the site;

(d) the detailed records of contractor's personnel and equipment;

(e) copies of quality assurance documents, test results and certificates of materials;

(f) list of notices of the employer's claims and contractor's claims;

(g) safety statistics, including details of any hazardous incidents and activities relating to environmental aspects and public relations; and

(h) comparisons of actual and planned progress, with details of any events or circumstances which may jeopardise the completion in accordance with the contract, and the measures being (or to be) adopted to overcome delays.

4.6.3 Cooperation

There is a great need for the A/E and the contractor to work in close cooperation and to understand the problems that the other has to face. Any advice, assistance and co-operation given to the contractor in his execution of the works are likely to benefit not only the contractor but also the A/E and the employer. Likewise any advice and assistance which the contractor can give to the A/E should benefit all parties. It is the common duty of the contractor's project manager and the A/E's representative to see that the works are executed in accordance with the specification and the drawings. Experience, commonsense and judgment are required in the exercise of this duty, and in the exercise by the A/E's representative of the powers delegated to him by the A/E under the contract.

Words, Phrases and Expressions 4.6

[1] capacity n. 身份

[2] the full powers vested in him by the contract documents 由合同文件授予他的全部权力

[3] in-house expertise 本单位专业知识

[4] statutory undertakings (水电通讯煤气等) 公用事业（单位）

[5] delegation of authority 将权限委托他人

[6] clerks of works 工程管理员，工程检查员（The clerk of works is paid for by the employer to providec onstant supervision, in order to ensure that the quality of materials and workmanship comply with the contract requirements. He is considered to act solely as an inspector, and has no authority to issue instructions (directions) to the contractor. However, as provided in some contracts, if instructions issued by the clerk of works are confirmed in writing by the Engineer within specified days of their

being issued, then such instructions shall be deemed to have been given by the Engineer. 工程管理员受雇于业主,履行日常监督职责,确保材料质量和施工工艺符合合同要求。一般认为,工程管理员任务仅限于检查,无权向承包商发出指示。然而,有些合同规定,如果工程管理员的指示在发出后规定的时间内经过工程师的书面确认,则此等指示将被视为由工程师所发出的。)

[7]　inspector n. 检查员
[8]　progress reports 进展报告(书)
[9]　workmanship 施工工艺(质量、水平)
[10]　to the extant that 在这一范围内,在这种情况下,只要

Notes 4.6

疑难词语解释

[1]　Figure 4.12 provides one method that can be adopted by an A/E when approaching a quantity surveyor.
【译文】 图 4.12 是建筑师/工程师在同工料估算师接洽时可用的一种方法。
[2]　The A/E should maintain a check on progress through regular reports submitted to him by the A/E's representative.
【译文】 建筑师/工程师应通过建筑师/工程师代表定期向其提交的报告控制工程的进展。
[3]　The A/E may at his own discretion delegate authority to the A/E's representative.
【译文】 建筑师/工程师可根据自己的斟酌将权限委托给建筑师/工程师代表。

4.7　Inputs to contract administration

Contract administration cannot proceed unless some items, whether internal or external to the contract, have been obtained, provided or made available to the employer or the A/E. The required items are referred to as inputs. Obviously, the contract documents are the most important input to the contract administration. In addition the contractor's monthly statements, claims, variations and many other performance related documentation are also inputs to contract administration.

4.7.1　Performance related documentation

The contractor's performance-related documentation includes:

☐　Documents of technical nature developed and supplied by the contractor under the contract, such as programme in bar chart or in network diagrams, calculations, computer programs and other software, drawings and models.

☐　Contractor progress reports. Sub-clause 4.21 under the *Red Book* requires the contractor to prepare and submit monthly reports to the engineer in six copies. Each report shall include charts and detailed descriptions of progress, photographs showing the status of manufacture and of progress on the site, records of the contractor's personnel and equipment, quality assurance documents, test results and certificates of materials, list of notices given under the contract, safety statistics and comparisons of actual and planned progress.

4.7.2 Work performance information

Work performance information, including the extent to which quality standards are being met, what costs have been incurred or committed, the contractor's invoices, etc., is collected as part of contract execution. The contractor's performance reports indicate which works items have been completed and which have not. The contractor must also submit invoices (sometimes called bills or requests for payment) on a timely basis to request payment for work performed. Invoicing requirements, including necessary supporting documentation, are defined within the contract.

Words, Phrases and Expressions 4.7

[1] Inputs to contract administration 合同管理的依据
[2] performance related documentation 履行合同过程产生的文件
[3] work performance information 工作结果信息
[4] performance reports 履行合同结果报告

4.8 Tools and techniques for contract administration

Tools and techniques for contract administration refer to some tangible things and/or defined systematic procedures adopted by the employer or the A/E to administer the contract to ensure that he and the contractor meet their contractual obligations and that their own legal rights are protected, including:

4.8.1 Variation control system

A variation control system defines the process by which the contract can be modified. It includes the paperwork, tracking systems, dispute resolution procedures, and approval levels necessary for authorizing variations.

4.8.2 Employer-conducted performance review

This is a review of the contractor's progress in executing the works and quality, within budget and on schedule, as compared to the contract. It can include a review of the contractor submitted samples, the contractor-prepared documentation and the inspections by the employer's personnel, testing conducted as specified in the contract, as well as quality audits conducted during the contractor's execution of the works. The objective of a performance review is to identify performance successes or failures, progress with respect to the contract, and contract non-compliance that allows the employer to quantify the contractor's demonstrated ability or inability to perform work.

4.8.3 Inspections

Inspections required by the employer and supported by the contractor as specified in the contract documentation, can be conducted during execution of the works to identify any weaknesses in the contractor's work processes or deliverables.

4.8.4 Communication skills

4.8.4.1 Factors affecting communication[15]

Communication problems in the construction industry stem partly from its scattered organization. Amount of information passing between the site personnel is large and creates the need for a well organized and effective communication network. However, the information conveyed may be difficult to understand, inaccurate or misleading.

The site personnel cannot always obtain the information they require when they want it. Estimates may be inaccurate, drawings incomplete or out of date and specification descriptions and engineer's instructions ambiguous.

Communications within small organisations are often good largely because there is extensive face-to-face contact between personnel. Larger firms tend to rely to a greater extent on written communications. Written communications provide a permanent record but can more easily lead to misunderstandings and delay in time.

4.8.4.2 Purposes of communication

Communication serves a variety of functions, all of which are important in contract administration and construction project management as a whole. The more commonly encountered activities have been identified as:

(1) Information is continually being exchanged between all persons connected with a building or civil engineering project.

(2) Performance of certain activities by appropriate personnel satisfactorily and within the required timescale is vital to the effective completion of the works.

(3) Social relationships can influence the smooth running of a contract.

(4) Expression of individual feelings permits an employee to make his views known on matters he considers important.

(5) Changes of the attitude of employees are often required to keep abreast of technological, legal and economic developments.

4.8.4.3 Organization of communication

Effective contract administration and project management as a whole depend on an efficient method of conveying instructions and information and of ensuring satisfactory feedback. On a building or civil engineering project, many diverse interests may be involved, such as those of consultants, resident engineer's site staff, main contractor, subcontractors, suppliers and the employer. Recognized channels of communication are needed to ensure that all the personnel concerned receive the information they require and when they need it.

A variety of communication channels may be used. For example, consultative arrangements between managers and operatives will facilitate the settlement of problems and improve working relationships.

Once the appropriate channels of communication is set up they have to operate effectively so that the information reaches the right people at the right time and in the manner required by the recipient.

4.8.4.4 Methods of communication

Communication methods in the organizations can be lateral, upward and downward. Lateral communication may take place between people of similar status and is primarily concerned with the exchange of information.

Upward communication provides a valuable source of feedback of useful information to management. The information can take various forms, such as submitting progress reports, making suggestions for improvements or work methods and seeking guidance on how to deal with certain problem areas.

Downward communication consists primarily of management passing instructions and information to the personnel involved. The opportunity is often taken to keep personnel adequately informed about overall progress, general guidance and policy matters. It is beneficial for employees to know how their particular activities fit into the firm's overall objectives and plans for the future.

Spoken communication provides an important connecting link on all building and civil engineering projects. It encompasses face-to-face conversation, direct telephone calls and so on. Face-to-face communication can be very effective if carefully considered and clearly expressed. People are inclined to communicate more freely in the absence of a permanent written record, but the latter can also be a disadvantage.

4.8.4.5 Project meetings

(1) Purpose of meetings

During the course of a contract, a variety of meetings will take place in site offices, on specific parts of the works and in suppliers' premises. Some may be called at short notice to resolve a problem on the site, while others will be formally arranged at regular intervals and are generally concerned with coordination and progress. The main objective of all meetings is to come to a decision, although supplementary aspects such as the exchange of information, generation of ideas and discussion of problems may also be important. Meetings can, however, fail to achieve these objectives through over-formality, ineffective chairmanship, failure to concentrate on key issues or an antagonistic attitude by one of the parties.

From time to time, it may be necessary to convene a formal meeting to discuss a specific matter having become important to the progress of the project.

(2) Site meetings

The most important meetings on a building or civil engineering project are the regular project meetings, sometimes referred to as site meetings or progress meetings. They are normally held at monthly intervals and provide the opportunity for a regular, comprehensive reappraisal of the project.

(3) Agendas

A formal agenda should be prepared for each project meeting to provide a sound basis for discussion at the meeting. It is usually formulated around a series of standard main headings, such as the major divisions in the bill of quantities, supplemented by relevant sub-headings, as illustrated in Figure 4.15.

AGENDA FOR SITE MEETING
To be held on 21 August 2006

1. Minutes of the last meeting
2. Matters arising and action taken
3. Weather report since last meeting: 2160 man/hours lost
4. Labour force on site: at date of meeting by trades
5. Questions on Architect's instructions issued to date
    Serial number of latest instruction (for verification). Verbal instructions requiring written confirmation
6. Daywork: date of last sheet passed to passed to Quantity Surveyors.
7. General Contractor's Progress Report
8. General Contractor's Report and questions on nominated subcontractors:
(i) Mechanical services,
(ii) Electrical services,
(iii) Structural frame,
(iv) Others.
9. General Contractor's Report and questions on nominated suppliers.
10. Comments from those in attendance.
11. Other terms. AOB.
12. Date of next meeting

To be distributed to: Employer, Quantity Surveyor, Consultants, Contractor, Clerk of Works, Subcontractors, Suppliers and Architect.

Figure 4.15 An Agenda of Site Meeting

(4) Minutes

The record of a meeting is commonly referred to as "the minutes". It is important that the record should be accurate and, if prepared by the A/E, the contractor should examine it carefully and challenge any inaccuracies in writing immediately, as shown in Figure 4.16. Do not wait until the next meeting.

Contractor's address
23 October 2005

Architect's address
Dear Mr. Wang,

<u>Re: The record of the site meeting</u>

Thank you for the record of the site meeting held on 21 October 2005. We would like to make the following comments:

Item 2.15 in Bill No.2 <u>Foundation to D Section</u>

Mr. Zhang did not state that this was 32 days behind schedule. It is, in fact, slightly ahead of schedule as now adjusted following your extension of time dated 16 October 2005.

We will appreciate it very much if you could correct the record at the next meeting. We are copying this letter to all those who attended the meeting that day.

Yours sincerely,

Liu Yugui
Project manager

Figure 4.16 Letter to Architect Questioning the Accuracy of the Record

If inaccuracies are allowed to go unchallenged, and the minutes need to be referred to subsequently, e.g., in a claims situation, it may be difficult if not impossible to prove that the minutes is in fact inaccurate. The minutes should be impartial. Figure 4.17 is a typical example.

MINUTES OF SITE MEETING No. 16
held on 28 October 2005

Job: Buford Industrial Park Phase 2, Land View Development Co., Ltd

Present: Mr. Wang Dagui   Land View Development Co., Ltd
         Mr. Liu Yugui   Beijing No.1 Construction Company
         Mr. Xiao Changle   Ditto

Ms. Pan Hongjie	Municipal Water Company	
Mr. He Yongge	Clerk of Works	
Ms. Sun Changchang	Beijing Vision Architects	

Serial	Description	Action
131	Minutes of last meeting: Agreed as correct	
132	Matters arising: Ref. Serial 124 Mr. Wangdagui confirmed that his board had agreed not to move the canteen.	
133	Labour force on Site: Foreman 1 Labourers 20 Bricklayers 5	
134	Delivery of materials: Mr. Liu Yugui reported that despite repeated telephone calls brick deliveries were still behind schedule from nominated suppliers. Architect to intervene.	

......

205 Door to Boiler House D.B.07: Mr. Liu Yugui said that his Works Engineer wished this to open on the opposite hand. Agreed. Architect to issue instruction.

206 Date of next meeting: 4 November 2005.

Distribution: as agenda.

Figure 4.17   A Sample Minutes of Site Meeting

(5) Implementation of decisions

At the meetings views are exchanged, proposals generated and decisions made. It still remains for the decisions to be implemented. The chairman of the meeting, possibly assisted by the secretary, will be responsible for ensuring implementation. The minutes will record who is to take the appropriate action and all participants will receive copies of the minutes. The action taken will be monitored at the next meeting under matters arising.

4.8.5   Information technology

The use of information and communication technologies can enhance the efficiency and effectiveness of contract administration by automating portions of the records management system, payment system, claims administration, or performance reporting and providing electronic data interchange between the employer and the contractor.

4.8.6   Progress reporting

Progress reporting both by the A/E and the contractor provides the employer with information about how effectively the contractor is achieving the contractual objectives.

4.8.7   Records management system

A records management system is a specific set of processes, related control functions, and automation tools which are consolidated and combined into a whole, as part of the project management information system. A records management system is used by the employer or the A/E to manage contract documentation and records.

The system is used to maintain an index of contract documents and correspondence, and assist with retrieving and archiving that documentation.

Documentation consists of the writings or records of persons who were present at events, written at the time or shortly after the time of the event. In many instances, it may be the only evidence in existence that reveals what actually occurred.

Good documentation is invaluable in resolving misunderstandings before they escalate into disputes. One party to a misunderstanding may have an incomplete or incorrect picture of the facts of an event or occurrence on the project. Good documentation of the true facts in the possession of the other party is very effective in clearing up the misunderstanding, thus avoiding a potential dispute before it starts.

4.8.7.1 Site records

The keeping of continuous and comprehensive site records provides an effective means of controlling and monitoring all activities on the site. They have a vital role to play in the assessment and settlement of disputes. They can take a wide variety of different forms and the following list embraces most of the more common records kept by the A/E's site staff.

(1) all correspondence between the resident engineer and the agent, including engineer's instructions, variation orders and approval forms;

(2) all correspondence between the engineer for the contract and the resident engineer, the employer and third parties;

(3) the minutes or notes of formal meetings;

(4) daily, weekly and monthly reports submitted by the engineer's site staff;

(5) equipment and labour returns, as submitted and corrected where necessary;

(6) work records such as dimension books, timesheets and delivery notes;

(7) daywork records, as submitted and corrected where necessary;

(8) interim statements, as submitted and including any corrections with copies of all supporting particulars and interim certificates;

(9) level and survey books, containing checks on setting out and completed work;

(10) progress drawings and charts and revised drawings;

(11) site diaries;

(12) laboratory reports and other test data;

(13) weather records;

(14) progress photographs; and

(15) administrative records, such as leave and sickness returns, and accident reports.

4.8.7.2 Correspondence

All letters, drawings and other documents should be recorded as they are received or dispatched, and all incoming documents should be date stamped. Oral instructions to the contractor should always be confirmed in writing and also telephone conversation where they convey instructions or important information. Copies of all correspondence, whether in the form of formal letters or handwritten notes, should be carefully retained, along with old diaries, notebooks, site books and similar data.

(1) The employer's brief and its development

Large public and private employers normally prepare a written brief before they engage the A/E for a project. It is well worthwhile spending considerable time and exercising great care in an effort to make the brief as precise and comprehensive as possible. It forces the employer to thoroughly think through his ideas about the project, which prevents the A/E from wasting time in considering matters on which the employer has already made up his mind.

Many A/Es undertake the subsequent design work in isolation from the employer. On submission of the completed proposals for approval, it may then take the employer a considerable time to understand them and they may contain certain details that could have been eliminated at an early stage if the employer had known what was being contemplated.

Some A/Es invite the employer's representatives to attend meetings of the design team. This practice certainly improves the employer's knowledge and understanding of the project and can quicken the decision-making process. However, some A/Es argue that it can inhibit designers in the performance of their work, lead to premature decisions and can be wasteful of time.

(2) Correspondence between A/E and employer

Extensive correspondence is bound to take place covering a wide range of matters, starting from the initial appointment through to design, contractual, financial and constructional aspects. The following examples will help to show how these specific matters could be dealt with. It is obvious that there is no single way of dealing with this type of correspondence and that it becomes very much a personal matter and each A/E has his own particular style. The letter as shown in Figure 4.18 deals with the employer's comments on the outline proposals for a project.

Land View Development Co., Ltd   5 July 2004

Dear Sirs,

Re: Buford Industrial Park

We thank you for your helpful comments on the outline proposals for the above project.

There has been no difficulty in incorporating most of your suggestions into our revised scheme. The only matter which has caused problems is the suggested relocation of the spillway channel.

We suggest that the best way to resolve this problem is for Mr. Zhang to visit you for a thorough discussion. We will call you shortly to arrange a suitable date and time for a meeting.

                                                                Yours faithfully

                                                                SAC Architects

Figure 4.18 A Letter Seeking for Comments from Statutory Undertaker

Figure 4.19 shows a letter that covers the employer's proposals with regard to a subcontract.

Land View Development Co., Ltd   6 September 2004

Dear Sirs,

Re: Buford Industrial Park

We refer to our discussion of 3 September 2004 concerning the proposed employment of Sparks for the sub-contract for electrical works on the above project and we confirm our advice to you.

(1) You have instructed us to obtain a tender from one firm only, namely Sparks, for the sub-contract works.

(2) We have advised you that, in our opinion, this firm is unsuitable for this class of work, and that tenders should be obtained from three other nominated firms.

(3) We have further advised you that in the event of the firm being nominated to do this work, the consequences could include additional expense, delays in progress and completion, and lower quality of work than anticipated.

(4) We shall, of course, carry out your instructions but we can take no responsibility should the outcome prove unsatisfactory, as the action taken is contrary to our advice. As soon as a sub-contract is signed we shall proceed to administer the contract provisions with diligence and impartiality.

Please consider the matter once again and let us have your final instructions as soon as possible.

                                                                Yours faithfully

                                                                SAC Architects

Figure 4.19 Employer's Proposals with Regard to a Subcontract

The letter written as in Figure 4.20 relates to a design change requested by the employer and the engineer takes the opportunity to include a timely warning.

Land View Development Co., Ltd,   20 September 2004

Dear Sirs,

Re: Buford Industrial Park

Thank you for your letter 16 September 2004 requesting us to alter the line and width of the approach road. The necessary design work is underway.

The alterations will require a certain amount of redrawing and rescheduling, although they are relatively minor.

We know, from our earlier discussions, that you appreciate the problems caused by quite small changes of design and the possible repercussions in terms of cost and programme time. The problems are likely to intensify in the later stages of design.

Yours faithfully

SAC Architects

Copy: Quantity Surveyor

Figure 4.20 A Letter Regarding Design Change

In Figure 4.21 the letter covers the important matter of a qualified tender and the recommended action.

Land View Development Co., Ltd   26 November 2004

Dear Sirs,

Re: Buford Industrial Park

Following today's meeting at which tenders for the above project were opened, we consider that it would be helpful to you if we listed the main points that we made at the meeting so that you may have the opportunity to consider them more fully.

(1) All tenderers were informed that the tendering procedure would be in accordance with the Instructions to Tenderers incorporated in the tender documents. Each tenderer understands the full implications and has the right to expect that, having expended a considerable amount of time and money on preparing tenders, these requirements will be strictly applied.

(2) A tenderer who amends the form of tender, inserts qualifications or submits a late tender is seeking to gain an unfair advantage. As you know, Mr. Liu Yugui was unwilling to withdraw his qualifications. If all tenderers had been given the opportunity to qualify their tenders as they thought fit, any common yardstick for assessing tenders would disappear.

(3) All other considerations apart, the adoption of a universally recognised system of tendering has a beneficial effect on the whole construction industry by keeping prices at a realistic level.

We therefore advise that the irregular tender submitted by Mr. Liu Yugui be rejected and that the normal checking procedure be applied to the next lowest tender. In the absence of any significant problems, and in conjunction with the quantity surveyor, we would expect to submit an acceptable tender total for your approval.

Yours faithfully

SAC Architects

Figure 4.21 A Letter Regarding a Qualified Tender

(3) Correspondence between A/E and contractor/subcontractors

On a substantial building or civil engineering contract a considerable amount of correspondence will take place between the engineer and the main contractor and subcontractors, ranging from contractual matters to technical matters. The letters shown in Figure 4.22 and Figure 4.23 illustrate commonly adopted forms of approach.

Guansha Construction Ltd   26 November 2005

Dear Sirs,

Re: Buford Industrial Park

The employer, Land View Development Co., Ltd, has instructed us to inform you that your tender of 19 November 2005 in the sum of ￥32,466,200 for the above project is acceptable and we are preparing the main contract documents for signature.

It is not the employer's intention that this letter, taken alone or in conjunction with your tender, should form a binding contract.

However, the employer is prepared to instruct you to commence siteworks and place orders for the materials required in the first month of your contract programme.

If for any reason the contract does not proceed, the employer's commitment will be strictly limited to payment for the operations listed above. No other work included in your tender must be carried out without a further written order.

Yours faithfully

SAC Architects

Copy: Quantity Surveyor

Figure 4.22 A Letter of Acceptance

Changlong Pipe Company   4 October 2005

Dear Sirs,

Re: Buford Industrial Park

Your tender for the supply of pipework for the above contract has been received. However, the tender cannot be considered in its present form as it does not satisfy the requirements of the contract. If you still wish to tender, please complete the Standard Form of Tender, a further two copies of which are enclosed, and return it to us not later than 18 October 2005.

Please note that the submission of the tender on your own office form or the insertion of your own special conditions, other than in the appropriate position on the Standard Form, will result in disqualification.

Yours faithfully

SAC Architects

Figure 4.23 A Acknowledgement Letter

(4) Communications between site personnel

A variety of communications will be required between various site personnel covering a wide range of matters. Some will originate from the A/E and will be mainly directed at the contractor and may be in the form of letters or instructions. Others will operate in the reverse direction, where the contractor is seeking information from or action by the A/E, while other communications may be internal ones between employers and their employees.

4.8.7.3　Reports

A report is primarily a summary of information and the principal method of conveying information on site matters to head office, the employer and other parties.

Daily reports by inspectors, supervising the constructional work on site, form an important part of site communications. These reports contain details of the work carried out, weather conditions, the number of contractor's employees engaged on the work being supervised, number and types of plant in use and hours worked, and details of any delays and their causes. Starts and finishes of activities will be noted. After processing, the reports should be filed and stored neatly and chronologically for ease of reference. Technical reports may be prepared on laboratory tests and special reports on specific problem areas.

4.8.7.4　Returns of labour and contractor's equipment

The contractor's labour and equipment returns constitute another commonly employed form of written record. The contractor is normally required to submit at prescribed intervals, such as monthly, the number and categories of labour and equipment engaged on the site. The A/E's site staff carry out checks on the information provided.

4.8.7.5　Drawings

Drawings provide a convenient and effective way of recording the progress of construction on the site. The type of information recorded includes the date and extent of construction (overall dimensions), the results of any tests and location of materials tested, and dates of approval of work. Record drawings may be purpose-drawn or the standard construction drawings suitably adapted to record construction progress. As the work proceeds, it is often necessary to revise drawings, normally on the original negatives and add dates and new references, to show alterations resulting from adoption of contractor's alternatives, different ground conditions and engineer's variations.

4.8.7.6 Laboratory tests

Laboratory reports and other test results are normally entered on standard forms and filed on a subject basis. Common tests include concrete cube strengths, earthworks density, compaction and moisture content, and analysis of bituminous products. On occasion, the information is more effectively presented diagrammatically as in the form of graphs for matters such as standard sieve analyses. Statistical analysis of data can encompass the determination of such parameters as range, standard deviation and coefficient of variation. The laboratory may also undertake the recording of rainfall, temperatures, wind speeds and tides?

4.8.7.7 Photographs

It is good practice to take photographs of the main features of the project from the same position at regular intervals to provide an excellent record of progress throughout the project. These photographs are often supplemented by photographs of particular features such as a rejected section of honeycombed concrete, irregular brickwork, bank slippage and extent of flooding resulting from exceptionally heavy rainfall.

4.8.7.8 Diaries

Diaries provide a complete narrative of the progress of the works and the activities of the A/E's site staff.

A diary provides a factual record of events on site, discussions with the staff of the Contractor's Representative and other personnel, instructions issued and weather conditions. Engineers and technicians must devote considerable amounts of time to the collection of appropriate information and its entry in their diaries. Rough notebooks and pocket tape recorders are useful for recording basic data and reminders. The resident engineer should check diaries regularly to ensure that they are up to standard.

Inspectors will need to record details of the deployment of plant and labour, movement of materials and progress of, and any problems associated with the work.

Assistant resident engineers or section engineers use the diaries to prepare a summary of the principal activities on both a weekly and monthly basis, which form a chronological review of the work in progress or completed.

4.8.7.9 Instructions for variations

Variations may be required to deal with variable site conditions, non-availability of materials or for other causes. On occasion an urgent decision has to be made by the A/E and he will probably instruct the variation verbally on the site or over the telephone, to be subsequently confirmed in writing.

Sub-clause 3.3 of the *Red Book* sets out: the Engineer may issue to the contractor from time to time instructions and additional or modified drawings necessary for the execution of the works and the remedying of any defects in accordance with the contract. If an instruction constitutes a variation the variation shall be handled as specified under clause 13.

The contractor shall comply with the instructions given by the Engineer or delegated assistant, on any matter related to the contract. Whenever practicable, their instructions shall be given in writing. If the Engineer or a delegated assistant: (a) gives an oral instruction, (b) receives a written confirmation of the instruction, from (or on behalf of) the contractor, within two working days after giving the instruction, and (c) does not reply by issuing a written rejection and/or instruction within two working days after receiving the confirmation, then the confirmation shall constitute the written instruction of the Engineer or delegated assistant (as the case may be).

These confirmations of verbal instructions require careful scrutiny to ensure that they do not contain inaccuracies, differences of emphasis or additional content. The example as given in Figure 4.24 shows how significant differences can easily arise between the contractor's confirmation of verbal instructions and the resident engineer's intended variation order.

CVI nr 14

The face of the concrete retaining wall adjoining the pump house is to be bush hammered to expose the aggregate. Payment to be made on a daywork basis.

Variation Order nr 116

The following additional work is to be carried out.
Subject to a satisfactory trial panel, measuring 2 m × 2 m, in a location to be agreed, the face of the southern boundary wall adjoining the pumphouse is to be lightly bush hammered to expose the aggregate. The bush hammering is to extend from 150 mm above finished surface level to 50 mm below the coping.
Payment is to be based on the bill rate for item 28/D.8

Figure 4.24 A Sample Variation Instruction

4.8.7.10 Samples for testing

It is advisable that a suitably completed form accompanies every sample that is sent to a laboratory for testing, and that a copy of the completed sample form and the test results are retained on the appropriate site file.

4.8.7.11 Delivery of materials

The contractor will have a set of standard letters relating to such matters as delivery of supplies of materials which have not left the supplier, tracing materials delayed in transit, and advising the supplier of damaged goods and requesting replacement. Figure 4.25 illustrates the type of letter that the contractor might send to the supplier in the latter case.

Precast Component Supplies Ltd.   8 April 2005
431 Madison Drive
Cherokee, GA 20045
Dear Sirs,

<u>Order No. 3001 - Precast Concrete Curbs</u>

We have to advise you that 18 of the curbs received on 13 March were badly damaged. We shall require the items to be replaced urgently and shall be obliged if you will notify us of the date of dispatch.

The hauler has been advised and we shall be pleased to receive your instructions with regard to the disposal of the damaged items.

Yours faithfully

Sound Road Builders, Ltd.

Figure 4.25 A Sample Expediting Letter

4.8.8 Payment system

Payments to the contractor are usually handled by the accounts payable system of the employer. The payment system includes appropriate reviews and approvals by the A/E, and payments are made in accordance with the terms of the contract.

4.8.9 Claims administration

Claims are documented, processed, monitored, and managed throughout the Contract life cycle, usually in accordance with the terms of the contract. If the parties themselves do not resolve a claim, it may have to be handled in accordance with the dispute resolution procedures established in the contract. These contract clauses can involve arbitration or litigation, and can be invoked prior to or after contract closure.

Words, Phrases and Expressions 4.8

[1] tools and techniques for contract administration 合同管理的工具与技术
[2] variation control system 变更控制制度
[3] tracking systems 追踪制度
[4] dispute resolution procedures 争议解决程序
[5] site meetings 现场会议
[6] regular project meetings 定期项目会议
[7] progress meetings 进展会议
[8] employer's brief and its development 业主设计任务书及其编制
[9] claims administration 索赔管理

Notes 4.8

疑难词语解释
[1] It often provides a valuable method of getting work done efficiently and quickly.
【译文】 它经常成为有效而又迅速完成任务的方法,很值得一用。

Chapter 5　Risk Management

5.1 Risk with construction projects

5.1.1　Definition of risk

5.1.1.1　Uncertainty

Uncertainty is synonymous with *doubt*, *dubiety*, *skepticism*, *suspicion* and *mistrust* and means lack of sureness about someone or something. It may range from a falling short of certainty to an almost complete lack of conviction or knowledge about outcome or result.

Uncertainty exists in the absence of information about past, present, or future events, values, or conditions. Uncertainty also exists when each possible outcome can be identified, but the probability of the occurrence can neither be determined nor assessed. There are degrees of uncertainty, meaning that some information about a construction project may be known, but there is not sufficient knowledge about it to provide certainty. There are three types of uncertainty. [29]

(1) Descriptive or structural uncertainty, is concerned with the absence of information relating to the identity of the variables which explicitly define a construction project.

(2) Measurement uncertainty. This is the absence of information relating to the assignment of a value to the variables used to describe the project.

(3) Result uncertainty. It occurs when the predicted results, and therefore their probabilities, cannot be identified.

5.1.1.2　Risk[29]

Risk is an uncertain event, action, condition, or a situation. The competitive world makes it inevitable for the participants in a construction project to take risks that may not be enjoyable.

Risk can be looked at from a number of different perspectives, such as:

(1) risk concerns future events, action or condition.

(2) risk involves change, such as in changes of the mind, opinion, actions, or places.

(3) risk involves choice, and all the uncertainty that choice itself entails.

Risk can be divided into two distinct types: speculative (or dynamic) risks and pure (or static) risks. Speculative risks are those having both profit and loss attached to them, while pure risks only have losses associated with them. Therefore, risks have both gain and loss associated with them.

It should be born in mind that risk in general, and loss in particular, is very dependent upon one's point of view.

It is certainly a misunderstanding that the gain of one party to a contract is certainly the loss of the other. In many cases there are no winners under a contract, as all the choices specified in a contact bring about some level of loss for all participants.

It is impossible to find a construction project which is not confronted with many different types and kinds of risks. Some of them may be important, some may not. The parties to a contract should be concerned with the risks created by the contract and the performance of it and the risks inherent in the construction activities.

It is noticeable that an event or action can be viewed as risk only if there is a chance of loss associated with risk.

A sure loss is not a risk, because it has a certainty of occurrence. In "certainty situations," the gains or benefits can be objectively treated straightforwardly against the losses or costs that exist.

In addition to chance, the idea of choice is important[29], as well, because there is no risk with an action, alternative offer, a method of construction, a procurement method, etc., if either party has not chosen to take, accept, or use it. Remember also, not choosing is also viewed as a choice.

5.1.1.3 Measurement of risk

In many cases the impact of a risk can be measured by combining the probability of its occurrence, denoted as p, and the magnitude of its consequences, denoted as c, as follows:

$$\text{Risk impact} = p \times c$$

Quantitative measurement of a risk permits a comparison of the magnitude of the various risks to which a construction project is exposed.

5.1.2 Causes of a risk

Risk, if it occurs, has a positive or a negative effect on at least one project objective, such as time, cost, scope, or quality (e.g., the time for completion as one of the both the employer and the contractor's objectives is specified in the Appendix to Tender).

A risk may have one or more causes and, if it occurs, one or more impacts. The risk event may be that the A/E may take longer than planned to issue further drawings, or a supplier has delivered defective goods. If either of these uncertain events occurs, there may be an impact on the cost, progress, or performance of the works. Risk conditions could include aspects of the project's or contractor's environment which may contribute to project risk, such as poor project management practices, lack of integrated management systems, concurrent multiple projects, or dependency on external participants who cannot be controlled.

The risks inherent in a construction project have origins in the uncertainty present in themselves and the project context as described in 1.6.

5.1.3 Risk management

Risk relates to threats against the success of a construction project or to opportunities in favour of its success. The threats may be accepted if they are in balance with the reward for taking the risk. The opportunities, such as acceleration which may be achieved by adding staff, if taken, help the project achieve its objectives.[16]

The parties to a contract should proactively and consistently manage the risks throughout the contract.

Risk management aims to increase the probability of and enhance effect of positive events, and decrease the probability and reduce negative consequences of events adverse to the project.

Careful and explicit planning for management of risks enhances the possibility of success.

Words, Phrases and Expressions 5.1

[1] descriptive uncertainty 说明不确定性
[2] structural uncertainty 结构不确定性
[3] measurement uncertainty 计量不确定性
[4] result uncertainty. 结果不确定性
[5] speculative risks 投机风险
[6] dynamic risks 动态风险
[7] pure risks 纯粹风险
[8] static risks 静态风险
[9] measurement of risk 风险的计量

5.2 Risk identification[16]

Risk identification determines which risks might affect the project and documents their characteristics. Participants in risk identification can include: project manager, project team members, risk management team (if assigned), subject matter experts from outside the project team, the employer, end users, other project managers, stakeholders, and risk management experts. On some occasions, the identification of a risk may suggest appropriate response to it. The process may be based on examining contract documents, reviewing the information gathered on context of the project, and analyzing various checklists and assumptions, etc..

Various diagramming techniques, such as cause-and-effect diagrams, system or process flow charts, and influence diagrams can help the identification.

5.2.1 Risk categories

The risks which may affect the project for better or worse, when identified, shall be described, documented, including their root causes and uncertain assumptions, and then organized into categories.

Risk categories shall be well defined and reflect risk response owners, (as defined below) common sources, or consequences of the risks inherent in construction projects and contracting. A categorization of the risks involved in construction projects is an illustration:[16]

(1) technical, quality, or performance risks;
(2) project management risks;
(3) organizational; and
(4) external risks.

Categorization of risks provides a structure that ensures a comprehensive process of systematically identifying risk to a consistent level of detail and contributes to the effectiveness and quality of risk identification.[16] Both the employer and the contractor can use a previously prepared categorization of typical risks. A risk breakdown structure, as shown in Figure 5.1, is one approach to providing such a structure, but it can also be addressed by simply listing the various aspects of the project. The risk categories may be revisited during the risk identification process. A good practice is to review the risk categories during the risk management planning process prior to their use in the risk identification process. Risk categories based on prior projects may need to be tailored,

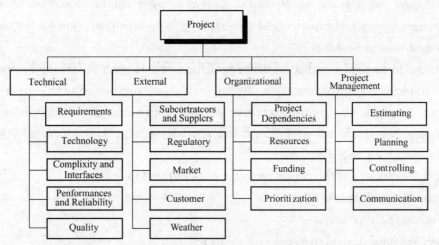

The Risk Breakdown Structure (RBS) lists the categories and sub-categories Within Which risks may arise for a typical project. Different RBSs will be appropriate for different types of projects and different types of organizations. One benefit of this approach is to remind participants in a risk identification exercise of the many sources from which project risk may arise.

Figure 5.1 Example of a Risk Breakdown Structure (RBS)[16]

adjusted, or extended to new situations before those categories can be used on the current project.

5.2.2 Identification of risks in a construction project

The risks inherent in a construction project can be viewed in terms of the impact which they may have, and thus two basic categories can be identified. The first category incorporates the risks which could lead to damage, physical loss, or injury and the second incorporates risks which could lead to lack or non-performance of the contract, delay in completion of the works and/or cost over-run of the work executed.

Defective design, materials or workmanship, force majeure, fire, human error and failure to take adequate precautions are the examples of the first category. Examples of the second category include late possession of the site, delay in receipt of information necessary for timely construction, changes in design, and variations to the original contract.

Attention, therefore, shall be given to the following areas of a construction project when risk identification is conducted:
(1) timing, such as time available for preparation prior to commencement, rate of the progress;
(2) nature of the works, for example, type of the works, size, complexity, location, market condition, readiness of the drawings and specifications, the designer's commitment to the contract;
(3) statutory requirements, such as changes in legislation;
(4) the employer's organizational structure, such as decision making procedure, anticipated variations, competency and experience of the employer's management team for the contract, nomination of subcontractors;
(5) procurement method (method of delivery of the works), such as fast track; and
(6) the employer's involvement in design and/or in contract control, such as the programme, progress, cost control and financial management, variation instruction control.

A survey was conducted in 1995 on the attitude of large U. S. construction firms toward risk and their risk management.[30]

The risk areas in a construction project were identified as follow.
(1) Permits and ordinances;
(2) Site access/right of way;
(3) Labor, equipment, and material availability;
(4) Labor and equipment productivity;
(5) Defective design;
(6) Changes in work;
(7) Differing site conditions;

(8) Acts of God(Force majeure);
(9) Defective materials;
(10) Changes in government regulation;
(11) Labor disputes;
(12) Safety;
(13) Inflation;
(14) Contractor competence;
(15) Change-order negotiations;
(16) Third-party delays;
(17) Contract delay resolution;
(18) Delayed payment on contract;
(19) Quality of work;
(20) Indemnification and hold harmless;
(21) Financial failure-any party;
(22) Actual quantities of work;
(23) Defensive engineering;

Words, Phrases and Expressions 5.2

[1]　risk identification 风险识别
[2]　diagramming techniques 绘图技术
[3]　cause-and-effect diagrams 因果(关系)图
[4]　system charts 系统图
[5]　process flow charts 流程图
[6]　influence diagrams 影响图
[7]　risk categories 风险分类
[8]　cost overrun 费用超支
[9]　defective design, materials or workmanship 有缺陷的设计、材料或施工工艺
[10]　permits and ordinances 许可证与条例
[11]　right of way(在他人土地上的)通行权
[12]　Acts of God (Force majeure)不可抗力
[13]　change order 变更指示(美国说法,等于 variation order 或 variation instruction)
[14]　financial failure 财务困境,财力不支
[15]　defensive engineering 保守的设计
[16]　risk response owner 风险应对负责人

Notes 5.2

[1]　On some occasions, the identification of a risk may suggest appropriate response to it. The process may be based on examining contract documents, reviewing the

information gathered on context of the project, and analyzing various checklists and assumptions, etc..

【译文】 有些场合，风险一经识别就会启发适当的应对办法。这一过程基本上可以靠审查合同文件与已收集到手有关项目环境的资料，以及分析各种各样核对表与假设等进行。

【解释】 注意，这里的 suggest 的意思是 bring (an idea, possibility, etc.) into the mind，使联想，使想起，启示，提醒，暗示。

［2］ The risks which may affect the project for better or worse, when identified, shall be described, documented, including their root causes and uncertain assumptions, and then organized into categories.

【译文】 会影响项目的风险，无论好坏，一经识别，就应以文字说明之、记载之，包括其最终原因，以及不确定的假设，并加以分类。

［3］ Categorization of risks provides a structure that ensures a comprehensive process of systematically identifying risk to a consistent level of detail and contributes to the effectiveness and quality of risk identification.

【译文】 风险分类成了一种方法，确保系统的风险识别过程全面，详细程度达到一致，有助于风险识别的效果与质量。

【解释】 structure 的意思是 way in which *sth*. is put together, organized, etc. 而 provide 的意思仍然是"形成"，"构成"，"是"。

5.3 Risk analysis

In this book, risk analysis refers to subjective and objective assessment of risks. For the purpose of this book only a brief description of risk analysis is given below.

5.3.1 Qualitative risk analysis[16]

Qualitative or subjective risk analysis involves prioritizing the identified risks for further action. Both the employer and the contractor can improve the project's performance effectively by focusing on high-priority risks. The priority of identified risks is assessed in terms of their measurements. as well as other factors such as the time frame and risk tolerance of the project constraints of cost, schedule, scope, and quality.

The time criticality of risk-related actions may magnify the importance of a risk. An evaluation of the quality of the available information on project risks also helps understand the assessment of the risk's importance to the project.

Qualitative risk analysis is usually a rapid and cost-effective means of establishing priorities for risk response planning, and forms the basis for quantitative risk analysis.

5.3.2 Quantitative risk analysis[16]

Quantitative or objective analysis is performed on risks that have been prioritized in the process of qualitative risk analysis as potentially and substantially impacting the project's competing demands. In the process the effect of those risk events is analyzed and a numerical rating is assigned to those risks. Various techniques such as Monte Carlo simulation and decision tree can be used to:

(1) quantify the possible outcomes for the project and their probabilities;

(2) assess the probability of achieving specific project objectives;

(3) identify risks requiring the most attention by quantifying their relative contribution to overall project risk;

(4) identify realistic and achievable cost, schedule, or scope targets, given the project risks; and

(5) determine the best project management decision when some conditions or outcomes are uncertain.

Availability of time and budget, and the need for qualitative or quantitative statements about risk and impacts, will determine which method(s) to use on any particular project.

Words, Phrases and Expressions 5.3

[1] prioritize *vt.* 确定(项目、目标等)的优先劣后、轻重缓急顺序(list or rate (as projects or goals) in order of priority)

[2] time frame 时间安排

Notes 5.3

[1] Qualitative or subjective risk analysis involves prioritizing the identified risks for further action, such as quantitative or objective risk analysis or risk response planning.

【译文】 定性即主观风险分析就是确定已识别风险的重要性大小顺序,以便将来采取行动,例如定量即客观风险分析或风险应对规划。

【解释】 involve 的意思仍然是"免不了",可译成"是"。

5.4 Allocation of risks

When risks are identified, assessed and analyzed they must be allocated to the various parties to plan suitable responses to them, in order to prevent the occurrence of their harmful consequences and thus reduce the risk. Such allocation is part of the risk management process, where the party to whom certain risks are allocated should be selected in accordance with certain principles.[13] The principles for allocation of risks in a construction project may simply revolve around the ability of a party to:

(1) make any arrangements which might be required to deal with the negative impact or any triggering factors relating to it;

(2) keep the risk under control or to influence any of its resultant effects;

(3) perform a task relating to the project, such as obtaining and maintaining insurance cover; and

(4) benefit from the project.

If risks are not allocated in a contract and a dispute arises between the parties to that contract concerning to whom a particular risk is allocated, then an arbitrator or a judge would most likely examine the following criteria for risk allocation and determine the dispute accordingly:

(1) which party could best foresee that risk?

(2) which party could best control that risk and its associated hazard or hazards?

(3) which party could best bear that risk?

(4) which party most benefits or suffers when that risk eventuates? [13]

Words, Phrases and Expressions 5.4

[1] allocation of risks 分配风险,风险分配

5.5 Risk response planning[16]

This is the process of developing options, and determining actions to enhance opportunities and reduce threats to the project's objectives. It follows the qualitative and quantitative risk analysis, and includes the identification and assignment of one or more persons (referred to as "risk response owner") to take responsibility for each agreed-to and funded risk response. Risk response planning addresses the risks by their priority, inserting resources and activities into the budget, schedule, and project management plan, as needed.

Planned risk responses must be appropriate to the significance of the risk, cost effective in meeting the challenge, timely, realistic within the project context, agreed upon by all parties involved, and owned by a responsible person. Risk responses reflect the trade-off of the parties to a contract between risk-taking and risk-avoidance.

Strategies in responding to a risk or their combination most likely to be effective should be selected and then specific actions developed to implement them. The strategies may either play primary role or a backup one. A fallback plan can be developed for implementation if the selected strategy does not prove fully effective. A contingency reserve is often allowed for time or cost. Finally, contingency plans can be developed, along with identification of the triggering conditions.

5.5.1 Strategies towards threats[16]

There are typically three strategies available for dealing with threats or risks that may have negative impacts on project objectives:

5.5.1.1 Avoidance

Avoidance of risk involves changing the project to eliminate the threat which a risk may pose, to isolate the project from the risk's impact, or to relax the project's objective which may be jeopardized, such as extending time of completion or reducing the project's scope. Some risks that arise early in the project can be avoided by clarifying requirements, obtaining information, improving communication, or acquiring expertise.

5.5.1.2 Transfer

Transferring liability for risk is most effective in dealing with financial risk exposure. and nearly always involves payment of a risk premium to the risk taker. Numerous tools can be used to implement this strategy, such as, insurance, performance bonds, warranties, guarantees, etc.. Contracts and/or subcontracts may also be used to offload specified risks to another party. For example, the main contractor may subcontract the curtain walling to a specialist firm.

5.5.1.3 Mitigation

Mitigation as a strategy involves reducing the probability of occurrence and/or possible negative impact of a risk to an acceptable level. Early proactive measure is usually more effective than aftermath responses. Well prepared tender, closely coordinated subcontractors and suppliers, and strict compliance with the applicable laws of the contract are examples for the contractor to mitigate the risks of delay and cost overrun. Where it is not possible to reduce probability, effort may be directed at the factors determining the severity of the possible negative impact. For example, adequate time reserve is provided for the critical activities to reduce the impact of delays in a critical activity.

Strategies for risk reduction which can be used by a contractor include:
- ☐ Review and revise tender and contract documents;
- ☐ Geo-technical information;
- ☐ Construct-ability reviews;
- ☐ Effective dispute resolution;
- ☐ Realistic time for completion;
- ☐ Need for contingency reserve;
- ☐ Planned communication;
- ☐ Pre-planning for permits/utilities/zoning;
- ☐ Use differing site conditions clauses;
- ☐ Recognize cost of design effectiveness;
- ☐ Delegate decision-making/authority to site.

5.5.2 Strategies towards opportunities[16]

There are three responses available to deal with risks with potentially positive impacts on project objectives.

5.5.2.1 Exploitation

This strategy may be taken to make an opportunity realize and to benefit from the positive impact of the risk, including assigning more competent personnel to the site to speed up the progress to win the bonus as promised in the contract and make the personnel available to another contract which can contribute more to the contractor's long-term business strategy.

5.5.2.2 Sharing

Sharing a positive risk involves allocating ownership to a party who is best able to capture the opportunity for the benefit of the project.

Certain principles of allocation of risks have been established and ought to be followed whenever possible. These principles are generally based on control over the occurrence of the risks and/or influence over the effect they cause if and when they eventuate. Only in certain circumstances can the allocation of risks be successfully based on the optimum ability of a party to perform a specific task related to the project or the inability of any party other than the employer who initiated the project to accept a certain risk.

There are exceptions to the freedom to allocate risks in accordance with the above principles and they depend on the prevalent legal rules, such as those which apply to exclusion clauses and limitation clauses in a contract. Table 5.1 illustrates the allocation of a few risks associated with a construction project.

Table 5.1 Allocation decisions regarding risks associated to a construction project

Risk area	Party to assume risk (risk owner)	Responses to risk
Site access	Employer	Advance planning and acquisition
Methods of construction	Contractor	Specific contract clause
Labor productivity	Contractor	Contractor's management
Site conditions	Employer	Geo-technical investigation and contract clause
Weather; Force Majeure	Shared (Employer assumes delay risk, contractor assumes cost risk)	Contract clause

5.5.2.3 Enhance

This strategy modifies the "size" of an opportunity by increasing probability and/or positive impacts, and by identifying and enhancing the key driving factors of the positive risks. The probability can be increased by facilitating or strengthening the cause of the opportunity and identifying and reinforcing its trigger conditions. The drivers of the negative risks can be dealt with in an opposite direction.

5.5.3 Strategy towards threats and opportunities[16]

The best strategy towards the threats and opportunities is acceptance when the risk occurs. This strategy can be taken where there is little possibility to eliminate all the risks present and the project team has decided not to change the project or is unable to identify any other suitable strategy. Passive acceptance needs no action. The project team can deal with the threats or opportunities as they occur. Provision of a contingency reserve is the most common active acceptance strategy, including amounts of time, money, or resources to handle the threats or opportunities.

5.5.4 Contingent response[16]

Some responses are designed for use only if certain events occur. For some risks, it is appropriate for the project team to make a response plan that will only be executed under certain predefined conditions, if it is believed that there will be sufficient warning to implement the plan. The triggers of a contingency response, such as missing intermediate milestones or gaining higher priority with a supplier, should be defined and tracked.

Words, Phrases and Expressions 5.5

[1] risk response planning 风险应对规划
[2] risk reduction 降低风险
[3] trade-off 权衡
[4] risk-taking and risk-avoidance 接受风险与回避风险
[5] fallback 撤退,后退(retreat, move or turn back)
[6] fallback plan 撤退计划,后退计划
[7] premium n. 额外代价,额外费用(addition to ordinary charges, wages, rent, etc.)
[8] contingency reserve 应急储备(The amount of *funds*, *budget*, or time needed above the *estimate* to reduce the *risk* of overruns of project *objectives* to a level acceptable to the *organization*. 为将达不到项目目标的风险降低到该组织可接受的程度而需要在资金或时间方面比估算多准备的那部分数额。)
[9] contingency plan 应急计划
[10] triggers 触发因素(Indications that a risk has occurred or is about to occur. Triggers may be discovered in the *risk identification* process and watched in the *risk monitoring and control* process. Triggers are sometimes called *risk* symptoms or warning signs. 风险业已发生或者即将发生的标示。触发因素可在风险识别过程中发现,并可在风险监视与控制过程中进行监视。触发因素有时称为风险症状或者警告信号。)
[11] triggering conditions 触发条件
[12] strategies towards threats 应对威胁的策略
[13] transfer 转移
[14] financial risk exposure 财务风险状况
[15] warranties 产品质量保证
[16] guarantees 保证书
[17] mitigate vt. 减轻(make or become less severe, intense, or painful)
[18] aftermath responses 事后应对措施
[19] exploit vt. 利用(utilize, especially for profit)
[20] sharing a risk 分担风险
[21] principles of allocation of risks 风险分配原则

[22] optimum ability 最大能力（原则）
[23] enhance vt. 增强（raise to a higher degree, as in value or quality）
[24] acceptance 自留
[25] contingent response 随机应变措施

Notes 5.5

[1] Avoidance of risk involves changing the project to eliminate the threat which a risk may pose, to isolate the project from the risk's impact, or to relax the project's objective which may be jeopardized, such as extending time of completion or reducing the project's scope.

【译文】 回避风险就是改变项目,消除某一风险可能施加的威胁,使项目不受风险后果的影响,或者降低可能受到损害的目标,例如延长竣工时间或缩小项目范围。

【解释】 involve 仍然是"免不了,需要,不可缺少"的意思,但译成汉文时,可译成"是"。

[2] Risk transference nearly always involves payment of a risk premium to the risk taker. Numerous tools can be used to implement this strategy, such as, insurance, performance bonds, warranties, guarantees, etc.

【译文】 转移风险几乎总要因补偿接受风险者而付出额外代价。实施这一策略的手段很多,例如保险、履约保证、产品质量保证、保证书等。

【解释】 involve 仍然是"免不了,需要,不可缺少"的意思,但译成汉文时,可译成"(需)要"。

[3] This is the process of developing options, and determining actions to enhance opportunities and reduce threats to the project's objectives.

【译文】 这是酝酿并提出办法以供选用,确定行动计划以便加强有利于项目目标的因素,削弱威胁项目目标的因素的过程。

【解释】 develop 的意思是 grow larger, fuller or complete,酝酿,孕育。option 的意思是 thing which is or may be chosen。

5.6 Contract strategy

In all construction work, the final cost cannot be identified accurately at the outset. This is due to a variety of "risks" associated with the work, such as unforeseen ground conditions, design changes, unusual weather conditions, inflation of material and labour costs, changes in international exchange rates etc.. This risk is shared in varying proportions between the employer and the contractor. The type of contract used for the procurement of the works has a very significant impact on the way the risks are shared between these two parties.

5.6.1 Cost reimbursement contract

All the risks are taken by the employer and none by the contractor. There is no incentive for the contractor to work efficiently - in fact there is an incentive to work inefficiently as the longer the work goes on and the higher the costs are, the greater is the reward to the contractor.

This contract strategy should only be used in emergency. An unsatisfactory arrangement will imply that no design or planning has been done prior to commencement of the works. This form of contract should be avoided if at all possible.

5.6.2 Remeasurement contracts

In general, the contractor will take more risk in this type of contract than with a cost reimbursement arrangement.

There is, however, little incentive for the contractor to adopt alternative cost saving methods of working, although in some contracts the contractor is offered the option of putting forward alternative designs. This will obviously increase the cost of bidding. The contractor can "weight" certain items in the bill to his advantage, and will include some element of cost to cover the risks he is taking.

5.6.3 Lump sum or fixed price contracts

The advantage to the employer is that he knows at the outset exactly how much the contract will cost him.

The employer takes none of the risk but, as a consequence of this, the price is likely to be higher than for other forms of contract—particularly if the employer has not fully defined the scope and detail of the project in complete drawings and specification. The contractor takes all the risk and has a high incentive to complete the works at the lowest cost to himself in the shortest possible time.

Quality may suffer—good site supervision by the employer or his engineer is essential.

Provision can be made within this form of contract for any rise (or fall) of material and labour cost. This transfers some of the risk from the contractor to the employer.

The contractor must make sure that the contract contains provisions for variations which may arise from unforeseen ground conditions, particularly if site investigation is suspected to be inadequate.

5.6.4 Contractual agreements[12][16]

The strategies described in 5.5 can be incorporated as warranties and/or conditions in the general or special conditions of contract which allocate the risks between the parties on a fair and equitable basis, having regard to such matters as insurability, sound principles of project management, and each party's ability to foresee, and mitigate the effect of, the circumstances relevant to each risk.

The allocation may be covered in the conditions of contract under the headings of indemnities, insurances, contractor's care of the works, the employer's risks, limitation of liability, force majeure, duty to minimise delay, and other items as appropriate, specifying each party's responsibility for specific risks, should they occur.

Words, Phrases and Expressions 5.6

[1] contract strategy 合同策略
[2] insurability n. 受保的可能性
[3] limitation of liability 责任限度
[4] indemnity n. 赔偿保障,补偿保障（an indemnity is a promise by one party to make good any loss suffered by the other party in respect of claims arising out of specified matters 保障是一方弥补另一方在事先规定的事物方面蒙受损失并提出要求时给予弥补的诺言）

Notes 5.6

There is no incentive for the contractor to work efficiently—in fact there is an incentive to work inefficiently as the longer the work goes on and the higher the costs are, the greater is the reward to the contractor.

【译文】 承包商没有提高工作效率的积极性,却有磨洋工的积极性,因为工作拖得时间越长,费用越高,承包商捞到的越多。

5.7 Allocation of risks under the *Red Book*

The *Red Book* is prepared on the basis of sharing of risks between the employer and the contractor. The principles of sharing of risks are similar to those described in 5.5.2.2.

5.7.1 Performance security

5.7.1.1 Security

Security is a contract under which one party, the surety, undertakes in the event of the default of a second party over the payment of a debt to, or the performance of some act in favour of, a third party, to be answerable to the third party for such default. It can be in the form of a demand guarantee or a surety bond.

A security is essentially a collateral contract for it cannot exist in the absence of the agreement of which performance is being guaranteed. To be enforceable it must be in writing, and if not under seal must be supported by consideration and formally accepted by the party to whom it is offered. In relation to construction contracts, guarantees need to be very carefully expressed so as to ensure that the surety is undertaking to guarantee due performance of those of the contractor's obligations which are crucial to the employer, the nature of which is frequently complex.

The provision of securities is sometimes a condition precedent of construction contracts. This is because of the complications surrounding the matter of consideration for a security, which usually exists between the surety and the potential defaulter rather than directly to support the obligation to protect the third party. Thus the insurance company which gives the guarantee or bond will usually be paid by the contractor and not by the employer. The amount of that payment will of course be reflected in the contractor's price and thus emanates from the employer.

The three-cornered arrangement of security and principal contracts (e. g. those between the employer and the contractor) makes the obligations of suretyship somewhat onerous and the law has evolved to afford some measure of protection for guarantors against acts which may prejudice their position. For example, concealment of material facts relating to, or the alteration of material terms within the principal contract may vitiate the guarantee, as may the failure of a party to the principal contract to preserve some security to which the surety is entitled.

The suretyship will end with the fulfilment of the obligations guaranteed, providing such performance cannot be set aside by legal process.

5.7.1.2 Performance security

Most conditions of contract require the contractor to undertake to provide a security for his due performance of the contract. This is usually provided by a bank or insurance company and entitles the employer to recover damages up to a prescribed amount (usually a percentage of the tender price) if the contractor does not fulfill all his obligations under the contract (for a sample see 3.2.10). The contractor shall execute the performance security at this stage as required by the employer. In some cases the satisfactory completion of this document is a prerequisite to the issue of the letter of acceptance by the employer.

The cost of such a security adds to the contract price. It must be borne in mind that in most forms of contract the following intrinsic securities for the performance of contractual obligations already exist:

(a) the contractor's equipment, when brought on to the site, shall be deemed to be exclusively intended for the execution of the works. The contractor shall not remove from the site any major items of contractor's equipment without the consent of the Engineer;

(b) the retention money (generally a percentage of the measured value of the work up to a stated limit) held by the employer;

(c) the materials supplied and work done by the contractor during the period preceding payment by the employer.

5.7.1.3 Differences between a demand guarantee and a surety bond

The major difference between the two forms of security is that under a demand guarantee the surety undertakes in the event of the contractor's default over the performance of his contractual obligations to pay the employer any sum or sums not exceeding in a specified amount, while under a surety bond, whenever the contractor shall be, and declared by the employer to be, in default under the contract, the employer having performed his own contractual obligations, the surety undertakes to promptly either remedy the default or take appropriate actions other than payment to the employer.

The surety under a surety bond may take following actions in favour of the beneficial, i.e. the employer:

(1) complete the contract in accordance with its terms and conditions; or

(2) obtain a tender or tenders from qualified tenderers for submission to the employer for completing the contract in accordance with its terms and conditions, and upon determination by the employer and the surety of the lowest responsive tenderer, arrange for a contract between such tenderer and employer and make available as work progresses sufficient funds to pay the cost of completion; but not exceeding the amount set forth in the bond; or

(3) pay the employer the amount required by employer to complete the contract in accordance with its terms and conditions up to a total not exceeding the amount of the bond.

5.7.2 Indemnities

A number of inherent characteristics of construction projects require that the liabilities arising from the duties and obligations of the parties to a construction contract should be covered by indemnities given by one party to the other, or provided in the form of policies. The characteristics, for example, may be:

(1) the contractor must complete the contract irrespective of whatever happens including accidents and other deterrents, other than a few circumstances as specified in the contract;

(2) construction involves a huge sum of money, frequently provided by banks, insurance companies and other financial institutions which require some form of guarantee as to the safety of their financing;

(3) numerous risks exist in a variety of categories; and etc..

5.7.2.1 Definition of indemnity

Indemnity is an undertaking by one party to hold another party free from pecuniary loss which might result from one or a number of specified occurrences and the quantum of which may be incalculable at the time the indemnity is given. An indemnity can be entered into orally and will be binding without specific acceptance by the indemnified, unlike a guarantee, which must be in writing and which must be formally accepted by the party to whom it is offered.

Indemnities are often incorporated in construction contracts. Usually, but not exclusively, they are given by contractors in favour of employers so as to protect the employer from pecuniary loss consequent upon claims made on him by third parties alleging injury as a direct consequence of the contract works. For such indemnities to be effective the indemnifier must be able to meet the financial consequences of any claim and consequently it is usual to require the indemnifier to take out insurance, in the joint names of the indemnifier and of the indemnified, to cover the occurrences referred to in the indemnity; this does of course require some evaluation to be made of the probable extent of any loss.

5.7.2.2 Indemnity clauses under the *Red Book*

(1) Sub-clause 17.1

Sub-clause 17.1 requires both the employer and the contractor to protect the other party from claims, including those from third parties, arising out of the contractor's execution of the works.[31]

The contractor shall indemnify and hold harmless the employer, the employer's personnel, and their respective agents, against and from all claims, damages, losses and expenses (including legal fees and expenses) in respect of:

(a) bodily injury, sickness, disease or death, of any person whatsoever arising out of or in the course of or by reason of the contractor's design (if any), the execution and completion of the works and the remedying of any defects, unless attributable to any negligence, willful act or breach of the contract by the employer, the employer's personnel, or any of their respective agents, and

(b) damage to or loss of any property, real or personal (other than the works), to the extent that such damage or loss:

(i) arises out of or in the course of or by reason of the contractor's design (if any), the execution and completion of the works and the remedying of any defects; and

(ii) is attributable to any negligence, willful act or breach of the contract by the contractor, the contractor's personnel, their respective agents, or anyone directly or indirectly employed by any of them.

The employer shall indemnify and hold harmless the contractor, the contractor's personnel, and their respective agents, against and from all claims, damages, losses and expenses (including legal fees and expenses) in respect of (1) bodily injury, sickness, disease or death, which is attributable to any negligence, willful act or breach of the contract by the employer, the employer's personnel, or any of their respective agents, and (2) the matters for which liability may be excluded from insurance cover.

These indemnities apply widely, but may not cover every type of claim[31]. In other words, there may be claims for which neither party is entitled to indemnity under this sub-clause. Claims for personal injury are to be borne by the contractor, if they are attributable to his execution of the works, and are not attributable to any act or negligence of the employer or employer's personnel. These claims may arise without any act or negligence by the contractor, and may be covered by the insurance specified in sub-clause 18.3.

The employer bears the cost of claims for personal injury which are attributable to any act or negligence of the employer or employer's personnel.

Claims for property damage are to be borne by the contractor to the extent that they are attributable to any act or negligence of the contractor or contractor's personnel, and by the employer to the extent that they are attributable to any act or negligence of the employer or employer's personnel.

Other claims for property damage may be covered by the insurance specified in sub-clause 18.3, subject to the last three paragraphs of sub-clause 18.1. [31]

Finally, the employer is required to bear the cost of claims in respect of the matters described as below:

(1) the employer's right to have the permanent works executed on, over, under, in or through any land, and to occupy this land for the permanent works;

(2) damage which is an unavoidable result of the contractor's obligations to execute the Works and remedy any defects; and

(3) any one of the employer's risks as listed in sub-clause 17.3, except to the extent that insurance cover is available at commercially reasonable terms.

Under sub-clauses 17.1 and 18.3(d)(ii), the employer indemnifies the contractor from claims in respect of "damage which is an unavoidable result of the contractor's obligations", but not in respect of any other damage which is a result of the particular arrangements and methods which the contractor elected to adopt in order to perform his obligations. The contractor should adopt appropriate arrangements and methods so as to minimise claims from third parties due to the performance of his obligations under the contract. [31]

(2) Sub-clause 17.5

It is noticeable that a separate sub-clause 17.5 is introduced to cover intellectual and industrial property rights.

This sub-clause provides appropriate protection to each party in respect of any breached of copyrights or of other intellectual or industrial property right. [31] If a third party makes a claim in respect of any of the matters mentioned in this sub-clause, the parties should each consider taking advice from a lawyer familiar with the applicable intellectual or industrial property right law. The sub-clause deals with many types of infringements, but does not attempt to cover all types. In other words, there may be claims for which neither party is entitled to indemnity under this sub-clause.

In particular, the sub-clause does not cover the situation where the employer is prevented by a competent court from operating the works by reason of an alleged infringement.

In this sub-clause, "infringement" means an infringement (or alleged infringement) of any patent, registered design, copyright, trade mark, trade name, trade secret or other intellectual or industrial property right relating to the works; and "claim" means a claim (or proceedings pursuing a claim) alleging an infringement.

5.7.3 Contractor's care of the works

Sub-clause 17.2 requires the contractor to take full responsibility for the care of the works and goods from the commencement date until the taking-over certificate is issued (or is deemed to be issued under sub-clause 10.1) for the works, when responsibility for the care of the works shall pass to the employer. If a taking-over certificate is issued (or is so deemed to be issued) for any section or part of the works, responsibility for the care of the section or part shall then pass to the employer.[31]

After responsibility has accordingly passed to the employer, the contractor shall take responsibility for the care of any work which is outstanding on the date stated in a taking-over certificate, until this outstanding work has been completed.

If any loss or damage happens to the works, goods or contractor's documents during the period when the contractor is responsible for their care, from any cause not listed in sub-clause 17.3, the contractor shall rectify the loss or damage at the contractor's risk and cost, so that the works, goods and contractor's documents conform to the contract.

The contractor shall be liable for any loss or damage caused by any actions performed by the contractor after a taking-over certificate has been issued. The contractor shall also be liable for any loss or damage which occurs after a taking-over certificate has been issued and which arose from a previous event for which the contractor was liable.

5.7.4 Insurance

Most standard forms of contract require the contractor to insure against specified liabilities, and the premiums for such insurances are included in the tender price. The heavier the liabilities placed on the contractor, the higher will be the tender price. The employer may insure against his liabilities under the terms of the contract, but some large organizations prefer to carry and finance the risk themselves. The employer should ensure that the contractor's insurances provide the protection required.

5.7.4.1 Insurance clauses under the *Red Book*

Clause 18 covers the insuring party's insurance responsibilities. Under this clause "insuring party" means, for each type of insurance, the party responsible for effecting and maintaining the insurance specified in the contract.

Clause 18 requires that the contractor, wherever he is the insuring party, effect each insurance with insurers and in terms approved by the employer before the date of the letter of acceptance.

Wherever the employer is the insuring party, each insurance shall be effected with insurers and in terms consistent with the details annexed to the particular conditions.

If a policy is required to indemnify joint insured, the cover shall apply separately to each insured as though a separate policy had been issued for each of the joint insured.

When each premium is paid, the insuring party shall submit evidence of payment to the other party. Whenever evidence or policies are submitted, the insuring party shall also give notice to the engineer.

The insuring party shall keep the insurers informed of any relevant changes to the execution of the works and ensure that insurance is maintained in accordance with clause 18:

The works, plant, materials and contractor's documents shall be insured for not less than the full reinstatement cost including the costs of demolition, removal of debris and professional fees and profit;

This insurance shall be maintained by the insuring party to provide cover until the date of issue of the performance certificate, for loss or damage for which the contractor liable arising from a cause occurring prior to the issue of the taking-over certificate, and for loss or damage caused by the contractor in the course of any other operations.

The insuring party shall insure the contractor's equipment for not less than the full replacement value, including delivery to site. For each item of contractor's equipment, the insurance shall be effective while it is being transported to the site and until it is no longer required as contractor's equipment.

Unless otherwise stated in the particular conditions, insurances under clause 18:
(a) shall be effected and maintained by the contractor as insuring party;
(b) shall be in the joint names of the parties, who shall be jointly entitled to receive payments from the insurers, payments being held or allocated between the parties for the sole purpose of rectifying the loss or damage;
(c) shall cover all loss and damage from any cause not listed in clause 17;
(d) shall also cover loss or damage to a part of the works which is attributable to the use or occupation by the employer of another part of the works, and loss or damage from the employer's risks, excluding (in each case) risks which are not insurable at commercially reasonable terms; and
(e) may however exclude loss of, damage to, and reinstatement of:
(i) a part of the works which is in a defective condition due to a defect in its design, materials or workmanship (but cover shall include any other parts which are lost or damaged as a direct result of this defective condition and not as described in the sub-paragraph (ii) below),
(ii) a part of the works which is lost or damaged in order to reinstate any other part of the works if this other part is in a defective condition due to a defect in its design, materials or workmanship,
(iii) a part of the works which has been taken over by the employer, except to the extent that the contractor is liable for the loss or damage.

The contractor shall effect and maintain insurance against liability for claims, damages, losses and expenses (including legal fees and expenses) arising from injury, sickness, disease or death of any person employed by the contractor or any other of the contractor's personnel.

In addition the insuring party shall insure against each party's liability for any loss, damage, death or bodily injury and damage to property which may occur to any physical property other than the works and contractor's equipment or to any person other than the contractor's personnel, which may arise out of the contractor's performance of the contract and occurring before the issue of the performance certificate.

5.7.4.2 Design risk

The accelerating technological advance and the availability of a wider range of materials imply that new design methods and concepts are rapidly replacing proven existing methods. At the same time, the increasing scarcity of oil, minerals and other energy resources means that projects are being sited at locations, some of which may have been avoided hitherto because of greater difficulty of development, possible exposure to natural catastrophes, or lack of hydrologic and other relevant data. Proper assessment of hazards may therefore be more difficult. [13]

It is appreciated that the A/E will normally alert the employer when his designs involve a "calculated risk" either in terms of the "state of the art" or because of insufficient data. But it is by no means certain that the employer will fully appreciate the effects of such a warning in terms of liability and insurance protection. Under the present system the employers have taken little direct interest in insurance, relying on the contractual pattern of risk-allocation and management. The employers may thus be unaware that in the changing technological setting of new large-scale projects this pattern will leave them unprotected in respect of very major exposures.

It should be noted that one of the risks defined in clause 17 is loss or damage to the extent that it is due to the design of any part of the works by the employer's personnel or by others for whom the employer is responsible. This is a central issue from which many disputes arise in settling insurance claims within the contract period. [13]

5.7.5 Employer's risks

The following risks shall be assumed by the employer under sub-clause 17.4 of the *Red Book*:

(a) war, hostilities (whether war be declared or not), invasion, act of foreign enemies;

(b) rebellion, terrorism, sabotage by persons other than the contractor's personnel, revolution, insurrection, military or usurped power, or civil war, within the country;

(c) riot, commotion or disorder within the country by persons other than the contractor's personnel;

(d) munitions of war, explosive materials, ionizing radiation or contamination by radio-activity, within the country, except as may be attributable to the contractor's use of such munitions, explosives, radiation or radio-activity;

(e) pressure waves caused by aircraft or other aerial devices traveling at sonic or supersonic speeds;

(f) use or occupation by the employer of any part of the permanent works, except as may be specified in the contract;

(g) design of any part of the works by the employer's personnel or by others for whom the employer is responsible; and

(h) any operation of the forces of nature which is unforeseeable or against which an experienced contractor could not reasonably have been expected to have taken adequate preventative precautions.

5.7.6 Consequences of employer's risks

If and to the extent that any of the employer's risks results in loss or damage to the works, goods or contractor's documents, the contractor shall promptly give notice to the engineer and shall rectify this loss or damage to the extent required by the engineer.

If the contractor suffers delay and/or incurs cost from rectifying this loss or damage, the contractor shall give a further notice to the Engineer and shall be entitled subject to sub-clause 20.1 to:

(a) an extension of time for any such delay, if completion is or will be delayed, under sub-clause 8.4, and

(b) payment of any such cost, which shall be included in the contract price. In the case of sub-paragraphs (f) and (g) of sub-clause 17.3, cost plus reasonable profit shall be payable.

After receiving this further notice, the Engineer shall proceed in accordance with sub-clause 3.5 to agree or determine these matters.

5.7.7 Limitation of liability[31]

When preparing tenders, tenderers will wish to assess their potential liability to the employer, and include in their prices some allowance for their risks. Sub-clause 17.6 is introduced for this purpose and to maintain a reasonable balance between the differing objectives of the parties, each of whom will wish to limit his own liability whilst being entitled to receive full compensation for default by the other party.

Under the sub-clause neither party is to be liable for certain types of loss.

The employer's liability is subject to sub-clause 14.14, that is, he shall not be liable to the contractor for any matter or thing under or in connection with the contract or execution of the works, except to the extent that the contractor shall have included an amount expressly for it (a) in the final statement and also (b) (except for matters or things arising after the issue of the taking-over certificate for the works) in the statement at completion.

The contractor's liability is limited to an amount which is to be stated in the particular conditions. Liabilities under the following sub-clauses are excluded from this limitation of liability:

(1) protection of the environment and payments for services, such as electricity, water and gas, and for the employer's equipment and free-issued materials actually used by the contractor,

(2) for indemnity against claims from third parties generally, and

(3) for indemnity against claims from third parties in respect of intellectual or industrial property rights.

Sub-clause 17.6 may be affected by the applicable law, which may limit a party's liability to a greater or lesser amount, and/or may limit the duration of its liability.

Although the general conditions do not specify any limit to the duration of the contractor's liability:

(1) under some common law jurisdictions, a period of liability may not begin until the employer ought reasonably to have been aware of the contractor's defective work.

(2) under some civil law jurisdictions, the contractor will be liable absolutely (i.e., without proof of fault) for hidden defects for ten years from completion (which is called decennial liability).

(3) unless the works include major items of plant, it may be inappropriate for the contract to limit the duration of the contractor's liability.

(4) if the works include major items of plant, it is usually appropriate for the contract to limit the duration of the contractor's liability for such plant; for example, to a stated number of years after the completion date stated in the taking-over certificate. After a few years' operation, it becomes increasingly difficult to establish whether any alleged defects are attributable to the plant's design, manufacture, manuals, operation, maintenance, or a combination of these and/or other matters.

Words, Phrases and Expressions 5.7

[1] insurance n. 保险((undertaking by a company, society, or the state, to provide) safeguard against loss, provision against sickness, death, etc., in return for regular payment)

[2] insure vt. 为……保……险(issue or obtain insurance on or for)

[3] insuring party 投保方,到保险公司支付保险费,取得保险单的当事人

[4] policy n. 保险单(Policy is a written contract of insurance between a policy-holder and his/her insurers which they intend to be legally binding. 保险单是受保人与其保险公司之间有法律约束力的书面合同。)

[5] claims vt. 索款,要求补偿,索补(To hold to be true against implied denial or doubt and to demand on the ground or right. 声明事实以澄清隐含的否认或怀疑,并据此或根据权利提出要求。)

[6] cover *n*. 保险范围(Cover is the extent of protection that insurance provides to the insured. The term *exceptions* is used to describe this cover by stating what is excluded. 是保险为受保人提供的保护范围。用术语"例外"以说明哪些内容排除在外的形式来说明这个保护范围。)

[7] cover *vt*. 给……保险(insure against risk or loss)

[8] effect *vt*. 取得,办理保险,购买保险(obtain insurance)

[9] insurers 承保人,保险公司(a person or company that insures or issues insurance)

[10] premium *n*. 保险费(amount or instalment paid for a insurance policy)

[11] insured *n*. 受保人(a person covered by an insurance policy)

[12] full reinstatement cost 完全重置费用

[13] full replacement value 完全重置价值

Notes 5.7

疑难词句解释

[1] Indemnity is an undertaking by one party to hold another party free from pecuniary loss which might result from one or a number of specified occurrences and the quantum of which may be incalculable at the time the indemnity is given. An indemnity can be entered into orally and will be binding without specific acceptance by the indemnified, unlike a guarantee, which must be in writing and which must be formally accepted by the party to whom it is offered.

Indemnities are often incorporated in construction contracts. Usually, but not exclusively, they are given by contractors in favour of employers so as to protect the employer from pecuniary loss consequent upon claims made on him by third parties alleging injury as a direct consequence of the contract works. For such indemnities to be effective the indemnifier must be able to meet the financial consequences of any claim and consequently it is usual to require the indemnifier to take out insurance, in the joint names of the indemnifier and of the indemnified, to cover the occurrences referred to in the indemnity; this does of course require some evaluation to be made of the probable extent of any loss.

【译文】 保障是一方为保障另一方在一件或若干件事先规定的事情发生时不受金钱损失而做出的承诺。保障的数额在给予保障时无法计算。保障可以口头给予,即使接受保障者并无具体接受表示也有约束力。这与保证不同,保证必须成文,且由保证书接受者正式承诺。

保障经常构成施工合同的一部分。保障一般由承包商向业主承诺,以便在第三方声称由于本合同工程的后果直接受到伤害而向业主提出补偿要求时保护业主不使其损失金钱。此种保障要产生效力,保障一方须有能力承担任何索赔的财务后果,如此一来,一般就要求保障者与受保障者联名,到保险公司为保障中提到的事件投保。这自然就要求对任何损失的可能范围进行某种评价。

[2] It is appreciated that the A/E will normally alert the employer when his designs involve a "calculated risk" either in terms of the "state of the art" or because of insufficient data. But it is by no means certain that the employer will fully appreciate the effects of such a warning in terms of liability and insurance protection. Under the present system the employers have taken little direct interest in insurance, relying on the contractual pattern of risk-allocation and management. The employers may thus be unaware that in the changing technological setting of new large-scale projects this pattern will leave them unprotected in respect of very major exposures.

【译文】 人们已经认识到，建筑师/工程师在其设计在技术水平方面或因为缺乏足够的数据而有"故意风险"时一般愿意告知业主。但是，这样做绝不表明业主一定完全明白这种警告对责任与保险提出的要求。在目前这种制度下，业主很少直接关心保险事宜，而是依赖合同在风险分担与管理方面的安排。如此一来，业主可能意识不到在新型大项目不断变化的技术背景中这种安排会使他们在遇到非常不利的情况时失去保护。

【解释】 exposures 意思是 being exposed to unfavourable things or conditions,可译成"非常不利的情况"。

[3] It should be noted that one of the risks defined in clause 17 is loss or damage to the extent that it is due to the design of any part of the works by the employer's personnel or by others for whom the employer is responsible. This is a central issue from which many disputes arise in settling insurance claims within the contract period.

【译文】 应当注意,第 17 条定义的风险之一是因业主人员或业主应负责的其他人对工程任何部分的设计而造成的损失或损害。这是在合同期间解决保险给付时引发许多争议的一个中心问题。

[4] Damage to or loss of any property, real or personal (other than the works), *to the extent that* such damage or loss arises out of or in the course of or by reason of the contractor's design (if any), the execution and completion of the works and the remedying of any defects, unless and *to the extent that* any such damage or loss is attributable to any negligence, willful act or breach of the contract by the employer, the employer's personnel, their respective agents, or anyone directly or indirectly employed by any of them.

【译文】 工程以外的任何不动产或动产的损坏或损失，只要是承包商对工程的设计、施工、竣工，弥补其中缺陷造成，或在其过程中或因其而发生，除非这些损坏或损失可归咎于业主、业主人员、业主人员各自的代理人或由其直接或间接雇用的任何人的疏忽、肆意行为或违反合同。

【解释】 注意,上句中第二个 *to the extent that* 已同 unless 合并翻译。

[5] The employer shall *indemnify* and *hold harmless* the contractor, the contractor's personnel, and their respective agents, *against* and *from* all claims, damages, losses and expenses (including legal fees and expenses) in respect of (1) bodily injury,

sickness, disease or death, which is attributable to any negligence, willful act or breach of the contract by the employer, the employer's personnel, or any of their respective agents, and (2) the matters for which liability may be excluded from insurance cover.

【译文】 业主应保障承包商、承包商人员与他们各自的代理人不对同以下各方面有关的所有索赔、赔偿费、损失与开销（包括法律费用与开销）承担责任，并始终不使其因此而受到损害：

(1) 可归咎于业主、业主人员、业主人员各自的代理人的疏忽、肆意行为或违反合同的身体伤害、小恙、疾病或死亡，以及 (2) 责任可能不在保险额以内的事物。

[6] The contractor shall *execute* the performance security at this stage as required by the employer.

【译文】 承包商在这一阶段应当按照业主的要求办理有约束力的履约保证。

【解释】 execute 的含义，请见第 3 章 Notes 3.2。这里译成"办理"。

[7] If a policy is required to indemnify joint insured, the cover shall apply separately to each insured as though a separate policy had been issued for each of the joint insured.

【译文】 如果要求有保险单保障联合受保人，则保险范围就应分别适用于每一个受保人，就如同向联合受保人中的每一个都签发了一个单独的保险单。

[8] This insurance shall be maintained by the insuring party to provide cover until the date of issue of the performance certificate, for loss or damage for which the contractor liable arising from a cause occurring prior to the issue of the taking-over certificate, and for loss or damage caused by the contractor in the course of any other operations.

【译文】 此项保险在签发完工证书日之前应始终由投保方持有，为承包商应负责的发生于签发接收证书之前，以及由承包商在其他任何作业过程中造成的损失或损坏保险。

【解释】 provide 在此句中的意思是"形成"，但是，在同名词 cover 结合之后，就没有必要翻译出来。

[9] Wherever the employer is the insuring party, each insurance shall be effected with insurers and in terms consistent with the details annexed to the Particular Conditions.

【译文】 只要业主是投保方，就应按照与专用条件附录中细节一致的条件到保险公司办理每一项保险。

[10] The employer should ensure that the contractor's insurances provide the protection required.

【译文】 业主必须确保承包商的保险足以保护(本工程)。

[11] A formal statement by a policy holder that an insured event has occurred coupled with a request for compensation.

【译文】 保险单持有人声明受保事件已经发生，连同赔偿要求的正式文件。

[12] evidence that the insurance described in this Clause have been effected.

【译文】 本条所述保险已经办理的证据。

[13] If and to the extent that any of the employer's risks results in loss or damage to the works, goods or contractor's documents, the contractor shall promptly give notice to the Engineer and shall rectify this loss or damage to the extent required by the Engineer.

(1) under some common law jurisdictions, a period of liability may not begin until the employer ought reasonably to have been aware of the contractor's defective work.

(2) under some civil law jurisdictions, the contractor will be liable absolutely (i.e., without proof of fault) for hidden defects for ten years from completion (which is called decennial liability).

(3) unless the works include major items of plant, it may be inappropriate for the contract to limit the duration of the contractor's liability.

(4) if the works include major items of plant, it is usually appropriate for the contract to limit the duration of the contractor's liability for such plant; for example, to a stated number of years after the completion date stated in the taking-over certificate. After a few years' operation, it becomes increasingly difficult to establish whether any alleged defects are attributable to the plant's design, manufacture, manuals, operation, maintenance, or a combination of these and/or other matters.

【译文】 如果任何业主风险，不管是哪一个，只要对工程、货物或承包商文件造成损失或损坏，承包商就应立即向工程师发出通知，并按照工程师的要求弥补这一损失或损坏。

(1) 在某些普通法管辖地区，只有在业主应在合理的时间内获知承包商有缺陷的工作的情况下才能计算责任期。

(2) 在某些民法管辖地区，承包商对于隐蔽的缺陷从竣工算起的十年内（叫做十年责任）负有不可推卸的责任（即无须提供缺陷证明）。

(3) 除非工程有大件设备，否则合同不宜限制承包商的责任期。

(4) 如果工程有大件设备，合同一般宜限制承包商为这些设备负责的期限；例如从接收证书载明的竣工日期开始在说明的年数内承担责任。在设备使用若干年后，就越来越难以确定所说的缺陷是否应归因于该设备的设计、制造、使用手册、运行、维护或这些方面同其他事项的共同作用。

[14] premium n. 保险费

Premium is the amount of money paid by the insured to the insurer as the consideration to the insurer for undertaking the insurance policy. It is calculated based upon the sum insured, the limit of indemnity, cover, exceptions, and any other conditions of the policy. For example, in the Contractors' (Construction) All Risks policy, the sum insured is the total of the value of the contract works, plus professional fees, the escalation in costs, and the costs for removing the debris. The premium shall be retrospectively adjusted on the actual value of the contract value when known.

【译文】　保险费是受保人付给保险公司的款项，这笔数额是对保险公司承担保险的报酬。保险费计算时要根据投保数额、补偿上限、保险范围、例外以及保险单任何其他条件。例如，施工一切险保险单，其中的投保数额是合同工程价值，加上专业服务费、费用涨价以及废弃物清除费的总和。保险费可以在明确了合同额真实价值之后进行事后调整。

Chapter 6 Contractor's Site Organization and A/E's Site Supervision

The successful tenderer has become the contractor when he is awarded the contract, which involves him in the pre-contract planning process before commencement of the works on the site.

This chapter examines the contractor's activities and deliverables which take place or are produced prior to and during progress of the works and the A/E's site supervision.

6.1 Arrangements prior to commencement

6.1.1 Procedures to be undertaken[25]

Prior to commencing the works on site, liaison will be necessary between the contractor, A/E, statutory undertakings and local authority. Initial contact will have to be made with key material suppliers and owners and nominated subcontractors (if known) together with plant hire firms likely to be participating in the early contract stages.

The time period for possession of the site is usually stated in the Appendix to Tender or in the preliminaries bill.

It is important for the prospective contractor's representative (agent, site manager) to be involved in the pre-contract planning as soon as possible after the award of the contract. This allows time for the contractor's representative to familiarize himself with the bills of quantities and the drawings and to take part in the preparation of the master programme. He at this stage will organize the setting up of the site including the preparation of the site layout plan. The short term programme for the project's early stages may also be prepared together with finalizing construction methods and requirements of the contractor's equipment.

The activities during the initial pre-contract period are mainly the pre-contract planning meeting, the registration of drawings, and the arrangements for commencing works.

6.1.2 Pre-contract planning meeting[32]

A meeting will be held to announce the award of the contract and to acquaint all concerned with its general background, which helps develop close coordination between the estimating, surveying and contract management teams and develop a team spirit.

Person	Responsibilities
Chief estimator	(1) Acting as meeting chairman (2) Handing over all tendering data to the respective personnel, e.g. Estimate summary and adjudication data Build-up of net bill rates Summary of subcontractors' and suppliers' quotations Method statements Preliminaries build-up
Contracts manager	(1) Organising the commencement of work by the programmed date (2) Appointment of contractor's representative and/or arrangements for transfer of site management personnel (3) Preparation or assistance in the preparation of the master programme in conjunction with the planning engineer and prospective contractor's representative (4) Finalizing the statement of construction methods and schedule of plant and equipment requirements (5) Preparation of the site layout plan (6) Preparation of expenditure budgets for labour, plant and preliminaries
Buyer	(1) Finalizing quotations with suppliers and subcontractors (2) Preparing schedules including coordinating material requirements with the master programme (3) Placing orders with material suppliers and own and nominated subcontractors. In certain organisations it may be considered a quantity surveying function to place subcontractors' orders (4) Preparing schedules of key dates for the delivery of materials
Chief quantity surveyor	(1) Checking and arranging for the signing of the contract documents (2) Appointment of surveying staff for the contract (3) Preparation of the contract valuation forecast and cash flow assessment (4) Distribution of copies of the bills of quantities (5) Preparation of forms of contract for own and nominated subcontracts in conjunction with the buying department
Office manager	(1) Preparation of master files for the contract (2) Distribution of all letters in relation to the contract (general office to act as central clearing house for contract mail) (3) Appointment of clerical staff for the contract to assist with the site administration (4) Arrangement of insurances (5) Serving of notices and payment of fees (6) Arrangement for office stationery and safety packs to be dispatched to site
Chief planning engineer	(1) Preparation of master programme for the contract This will involve a review of the tender programme and current drawings available. Liaison with the contracts manager or contractor's representative regarding the sequence of work, key dates and labour resourcing (2) Appointment of staff for undertaking short term planning procedures on site
	The above items indicate areas of responsibility allocated to each of the company departments. Responsibility for planning and cost control may however be undertaken by a production control section, as is often the case within the large construction firms

Figure 6.1 Responsibilities of the Persons Involved in Pre-contract Planning[32]

The chairperson should be fully conversant with all decisions made during the preparation of the tender. The meeting may be used as a vehicle to pass over the tender documents from the estimating to the contracts department.

Figure 6.1 covers the responsibilities of the attendants as to the pre-contract planning.

6.1.3 Custody of drawings

It is of importance to the contractor that records are kept of the receipt of all contract documentation as at some later stage contractual claims may arise. From the initial stage a drawing register must be kept to record the issue date of drawings received from the A/E and other consultants.

Communication channels must be set up which ensure that all drawings are sent to the contractor's head office for distribution by the office manager. In this way a full set of up to date drawings is permanently held for reference at head office.

6.1.4 Arrangement for commencing works[25]

Careful planning at this stage can result in monetary and time savings during the contract. Consideration should be given to the following items in order to ensure a smooth start to the contract.

(1) Preparation of the site layout plan
(2) Temporary site services
(3) Licenses, permits and notices
(4) Requirements of small tools and plant

A pre-contract checklist covering essential items is used within certain of the large construction companies. This provides an important guide to new project management staff.

Words, Phrases and Expressions 6.1

[1] plant hire firms 设备租赁公司
[2] possession of the site (业主向承包商) 移交 (施工) 现场
[3] pre-contract planning meeting 开工前规划会议
[4] develop a team spirit 培养团队精神
[5] chief estimator 总估算师
[6] contracts manager 合同经理
[7] buyer 采购员
[8] chief quantity surveyor 总工料估算师
[9] office manager 办公室总管
[10] chief planning engineer 施工组织设计总工程师
[11] finalizing quotations with suppliers and subcontractors 与供应商和分包商最终确定其报价
[12] central clearing house for contract mail 集中收发与合同有关信件的处所
[13] distribution of all letters in relation to the contract 分发与合同有关的信件
[14] clerical staff (全体) 办事员
[15] serve *vt*. 送达 (deliver (a summons, etc)) to the person named in it
[16] conversant *adj*. 具有关于……的知识，懂得，熟谙 (conversant with, having the knowledge of)

Notes 6.1

疑难词语解释

[1] serving of notices

【译文】 递送通知

[2] conversant with all decisions made during the preparation of the tender

【译文】 熟悉编制标书期间所做的所有决策

6.2 Contractor's organizational structure[32]

The contractor's organizational structure for managing the projects shall match the division of the responsibilities of his senior management and site management and facilitates the site management.

The contractor usually has two main types of staff perform various technical and managerial functions. The technical functions usually include:

(1) engineering: such as designs of temporary and permanent structures and the planning and programming of work;

(2) estimating and tendering: preparation of specification and pricing of the bills of quantities, interpretation and application of contract documents and negotiation with the A/E;

(3) supervision and reporting: of site operations, monitoring of progress, preparation of reports, liaison with the A/E, purchasing of materials and equipment, and preparation of monthly valuations;

(4) quality control, investigation and research and development: such as site investigations and geotechnical processes;

(5) operation of central plant and transport depots with workshops and repair facilities, routine inspection of plant and equipment, and purchases and sales of plant; and

(6) staff education and training.

The non-technical staff deals with a wide range of activities, including:

(1) secretarial and legal matters;

(2) finance, accounts, audits, payments, cash and payroll checks;

(3) insurance, licenses and taxation returns;

(4) orders, monitoring deliveries and checking invoices;

(5) cost records and analyses;

(6) plant and transport records and registers;

(7) general correspondence and records;

(8) labour relations; and

(9) staff training and development.

6.2.1 Company's organizational structure

The business of construction is basically project-based and it is obvious that the organizational structures of construction firms differ from that of operations-based organizations.

Nowadays the state-owned construction companies in most parts of China have been made shared and matrix organized, as shown in Figures 6.2.

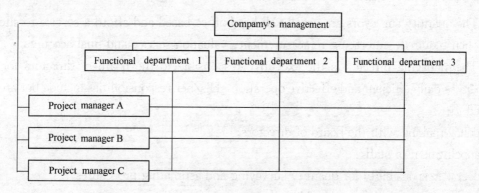

Figure 6.2 Matrix Organization

6.2.2 Head office Organisation

Figure 6.3 indicates the senior management structure above contractor's representative, who spends his full time on site, together with the area of responsibility allocated to each of the directors of a shared company. All major contracts are normally under the direct control of the contracts director and he is assisted by the contracts manager in the control of the contracts. Minor contract up to a value of, say, ￥2,000,000 are the responsibility of a further director controlling a number of foremen each engaged on a contract.

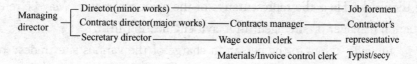

Figure 6.3 Staff under the Managing Director

6.2.3 Responsibilities of contractor's head office management[32]

The head office and site management personnel of construction firms are given a number of similar job titles and assigned similar responsibilities.

For example, a shared construction firm usually has the managing director take the responsibilities for the liaison between head office and the construction projects covering quantity surveying services, planning and various aspects of financial control. Estimating, surveying and planning services are under his direct control. Figure 6.4 indicates the probable staff under the control of the managing director.

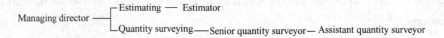

Figure 6.4 Probable Staff under the Managing Director

The quantity surveyors are usually all head office based and afford a service to sites as and when required, embracing remeasurement, valuation service and final accounts.

The managing director is the executive responsible to the board of directors for the company's daily efficient and effective operation. His/her responsibilities include but are not limited to:

(1) policy-making with the board of directors,
(2) appointment of staff,
(3) overall responsibility for quantity surveying and estimating procedures,
(4) financial control aspects-monthly cost/value comparisons,
(5) progress and cost reports - liaison with the contracts director,
(6) planning procedure at the pre-contract and contract stage,
(7) obtaining work-contact with employers, and
(8) investment and raising funds.

Contracts manager is a person responsible for organising a number of contracts at the same time. Each site will be controlled by a contractor's representative. The contracts manager is generally head office based and visits all the sites under his control periodically. He normally works in a defined region or area and is responsible for the coordination of all contracts under his control.

Contractor's representative is the person responsible on the site for controlling and organising the work. He is the representative of the main contractor. May be referred to as the construction manager or site agent in a civil engineering firm.

General foreman refers to the person in charge of the various site trades, responsible for coordinating all the trades. On the smaller contract the general foreman may act as the contractor's representative if he is the representative of the contractor on site.

The title of *project manager* may be used for the person in charge of the larger type of project. He will be in control of a number of contractor's representatives employed on the various sections of the works coordinated by the project manager. More usually, however, the title is reserved for a manager under whom a number of firms work together. On overseas contracts these may include design firms.

The person in charge of a particular trade, such as a foreman joiner, bricklayer, steelfixer or plasterer, called *trades foreman*. The trade foreman in charge of the labour gangs is referred to as a *ganger*. He receives daily instructions from the general foreman, or in the case of a smaller contract, the contractor's representative.

6.2.4 Organisational structure for major projects[32]

Liaison with construction sites is done through the contracts manager responsible for coordinating materials, plant and labour between the contracts, under his control. Weekly progress reports are prepared at site level and the contracts manager presents a monthly progress report to the contracts director.

Contract planning during the progress of the work is undertaken by the contracts manager in consultation with the contractor's representative. Short term planning procedures are the responsibility of the contractor's representative. Contractor's equipment is largely hired by arrangement with the contracts director.

Figure 6.5 and Figure 6.6 indicate the organisation structure for a building works in the U.K. and in China, respectively.

A director in the head office will be given overall responsibility for the contract, which must be made clear to the employer and the A/E to enable the employer and/or the A/E to have a direct link with the contractor's top management regarding any major problems during the contract. All senior management reports will be sent to the contracts director and he will visit the site at regular intervals during the progress of the works.

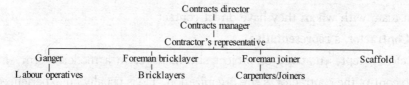

Note: Contracts director is responsible for overall financial control, cumulative/monthly cost/value comparisons; contracts manager is in charge of monthly reports, arranging site meetings and coordination between contracts(projects); responsibilities of the contractor's representative include weekly progress reporting, short term planning and coordination of labour/subcontractors. The services with which the head office provides the contract include estimate analysis, contract budgets, wages and clerical procedures; quantity surveying (remeasurements, valuation and final account); planning and cost and bonus services.

Figure 6.5 Organisational Structure for a Building Works in the U.K.[32]

The contract may be controlled by a project manager under the direct control of the contracts director.

As the contract will involve various phases or separate sections, each phase will be controlled by a contractor's representative or site agent responsible directly to the project manager.

As an alternative the contract may be under the control of a project manager responsible to a contracts manager. The contracts manager will be responsible for coordination between the head office, site and the A/E.

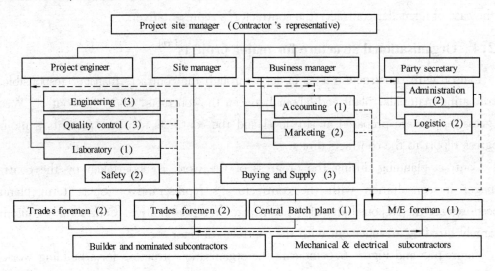

Note: The parenthesized figures represent the number of staff.

Figure 6.6　Organizational Structure of a Building Works in Beijing

6.2.5　Site personnel

The contractor's site management personnel assumes various duties which are described briefly below, together with an indication of their relationships with other members of the construction team with whom they have direct contact.

6.2.5.1　Contractor's representative

Contractor's representative, or project site manager in a modern term, is normally appointed to control the contractor's site organization. He is usually an experienced engineer and the contractor generally gives him wide discretionary powers. In addition to sound engineering and contractual experience, he must possess good qualities of leadership and integrity. The main duties of the contractor's representative's are to ensure that the works are administered effectively and that construction is carried out economically, in accordance with the contract documents and that they satisfy the requirements of the A/E's representative.

Sub-clause 4.3 of the *Red Book* requires the contractor, prior to the commencement date, to submit to the engineer for consent the name and particulars of the person he proposes to appoint as contractor's representative. The contractor shall not, without the prior consent of the engineer, revoke the appointment of the contractor's representative or appoint a replacement.

The contractor's representative shall give his whole time to directing the contractor's performance of the contract. He shall, on behalf of the contractor, receive instructions from the engineer.

The contractor's representative may delegate any powers, functions and authority to any competent person, and may at any time revoke the delegation. Any delegation or revocation shall not take effect until the Engineer has received prior notice signed by the contractor's representative, naming the person and specifying the powers, functions and authority being delegated or revoked.

The contractor's representative shall be fluent in the language for communications defined in the contract.

6.2.5.2 Staff of contractor's representative[15][32]

The contractor's representative is supported by both technical and non-technical staff with their numbers and duties dependent on size of the works. For instance, the personnel might comprise an engineer, general foreman, plant and transport foreman, cashier, timekeeper and storekeeper. Larger contracts require proportionately larger staffs. The allocation of duties will be influenced by many factors, including the locality and nature of the work, the amount of assistance from head office and the experience and capability of the available personnel. The allocation of duties should seek to ensure smooth and effective communication and the ability to introduce checks at critical points.

Figure 6.7 illustrates a contractor's typical site organization based on an example in *Civil Engineering Procedure*[7], although it will be appreciated that there is no one universal system.

6.2.5.3 Section managers

Section managers, also called *sub-agents*, are often appointed to control the various work sections of a large project, and they, in their turn, may have a number of section engineers responsible to them for the supervision of the actual operations. The direct control of labour and operation of plant and transport is usually undertaken by various grades of foremen.

6.2.5.4 General foreman

The general foreman or works manager directs the day-to-day distribution of labour to particular operations under sectional or trade foremen, to supervise the supply of materials and stores and disposition of plant, and to activate the necessary site communications.

6.2.5.5 Site services manager

The site services manager/engineer controls a number of departments providing services needed for the effective execution of the project. These services often include a concentrated batching plant, steel bending yard and plant and transport department.

6.2.5.6 Chief engineer

The chief engineer is responsible to the contractor's representative for the accuracy of the works through section engineers. He needs to check the coordination of drawings received from the A/E and then issue them to the relevant personnel in the appropriate

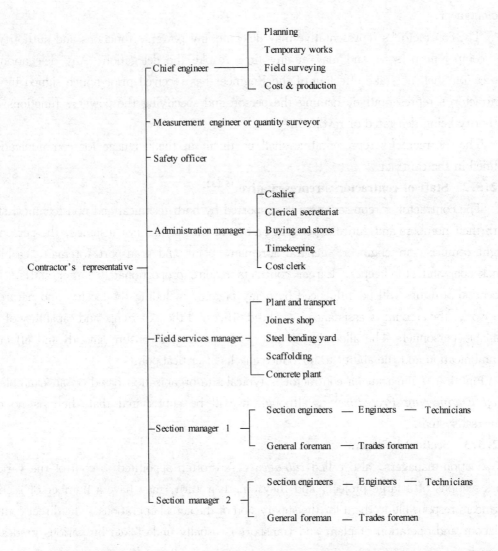

Figure 6.7 Chart of a Contractor's Typical Site Organization[7]

sections of the project. He also carries out any local designs that may be required, particularly those relating to temporary works, and gives general technical guidance where necessary to personnel on the site.

6.2.5.7 Production control and costing engineer

He is responsible to the chief engineer and keeps routine progress records and costs and normally operates through departments controlled by sub-agents. He is often responsible for the routine measurement of work on site.

6.2.5.8 Administration manager

This staff member is the site office manager and is responsible to the contractor's representative for the efficient administration of the non-technical personnel. He usually prepares detailed lists of duties for the guidance of clerical staff and devises checking and counter-checking procedures for cash transactions and stocktaking. He needs to be familiar with the standing instructions of the head office and to ensure compliance with them.

His department also controls the payment of wages through time-keepers and cashiers and the purchasing and checking of receipt of materials and components, in addition to the checking of accounts, insurance, safety precautions, site welfare and other matters relating to labour relations.

6.2.5.9 Section engineers

They are usually engineers with experience of both design and field work who, although ultimately responsible to the A/E, often report to sub-agents on the accuracy and control of the building or civil engineering works. Each section engineer will liaise with his foreman to plan the work of his section and report on matters of detail to the measurement engineer or the quantity surveyor engaged on the project.

6.2.5.10 Measurement engineer or quantity surveyor

On large contracts a measurement engineer or quantity surveyor may be employed to take measurements and check the quantities and value of completed work.

6.2.6 Site management[15]

Many contractor's representatives have to make many *ad hoc* decisions because of the large number of inputs on to a construction site the diversity of conditions from one site to another. They often have to take decisions based on personal knowledge and experience, without reference to senior personnel at head office.

Poor management is reflected in increases in total costs of various inputs to the construction process. The cost of subcontractors' work can rise because of poor planning, resulting in a delayed start on site and the subsequent submission of claims against the main contractor. The cost of plant can be increased because of low usage and inadequate maintenance.

The majority of project managers on building or civil engineering works are qualified professional engineers. The major task of a civil engineer in charge of a site was generally considered to be the satisfactory execution of the works. However, the employer's other requirements, such as construction to a time and cost budget, have increased in importance and this has necessitated greater management control over resources.

Nowadays civil engineers and even architects generally display more interest in management than before, partly because the increasing complexity of building and civil engineering works requires more involvement on their parts. They are expected to take a wide range of responsibilities on site, including the measurement and valuation of completed work. However, the larger contractors are increasingly engaging specialists, such as quantity surveyors and accountants, on construction sites to perform functions previously undertaken by civil engineers. Civil engineers are recognizing the need to acquire and demonstrate expertise in these areas, and they are now included in the curriculum of many architectural or civil engineering degree courses.

6.2.7 Human resources management[15]

Effective management of human resources is an important aspect of the duties of the contractor's representative.

Some contractors have established personnel or human resources departments, responsible for recruiting, employing and developing human resources.

One of the more important strategic tasks of human resources management is to continually analyze and reappraise the organization's operations. This will assist in making decisions on suitable work structures and formal roles and relationships, to allocate responsibilities and define levels of authority. The human resources manager can contribute to the development of manpower forecasting and budgeting techniques. He can also play an important role in identifying the strengths and weaknesses of the organization and assessing the effects of social, legal, economic and other changes.

Words, Phrases and Expressions 6.2

[1] organization(al) structure 组织结构
[2] senior management 高层管理人员
[3] site management 现场管理, 现场管理人员
[4] technical and managerial functions 技术与管理职能
[5] project-based 以项目为…… *The business of construction is basically project-based*. 建筑业基本上以项目为业。
[6] matrix organization 矩阵组织
[7] head office organisation 公司总部组织
[8] shared company 股份公司
[9] contracts director (负责)合同(的)董事,(负责)项目(的)董事
[10] foremen *n*. 领班, 班长, 队长
[11] head office management 公司总部管理人员
[12] job titles 岗位头衔
[13] board of directors 董事会
[14] policy-making 制订方针
[15] general foreman 总管, 总负责人
[16] trades foreman 工种班长, 工种领班
[17] site personnel 现场人员
[18] afford *vt*. 给予, 供给(give, supply sth.)
[19] site investigations 现场勘测
[20] geotechnical processes 岩土工艺

[21] operation of central plant 管理集中生产设备
[22] taxation returns 税务报表
[23] labour relations 劳资关系
[24] plant and transport foreman 设备与运输负责人，设备与运输领班
[25] section manager 单项工程负责人，栋号负责人
[26] section engineer 栋号工程师
[27] various grades of foremen 各种级别领班
[28] field services manager 现场设施经理，现场设施负责人
[29] concentrated batching plant 集中搅拌站
[30] steel bending yard 钢筋加工场
[31] production control and costing engineer 生产与费用控制员
[32] administration manager 行政经理
[33] measurement engineer 工程量计量员
[34] *ad hoc* 为此目的安排的，专门，特别((Latin) arrange for this purpose)
[35] *ad hoc* decisions 特别做出的决定

Notes 6.2

疑难词语解释

[1] The quantity surveyors are usually all head office based and afford a service to sites as and when required, embracing remeasurement, valuation service and final accounts.

【译文】 工料估算师通常在总部办公室工作，在现场需要时为其服务，包括计量、估价与结算。

[2] make many *ad hoc* decisions

【译文】 做出许多特别决定

[3] He usually prepares detailed lists of duties for the guidance of clerical staff and devises checking and counter-checking procedures for cash transactions and stocktaking. He needs to be familiar with the standing instructions of the head office and to ensure compliance with them.

【译文】 他通常编制详细的职责名单用于指导办事人员工作，并设计现金往来与物品领用的核对与复核程序。他需要熟悉总部的既定方针，并确保遵守之。

[4] orders, monitoring deliveries and checking invoices

【译文】 订货、监视货物的送达情况并核对发票

[5] planning 和 programming 的区别

【解释】 programming 侧重于"制订进度计划和施工组织设计"，技术性很强，而 planning 包括工程的所有的方面。

[6] wide discretionary powers

【解释】 根据具体情况加以斟酌和权衡自行处理问题的权力，与"按照明文规定处理问题的权力"相对。

6.3 Site layout[15]

A site layout plan should be prepared showing the proposed locations of all facilities, accommodation and plant to secure optimum economy, efficiency and safety during construction. A tidy site is the outward symbol of an efficient organization. The following aspects deserve particular attention.

6.3.1 Access

The requirements vary depending on the type of building or civil engineering works and the stage of construction. Access from the public highway should desirably be duplicated, with short direct routes and one-way traffic to ensure a smooth flow of vehicles. Temporary access ways may be constructed of hardcore, sleeper, concrete, proprietary track or transportable mats for mechanical plant. Any permanent works should be used when sitting temporary roads or hard-standings.

Permission must be obtained from the local authority for access over or encroachment onto public footpaths. The police must be notified if roads are to be closed or diverted. It may be necessary to erect lighting, to maintain rights of way and install pedestrian walkways or vehicle tracks over trenches.

6.3.2 Materials storage and handling

The principal objective is to minimize wastage and losses arising from careless handling, poor storage or theft, and to eliminate double handling or unnecessary transportation of materials and components. Suitable stores and compounds must be provided for tools and equipment, plant spares, and breakable or expensive materials and components. Newly erected ancillary buildings can often be used for storage purposes and this can be facilitated by building them early in the construction programme. The suitable space shall be allocated for subcontractors' huts and materials.

Security measures must be given high priority, including locked buildings, substantial fences and gates, careful location of checkers' huts, the possible installation of a weigh-bridge, effective procedures for receipt and issue of stores, fire precautions, and the employment of a night watchman, guard dog or visiting patrolman.

Efficient distribution of the offices, stores and labour force, and an intelligently arranged programme, can ensure that materials are handled as little as possible and used in the order of arrival on site.

6.3.3 Administration building and other facilities

The administrative offices of both the contractor and the A/E's staff shall be ideally sited overlooking the works yet reasonably free from the noise and dust emanating from site operations. The requirements of privacy for discussion and accommodation for site meetings must be taken into account.

Welfare facilities have become an increasingly important item on construction site, necessitating adequate provision of mess-rooms or canteens, drying or changing rooms, and toilets.

6.3.4 Equipment, workshops and services

The choice of the most suitable type of equipment is a matter of major significance. There needs to be a good balance of plant and manpower for each site operation.

The number and size of workshops must be determined and their location chosen so that ready access to construction work is obtained without causing congestion of the site. The lines of both new and existing services must be considered when sitting temporary buildings and roads. The installation of temporary diesel pumps, water and compressed air mains, electric power and telephone lines, and other services requires negotiation with the relevant authority and coordination with the general scheme of work.

Words, Phrases and Expressions 6.3

[1] site layout 现场布置
[2] site layout plan 现场布置平面图
[3] A tidy site is the outward symbol of an efficient organization. 整洁的现场是有效组织的外在象征。
[4] storage and handling 存储与搬运
[5] double handling 二次搬运
[6] hardcore n. 硬质
[7] sleeper n. 枕木
[8] proprietary track 专用轨道
[9] transportable mats 移动式垫板
[10] liaision n. 联络
[11] transport depot 运输仓库

6.4 Planning and monitoring[15]

When the contractor has accepted the contract, he must commit all the required resources to it. He must review the information available to him for pre-contract and contract planning and establish his plans. Decisions must be made relating to the financing of the contract, the appointment of subcontractors and suppliers and the availability of information requirements and resources. Budgets, plans of action and programme must be prepared. A cash flow assessment will indicate the funds which may be required to finance the contract, and the necessary loans can be raised.

The prospective contractor's representative should become involved in the planning activities as these are, in effect, an attempt to formulate the method he will adopt.

6.4.1 Objectives of programming

In order that building or civil engineering works shall be carried out efficiently, they shall be carefully and properly planned in the first instance. Decisions shall be made on construction methods, temporary works and contractor's equipment, labour, material and transport requirements, all set against time. Full consideration shall be given to alternative methods and the effect of each planned activity on the others.

The works shall be broken down into a series of operations in the programming, from the temporary and preliminary operations throughout the completion of the permanent works. The programme should ideally be discussed and agreed by all concerned before work starts, to avoid confusion, delays and increased costs.

The programme can be in three broad classifications.

(1) The A/E and the tenderer shall prepare their own *outline programmes*. The one prepared by A/E is included with his report to the employer, while that of the tenderer is submitted with his tender.

(2) The contractor shall prepare a master programme as soon as possible after he receives instructions to commence on the site. It shows the periods during which the individual work sections are to be carried out in order to achieve the correct sequence of operations and completion of works within the contract period. The programme will also indicate the dates by which detailed drawings will be required, the dates when the various sections of work will be completed, ready for use, or for the installation of plant by other contractors. It also provides valuable information relating to material, labour and plant requirements.

(3) *Detailed programmes* cover each section of the works and they dovetail into the master programme.

The programme must be realistic and capable of fulfillment. The A/E must examine the programme thoroughly with a view to detecting any errors in the logic or basic assumptions underlying the sequence and linkage of the various activities shown.

6.4.2 Progressing

Progressing consists of taking systematic steps to ensure that the programme is followed as closely as possible. Unless this monitoring procedure is adopted, the value of the programme will be largely lost. Lack of progress will entail some modification of the programme. It is essential to introduce an adequate recording procedure to ensure that a thorough, accurate and regular check is made of the work executed under the work section heads or operations contained in the programme.

Words, Phrases and Expressions 6.4

set against time 根据时间安排

6.5 Resource scheduling[15]

A network cannot normally, on its own, identify the volume of resources needed at any given point of time in which the works is under way. When the network is initially prepared, no account is taken of the resources which will be available. The start of an activity is usually assumed to be dependent upon the completion of previous related events, and not on the availability of suitable operatives and contractor's equipment. As a matter of fact, no works is carried out with unlimited resources and it is necessary to have regard to resource limits in meeting the employer's requirements.

Resource scheduling is an essential follow-on step from the initial planning of the network. The results of time analysis are used to determine priorities when different activities compete simultaneously for the same limited resources. Scheduling decisions will be needed to remove the bottlenecks. For example, it may be necessary to employ additional subcontract labour over a critical period or to provide for no-critical activities to be delayed in favour of those which have less float.

Words, Phrases and Expressions 6.5

[1] resource scheduling 制订资源使用计划
[2] follow-on step 后继步骤
[3] critical period 关键时段
[4] no-critical activities 非关键活动,非关键工序
[5] float n. 浮动时间,时差,机动时间

Notes 6.5

疑难词语解释
[1] For example, it may be necessary to employ additional sub-contract labour over a critical period or to provide for a no-critical activities to be delayed in favour of those which have less float.
【译文】 比如,也许在某一时段必须增雇分包劳务,或者推迟非关键活动,为机动时间少的活动创造有利条件。
【解释】 provide for 在这里的意思是"为……准备条件"。
[2] The start of an activity is usually assumed to be dependent upon the completion of previous related events, and not on the availability of suitable operatives and

contractor's equipment.

【译文】 一般假定任何一项活动的开始是根据紧在其前的有关事件的完成而确定的,而非由是否有合适的操作人员和施工机具而定。

6.6 Costing and accounting arrangements[15]

6.6.1 Site staff

In conjunction with the foreman and gangers, he allocates times to the various project operations for costing purposes.

With regard to purchasing, some contractors negotiate and place all orders through head office buyers, while others delegate to regional offices controlling a number of contracts or directly to a specific contract where it is of sufficient size.

The *storekeeper* checks and takes into custody all materials received, and subsequently issues them under a recognized system of authorization. On some projects the storekeeper is also responsible for maintaining the plant register.

6.6.2 Accounting arrangements

The functions of an accounts system is listed as follow:
(1) stating expenditure to date, with commitments and liabilities incurred in red, in a form useful to the contractor's representative;
(2) summarizing relevant information to aid management in assessing the present and future financial state of the contract;
(3) providing up to date accurate financial statements at regular intervals and realistic estimates at interim periods; and
(4) preparing company accounts and returns in conformity with statutory requirements.

6.6.3 Cash flow

When an employer contemplates investing funds in a large project spread over several years, he wants to know the sum of money involved and the timing of payments. The employer needs to make arrangements with his banker or other financial source for the sums to be available as required. If the employer expects to make an eventual profit on the project, forecast shall be made of the extent and timing of revenue receipts as part of the project appraisal in order to assess the commercial viability of the project.

The contractor will need to monitor the costs very closely. At site level, the cost control often encompasses standard costs and variance analyses normally on a weekly basis. Head office management usually prepares cost/value which are statements of the total costs and values relating to each works normally updated at each monthly valuation.

Words, Phrases and Expressions 6.6

[1] costing and accounting arrangements 成本与会计核算安排,成本与会计核算工作
[2] site staff 现场人员
[3] gangers n. 班组长
[4] recognized system of authorization 正式的审核制度
[5] plant register 设备登记册
[6] cash flow 现金流
[7] timing of payments 支付的时间安排

Notes 6.6

疑难词语解释

[1] commitments and liabilities incurred in red
【译文】 已安排用途的未来支出和债务
【解释】 commitments 的意思是"虽然尚未实际支出但已经安排了用途的未来支出"。西方国家,账目中支出者用红颜色数字记载,所以 commitments and liabilities incurred in red 应当理解为"已经安排了用途的未来支出和债务"。

[2] financial statements at regular intervals and realistic estimates at interim periods
【译文】 定期财务报表,以及各期间的反映实际情况的估算书

[3] preparing company accounts and returns in conformity with statutory requirements
【译文】 按照法律要求编制公司账目与报表

[4] With regard to purchasing, some contractors negotiate and place all orders through head office buyers, while others delegate to regional offices controlling a number of contracts or directly to a specific contract where it is of sufficient size.
【译文】 在采购方面,某些承包商通过总部议价并下所有订单,而另外一些承包商则将下定单任务委托控制若干合同的地区办事处,或在合同规模足够大时直接委托给该具体合同。

6.7 Usage and costing of contractor's equipment[15]

6.7.1 Requirements of contractor's equipment

Contractor's equipment has to withstand severe weather conditions and exceptional stresses. Hence it needs to be tough, durable and resilient, easy to dismantle and reassemble on overhaul and capable of movement from one site to another.

When selecting his equipment, the contractor should give special consideration to the following criteria:

(1) It must be the right type for the work in hand.

(2) It must be able to carry out the task efficiently, economically and, if necessary, continuously.
(3) There must be sufficient work on the site to keep it occupied.
(4) There must be sufficient operatives and ancillary equipment to maintain optimum output.

For example, an excavator must have sufficient earthwork to ensure optimum use and it must be adequately serviced by vehicles running between the site and the tip. Contractor's equipment must be maintained in full repair and should be kept on site for as short a period as possible. This applies particularly to equipment or plant hired by the contractor since he will to pay for it whether it is in use or not.

The following factors should be considered when selecting equipment for a works:
(1) workload to be undertaken;
(2) time allowed in the programme for the work;
(3) its capabilities and output and the various tasks which it can perform;
(4) transportation costs involved; and
(5) maintenance requirements and facilities.

Other factors may include the nature of the works, site conditions, access, obstructions, boundaries, adjoining buildings and noise limitations.

6.7.2 Costs of contractor's equipment

A contractor should make the decision as to purchase as opposed to hiring, based upon an economic analysis of the alternatives. Included in this analysis will be such factors as availability of cash for a down payment or outright purchase, current interest rates on loan or installment purchase plans, current hire rates, tax benefits, and expected use and anticipated life of the equipment. Ownership of the equipment can be a drain on resources if the equipment is underused, subject to frequent breakdown or become obsolete.

As a general rule, large and expensive items of the equipment are best hired as required, usually for relatively short periods and achieving a high rate of output. General items of contractor's equipment which have a high usage factor are usually better owned. However, on occasions, the duration of a contract and the amount of work involving the use of contractor's equipment may create a situation where it could be advantageous to purchase all new contractor's equipment and dispose of it on completion of the contract.

Words, Phrases and Expressions 6.7

overhaul 大修

Notes 6.7

疑难词语解释

Ownership of the equipment can be a drain on resources if the equipment is underused, subject to frequent breakdown or become obsolete.

【译文】 如果设备使用不足、常出故障或技术陈旧，则拥有设备就会使（企业的）资源逐渐耗尽。

【解释】 subject to 在这里的意思是"动不动就……，易受……的；经常……的"。drain 的意思是"财富不断外流，逐渐流光"。

6.8 Safety aspects[15][32]

Construction site often create potentially dangerous situations and over 1,200 persons are killed on them in China each year. The accidents on the site can be heard of every day.

Some activities are more vulnerable than others. For example, steel erection and demolition accounts for many fatalities and serious accidents. Many site injuries results from operatives falling from structures or being hit by falling objects. Many others are caused by the misuse of mechanical plant and site transport.

There are a number of legislations in connection with construction site operations providing the legislative framework within which to promote, stimulate and encourage high standards of health and safety at work. It aims to persuade all those involved in the construction industry, including employees, to promote an awareness of safety during the construction process, and to do all that is necessary to avoid accidents and occupational ill-health.

The main objectives of the legislations are:
(1) to secure the health and safety and welfare of persons at work;
(2) to protect persons other than those at work against risks to health and safety arising out of work activities;
(3) to control the keeping and use of explosives and highly flammable or otherwise dangerous substances and generally prevent the unlawful acquisition, possession and use of such substances; and
(4) to control the emission into the atmosphere of noxious or offensive substances.

Employers have detailed responsibilities:
(1) to develop systems of work which are practicable, safe and have no risk to health;
(2) to provide plant to facilitate this duty, and this general requirement is to cover all plant used at the workplace;
(3) to provide training in the matter of health and safety; employers must provide the instruction, training and supervision necessary to ensure a safe working environment;
(4) to provide a working environment which is conducive to health and safety; and
(5) to prepare a written statement of safety policy and to establish an organizational framework for carrying out the policy; the policy must be brought directly to the attention

of all employees.

However, employees also have the following specific duties:
(1) to take care of their health and safety and that of other persons who would be affected by acts or an omission at the workplace; and
(2) to cooperate with the employer to enable everyone to comply with the statutory provisions.

Words, Phrases and Expressions 6.8

[1]　vulnerable *adj.* 敏感的,易受伤害的,易损的(that is liable to be damaged, not protected against attack)

[2]　accounts for *vi.* ((used with *for*) explain the cause of; serve as an explanation of; answer concerning one's conduct) *steel erection and demolition accounts for many fatalities and serious accidents* 钢结构安装与拆除是许多死亡和严重事故的原因

[3]　awareness *n.* 意识(knowledge or realization of) *promote an awareness of safety* 提高安全意识

[4]　develop systems of work 建立工作制度

Notes 6.8

疑难词语解释

[1]　provide a working environment which is conducive to health and safety
【译文】　形成有利于健康与安全的工作环境

[2]　prepare a written statement of safety policy and to establish an organizational framework for carrying out the policy; the policy must be brought directly to the attention of all employees
【译文】　编写一份有关安全方针的书面声明,并落实执行这一方针的组织保证;这一方针必须直接同所有员工见面

[3]　Some activities are more vulnerable than others
【译文】　有些活动同其他活动相比,容易出事故

[4]　It aims to persuade all those involved in the construction industry, including employees, to promote an awareness of safety during the construction process, and to do all that is necessary to avoid accidents and occupational ill-health.
【译文】　其目的是劝说所有置身于建筑业的人,包括雇员,都在施工过程中提高安全意识,都要付出所有必要的努力避免事故与职业病。

6.9　A/E's site supervision

6.9.1　A/E's site supervision duty

It is the primary duty for the A/E to ensure that the contractor executes, completes the works and remedies any defects as required in the contract. The primary function of the resident engineer's site staff is to supervise the contractor's activities. Although there are many occasions when the contractor generates action, for example, by requesting approval under a sub-clause of the conditions of contract to place concrete or to put out of view a work, the most routine supervision is undertaken by the engineer's site team acting on their own initiative.

The resident engineer can establish procedures for regular inspections and encourage his staff to maintain a close watch on the contractor's operations, but much will depend on the enthusiasm and diligence of the team. The individuals making up the team work together as a closely knit group. The resident engineer has to ensure that they are provided with a satisfactory working environment to achieve the best results.

6.9.2 Organization of A/E's site supervision[15]

An organisational chart is shown in Figure 6.8 to cover the engineer's site team for a highway contract. The resident engineer is backed up, in this case, by three assistant engineers each having 'line management' responsibilities for a major part of the project.

The main responsibility of assistant engineers is to manage the team's daily activities. An assistant engineer ensures that staff is effectively deployed, adequate records maintained and all necessary measurements agreed and recorded. He oversees routine communications with the contractor, liaises with statutory undertakers and other third parties, supervises any new design work and modifications to the contract drawings and controls the arrangements for approval of work on site.

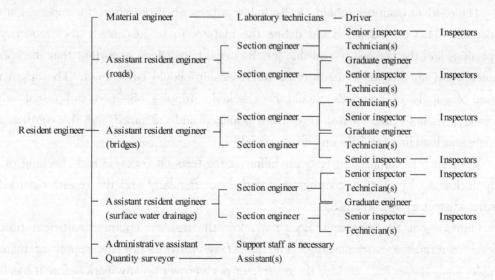

Figure 6.8 Organization Chart of Resident Engineer's Wite Team

The section engineers report to the assistant engineers. Duties may be allocated to them on the basis of location or function, depending on the nature and phasing of the works. The section engineers work at the bottom level of site supervision. They should be fully aware of what is taking place on the site within their specific scope of interest and should ensure that adequate records are kept and matters of significance reported to the assistant engineer. They shall make sure that the processes of inspection and supervision operate effectively.

Each section engineer normally leads a group of personnel made up principally of technicians and inspectors. Engineers and technicians concentrate on such activities as checking calculations and setting out, carrying out survey work and assessing performance, while inspectors, who are normally recruited from trade operatives, are mainly involved with the supervision of workmanship and the method aspects of the specification.

It is a common practice to find some specialists among the engineer's site staff such as a quantity surveyor and materials engineer. The quantity surveyor is concerned with the measurement and valuation of the contractor's work and is sometimes referred to as a measurement engineer on a civil engineering contract. The materials engineer will be involved in the testing of materials and components and reports on the tests and their implications.

Apart from involving the engineer's site staff in inspecting of work and materials to ensure compliance with the specification, the contractor will also institute his own quality control procedures, in addition to comparing progress with the approved programme, as quality assurance has now become an important activity.

6.9.3 Resident engineer's duties[12] [15]

The resident engineer should ideally follow a prescribed procedure for supervision and inspection. This procedure should define the matters to be inspected, the frequency of inspections and the persons responsible for the inspections. It is important that the correct materials and an acceptable quantity of workmanship should be secured. The origin and source of supply of materials should be checked, samples obtained and tested where necessary, manufacturer's test certificates obtained and examined and the contractor's storage and handling facilities on site approved.

Tests such as compaction tests on filling, cube tests on concrete and checking of the film thickness of protective coatings are carried out regularly and the results recorded to ensure satisfactory workmanship.

On occasions the contractor does not allow the resident engineer sufficient time to inspect materials or workmanship. The *Red Book* provides for the supply of material samples (sub-clause 7.2) and for the contractor not to cover up any work before it has been approved and the measurements agreed (sub-clause 7.3). It is advisable to remind the contractor of these provisions before work is started in order that he shall allow adequate time in his programme for inspections and measurements.

6.9.4 Implementation of specification[12] [15]

6.9.4.1 Supervisory implications

Most of the A/E's principal supervisory functions are directly relevant to the specification, which are briefly described as follows.

(1) Precautious measures

The A/E's site staff can, with advantage, use their knowledge of the site and the design to identify possible problem areas and to alert the contractor to them and to suggest methods by which the problems can be overcome. The objective is to assist in the smooth running of the contract by avoiding the execution of abortive work.

(2) Overseeing site operations

The A/E's site staff and, in particular, the inspectors keep a close watch on the standard of workmanship, having regard to the specification requirements and generally accepted standards of good practice. Some specification requirements are more easily enforced than others. It is also possible that equally satisfactory results can be achieved by using alternative methods to those prescribed in the specification. Any deviations from the specification require prior approval and careful scrutiny by the engineer.

Supervision of each and every activity on a large site to ensure compliance with the specification may prove very difficult or even impracticable because of the limitations on the resources available. Furthermore, the contractor is not obliged to notify the resident engineer of the location and timing of every individual operation. Although he is required under sub-clause 7.3 of the *Red Book* to give notice of any work which is to be covered up or put out of view.

There are usually provisions in the contract documents for the giving of advance notice of important activities such as concrete pours (such as the *Red Book*'s sub-clause 8.3) and operations to be undertaken outside normal hours (such as the *Red Book*'s sub-clause 6.5). The extent of supervision also varies with the type of work. For instance, compaction of extensive areas of fill, and structural concrete pours all require full-time supervision, while other operations, such as pipe laying, and placing of steel reinforcement, can be covered by periodic visits.

(3) Checking performance against standards

The A/E's site staff actively checks if the work complies with the standards prescribed in the specification. The staff checks on its own many items, such as surface tolerances on road construction, and concrete faces after striking of formwork. Materials and components are inspected on delivery to the site and samples of completed work are sent to the laboratory for testing where appropriate. Pipe runs must be checked for satisfactory lines, levels and joints before the trenches are backfilled, and formwork and reinforcement approved before concrete is placed.

(4) Approving and monitoring contractor's proposals

The contractor, in the early stages of the works, submits to the A/E for his approval, details of his methods of working, sources of materials and suppliers of components, in some cases the engineer requires the construction of sample panels of materials, such as bricks and concrete finishes, for approval prior to starting the work on the site. In other cases the A/E wishes to inspect the works where products are produced, such as gravel quarries and fabricating yards. The contractor is responsible for ensuring that all workmanship and materials satisfy the requirements of the contract.

6.9.4.2 Compliance with specification

The A/E's site staff has from time to time to make judgements as to compliance of the contractor's methods, materials and workmanship with the specification, or in the absence of a detailed statement of performance, accordance with the normally accepted standards of good practice. A number of approaches are available for doing so, depending on how the requirements have been expressed.

(1) The provisions in the specification may be extremely rigid. For example, the hooked ends of bar reinforcement shall each have an internal diameter of curvature and a straight length beyond the semicircle of at least four times the diameter of the bar.

(2) Some specification provisions are in the "deemed to satisfy" category, such as where the supplier of a specific material or component is listed which effectively pre-determines compliance and eliminates the need for individual assessment.

(3) Certain tolerances may be permitted such as a compacting factor of 0.82 to 0.85 for vibrated concrete in reinforced concrete floors, beams and slabs. This establishes a range of compliance against which individual results are assessed.

(4) Where large quantities of components are involved such as a structural concrete, it is customary to work to a statistically based standard of compliance, thus determining the degree of consistency of the end product instead of concentrating on individual samples.

(5) A specification requirement may be accompanied by a discretionary power of acceptance of an alternative based on the supervisor's own judgememt. This approach can sometimes cause problems as the engineer may not always approve the alternative product which the contractor selected and allowed for in his tender.

(6) A specification may include some items of work that are done "as required by the engineer" or "to the satisfaction of the engineer".

Compliance then becomes a matter for the supervisor's discretion, subject only to the customary standards and practice of the construction industry.

The circumstances may arise where a proposal by the contractor meets the requirements of the specification but does not produce the optimum result. Supervision is concerned with two criteria: adequacy of the permanent works and compliance with the specification, and a conflict can arise between the two on occasions. Hence compliance with a particular requirement of a specification will be a necessary condition for the acceptance of a contractor's proposal but it may not be a sufficient condition.

6.9.4.3 Problems in implementing specifications

Problems sometimes arise in the implementation of specifications because of mistaken, inconsistent or disputable interpretation. The document may, for example, contain excessively long paragraphs with the wording difficult to understand.

A specification should be complete, authoritative, up-to-date, to the point, specific to the contract in question and readily understood. Standard specification can be used as a guide, but great care should be taken to ensure that it contains no irrelevant clauses, covers all the features of the contract in hand. Difficulty can be experienced by the engineer's site staff when attempting to implement an unsatisfactory specification which will inevitably lead to unnecessary and time consuming disputes.

Notes 6.9

疑难词语解释

The most routine supervision is undertaken by the engineer's site team acting on their own initiative.

【译文】 最经常的监督由工程师现场人员根据自己的见解承担。

Chapter 7　Measurement, Valuation and Payment

It is the A/E's a primary responsibility to ensure that the contractor gets paid for the work executed by him and the employer gets the work for which he is paying. A major portion of construction contracts are of measure and value type and the contractor is paid for based upon the actual ascertained quantity and the rates tendered by him beforehand, on one hand. Variations are inevitable, on the other hand. Therefore, it takes the A/E a great portion of his time to measure the work executed by the contractor and to value the varied work.

The topics dealt with in this chapter include method of measurement, taking measurements on site and keeping of essential records, evaluation of work on a daywork basis, the making of interim and final valuations, and valuation of variations.

7.1　Method of measurement

7.1.1　Standard methods of measurement

The *Red Book* creates a remeasurement type of contractual arrangement. Thus the contractor is paid at the rates in the priced bill of quantities accepted by the employer for the actual quantity of work completed.

In many parts of the world methods of measurement for construction contracts have been formulated and standardized. For example, the *Standard Method of Measurement of Building Works* and the *Civil Engineering Standard Method of Measurement* in their regularly updated editions have been used as the principal codes in the United Kingdom, commonwealth countries and Hong Kong for building and civil engineering contracts, respectively. Figure 7.1 shows the grouping of items of work in SMM7, Seventh Edition Revised 1998, to erect a building.

       A Preliminaries/General conditions
       C Existing site/buildings/services
       D Groundwork
       E In situ concrete/Large precast concrete
       F Masonry
       G Structural/Carcassing metal/timber
       H Cladding/Covering

J Waterproofing
K Linings/Sheathing/Dry partitioning
L Windows/Doors/Stairs
M Surface finishes
N Furniture/Equipment
P Building fabric sundries
Q Paving/Planting/Fencing/Site furniture
R Disposal systems
S Piped supply systems
T Mechanical heating/Cooling/Refrigeration systems
U Ventilation/Air conditioning systems
V Electrical supply/power/lighting systems
W Communications/Security/Control systems
X Transport systems
Y Mechanical and electrical services measurement

Figure 7.1   Group Titles of Items of Work in a Building

The Ministry of Construction, in China, establishes the principles of measurement for construction works in the National Standard GB 50500—2003 *Code of valuation with bill quantity of construction works*. The code requires that the building and civil engineering works using public money be estimated as specified in the code as of 1 July 2003. According to the code it is optional for private and foreigner financed works. The authorities under the local construction commissions or construction industry administration bureaus in China have standardized methods of measurement, based upon the principles, for estimating, tendering for, and valuing, etc. the building and civil engineering works in their localities. The methods of measurement are similar to, but not completely identical to the SMM7 or CESMM6.

The code GB 50500—2003 covers following topics:
1. General,
2. Definitions,
3. Preparation of bills of quantities,
4. Estimating procedure and method, and
5. Formats and guidelines for preparation of bills of quantities.

Appendix A: Items of work of building works and their measurement

Appendix B: Items of work of fitting out and finishing and their measurement

Appendix C: Items of work of mechanical/electrical installation and building services and their measurement

Appendix D: Items of work of municipal works and their measurement

Appendix E: Items of work of landscaping and their measurement

The codes prescribe the manner in which items of work are to be classified and provide schedules of items of work with their descriptions and estimated quantities. This procedure ensures a reasonable standard of uniformity in the approach to the measurement of the items of work.

The preparation of bills of quantities for building and civil engineering works in most cases follows these codes.

The descriptions of items of work for permanent works generally identify the component of the works and not the tasks to be carried out by the contractor. The work classifications used in the codes provide a minimum detail of description and itemization on which contractors can rely, but encourages greater detail in non-standard circumstances, particularly where cost-significant aspects are involved.

7.1.2 Fixed and time-related cost items[15]

The costs arising from many building or engineering operations are not proportional to the quantity of the resulting permanent work, but related to the contractor's general arrangements and methods of construction, and referred to as method-related costs. There are two basic types of method-related costs: time-relatedcosts and fixed costs. The cost of bringing an item of contractor's equipment onto a site and its subsequent removal is a fixed cost, not related to quantity or time, while that of its operation a time-related cost.

The A/E specifies certain requirements by entering preliminaries and/or general items in the bill of quantities covering such items as accommodation, buildings, services, contractor's equipment, and equipment for use by and attendance upon the A/E's site staff; testing of materials and works; supervision, and temporary works. The charges which the tenderers make to cover the items are not directly related to the quantities of permanent work.

Then tenderers enter prices for the items based upon their estimates of the cost of each operation, together with amounts-sometimes proportioned over all the items to cover overheads, site offices, head office expenses, financing, insurance, general contractual liabilities, profit, contingencies and so on.

Alternatively some of these items may be in a section of the bill on preliminaries. If the bill of quantities is based on the CESMM tenderers may enter both fixed and time-related items in the method-related cost section of the bill, and include in the rates entered against the items of permanent work only those elements which are quantity-related.

It is not a secure for the contractor to recover the cost of bringing a tower crane onto the site and its hire, operation and subsequent removal by entering a lump sum against an item in the preliminaries, where provided, or by making an allowance in the cost per cubic metre of the various work sections for which the tower crane is to be used. It is believed that method-related cost items provide a better way of representing the contractor's site operation costs, such as the provision of site accommodation and temporary works, and the setting up of labour gangs, sometimes referred to as "site mobilization".

The temporary works can involve a wide range of diverse operations, such as traffic diversion and regulation, access roads, cofferdams, pumping, dewatering, access and support scaffolding, piling shafts, pits and hard-standings.

7.1.3 Measured items of work[15]

The contractor must clearly and fully identify the items of work entered by the A/E in the bills of quantities and distinguish between the method-related items and the items of work to be measured, as prescribed in the codes of measurement, of which quantities are related to the length of time to be spent on them. Against the time-related items of work the contractor shall enter their rates together with full descriptions, including time and cost elements, to facilitate their subsequent costing.

The method-related items are not subject to measurement, although the contractor will be paid for these sums (prices) in interim payment certificates in the same way as he is paid for measured items of work.

The prices entered against method-related items will appear in the final account as well, and will not be changed as a result of the quantity of work executed actually by the contractor having been found different from that in the bill of quantities provided that the A/E has not instructed any variation. The contractor is not bound to execute the items of work using the methods or techniques listed by him for the method-related items, but he will nevertheless be paid as though the techniques indicated had been adopted. For example, if the contractor inserted prices for a concrete batching plant and subsequently used ready-mixed concrete, the appropriate interim payments will be distributed over the quantity of concrete placed.

The prices against the method-related items and the rates against the items of work to be measured entered by the contractor are also referred to as "fixed charges" and "variable charges" reflecting the fixed cost and variable cost to the contractor, respectively.

It is not difficult for the contractor's estimator to price the most items of work billed as prescribed in the codes of measurement. However, he may feel confused when pricing manholes of which types, sizes and connections vary widely. The enumerated approach to their measurement prescribed by CESMM6 is not sufficiently sensitive to their actual cost.

Words, Phrases and Expressions 7.1

[1] actual ascertained quantity 双方实际确认的(工程)数量
[2] rate 或 unit price n.分项工作单价
[3] price 或 lump sum n.总和价

[4] evaluation of work on a daywork basis 按点工估算工程(工作)价值,按计日工作计价
[5] method of measurement 计量方法
[6] Standard Method of Measurement of Building Works 建筑工程标准计量方法
[7] Civil Engineering Standard Method of Measurement 土木工程标准计量方法
[8] Carcassing metal 骨架用钢(包括结构用钢),钢骨架(包括钢结构)
[9] Cladding/Covering 维护/覆盖(面层)
[10] Linings/Sheathing/Dry partitioning 衬板/护墙板,外墙板/干隔断
[11] Building fabric sundries 建筑物骨架配件
[12] Disposal systems 垃圾处置系统
[13] National Standard GB 50500—2003 Code of valuation with bill quantity of construction works. 建设工程工程量清单计价规范 GB 50500—2003
[14] lump sum 总数,一次总付之款(one payment for a number of separate sums that are owed)
[15] lump sum items 总和项(不宜译成"包干项")
[16] method-related items 与(施工)方法有关的分项(工作)
[17] measured items of work 计算数量的分项(工作)
[18] time-related items 与时间有关的分项(工作)

Notes 7.1

charge 与 cost 的区别

charge 指卖方向买方索取的数额,而 cost 是卖方生产过程中需要的人工、材料、施工机械使用、监督、管理等费用。一般说来,charge 的数额大于 cost。另外,在工程合同里,charge 常用来指政府或其他凭权力收取的费用,例如公路管理局收取的养路费,环保局收取的环保费。这种费用不一定是承包商生产过程需要的。而 cost 一定是承包商生产过程中需要的,如人工、材料、施工机械使用、监督、管理等等费用。

7.2 Measurement of work executed and cost records

7.2.1 Measurement of work executed

Where the A/E requires any part of the works to be measured, he shall give reasonable advance notice in writing to the contractor, who shall either attend or send a qualified representative to assist in making the measurement. If the contractor fails to attend the measurement or send a representative, then the measurement made by the A/E or his representative shall be taken as being the correct measurement (sub-clause 12.1 of the *Red Book*). It is important that any delegation of the A/E's duties under this sub-clause, and any limitations placed on them, shall be fully particularized and notified in writing to the contractor.

7.2.2 Method of measurement

Sub-clause 12.2 of the *Red Book* states: except as otherwise stated in the contract and notwithstanding local practice:

(a) measurement shall be made of the net actual quantity of each item of the permanent works, and

(b) the method of measurement shall be in accordance with the bill of quantities or other applicable schedules.

These two sub-paragraphs describe what is typically referred to as the "method of measurement" applicable to the works. This method relates primarily to what quantities are to be applicable to the evaluation, rather than to the measuring techniques (although they may also be described), and plays an important part in the whole evaluation of the contract price. This method (or principles) of measurement may comprise:

☐ principles for measurement which are specified in a preamble to the bill of quantities,

☐ a publication which specifies principles of measurement and which is incorporated (by reference) into the bill of quantities, or

☐ for a contract which does not contain many or complex items of work, principles included in each of the item description in the bill of quantities.

Each item of the works to be measured in accordance with such principles or method of measurement, which take precedence over the general principle described in sub-paragraph (a) of sub-clause 12.2.

It may be unfair to assume that an international contractor is familiar with all aspects of local practice on these matters.

It is important to verify, before the bill of quantities is issued to tenderers, that it includes the correct quantities and item descriptions of the work defined in the drawings and specification.

Generally, tenderers will have limited opportunities to verify the correctness of the bill of quantities which they receive.

However, during the measurement of the completed works, omissions in the original bill of quantities may be discovered, or be alleged by a party, in such a case, a dispute may arise as to whether an additional bill item will be required, or as to whether the work is covered by another item in the bill of quantities. Although resolution of the matter may need to take account of various provisions in the contract and of the applicable laws, it is suggested that the effect of clause 12 of the *Red Book* would typically be as follows:

☐ if the bill of quantities includes (either incorporated by reference or specified) principles of measurement which clearly require that an item of work be measured, and if the bill of quantities contains is no such item, then an additional bill item will be required in order to satisfy the requirements for measurement in accordance with such principles.

☐ if the bill of quantities includes (either incorporated by reference or specified) principles of measurement which do not clearly require that a particular item of work be measured, and if the work was as described in the contract and did not arise from a variation, then measurement in accordance with such principles does not require the addition of a new bill item.

☐ if the bill of quantities does not includes principles of measurement for a particular item of work, and the work was as described in the contract and did not arise from a variation, then measurement in accordance with such principles does not require the addition of a new bill item.

7.2.3 Evaluation

Evaluation of the executed work shall be made as stated under sub-clause 12.3, if the *Red Book* is used for a contract. The sub-clause states:

Except as otherwise stated in the contract, the Engineer shall proceed to agree or fairly determine with due consultation with the parties, having regard to all the circumstances, the contract price by evaluating each item of work, applying the measurement agreed or determined in accordance with the sub-clauses 12.1 and 12.2 and the appropriate rate or price for the item.

For each item of work, the appropriate rate or price for the item shall be the rate or price specified for such item in the contract or, if there is no such item, specified for similar work.

Any item of work included in the bill of quantities for which no rate or price was specified shall be considered as included in other rates and prices in the bill of quantities and will not be paid for separately. However, a new rate or price shall be appropriate for an item of work if:

(a) (i) the measured quantity of the item is changed by more than 10% from the quantity of this item in the bill of quantities or other schedule,

(ii) this change in quantity multiplied by such specified rate for this item exceeds 0.01% of the accepted contract amount,

(iii) this change in quantity directly changes the cost per unit quantity of this item by more than 1%, and

(iv) this item is not specified in the contract as a "fixed rate item"; or

(b) (i) the work is instructed under clause 13,

(ii) no rate or price is specified in the contract for this item, and

(iii) no specified rate or price is appropriate because the item of work is not of similar character, or is not executed under similar conditions, as any item in the contract.

Each new rate or price shall be derived from any relevant rates or prices in the contract, with reasonable adjustments to take account of the matters described in sub-paragraph (a) and/or (b), as applicable. If no rates or prices are relevant for the derivation of a new rate or price, it shall be derived from the reasonable cost of executing the work, together with profit, taking account of any other relevant matters.

7.2.4 Break-even analysis

The statements made in sub-clause 12.3 can be elaborated in a manner as follows.

It is common that the rates entered by the contractor against some items of work in the bills of quantities relate to fixed overheads and/or other preliminaries and general items which do not increase pro rata with the quantity of an item of work. For example, if the construction, maintenance and reinstatement of a temporary drainage channel were priced in the bill of quantity at a rate per week, doubling the length of time which is required to construct the drainage channel does not double the cost to the contractor and the rate for the extended time requires adjusting.

Sub-paragraph (a) of sub-clause 12.3 specifies four criteria applicable without reference to clause 13, and a new rate shall only be appropriate if all four criteria are satisfied. The first two criteria relate to the change in quantity; the third criterion relates to its effect on costs; and the fourth criterion allows adjustment of some items to be precluded. [31]

(i) The measured quantity, denoted by Q_M, of the work must be less than 90%, or more than 110%, of the quantity stated in the bill of quantities, denoted by Q_B.

This criterion is consistent with the principle that Q_B is only an estimate, as stated in sub-clause 14.1(c).

(ii) When the difference in quantity Q_M-Q_B is multiplied by the rate per unit quantity stated in the bill of quantities, the product must be more than 0.01% of the Accepted Contract Amount. This criterion is specified in order to avoid adjusting a rate if the adjustment will have little effect on the final contract price.

(iii) The difference in quantity (namely the change from Q_B to Q_M only and excluding other matters) must have affected the "cost per unit quantity", which is the cost incurred executing the work covered by the item, divided by the quantity of the item as measurable in accordance with the applicable method of measurement. This criterion relates firstly to the cost actually incurred, denoted by C_{qm}, in the execution of the measured quantity Q_M; and secondly to what the cost would have been C_{qb} if the measured quantity had been equal to Q_B. Under this criterion, the cost actually incurred divided by the measured quantity, that is C_{qm}/Q_M, must be less than 99%, or more than 101%, of the cost per unit quantity, that is C_{qb}/Q_B, which the contractor would have incurred if he had executed Q_B. Note that C_{qb} is the cost which the contractor would actually have incurred if $Q_M = Q_B$; and not (for example) his foreseen or estimated cost, or any amount included in (or otherwise derived from) the bill of quantities. This criterion is specified in order to avoid adjusting a rate by less than 1%. As noted below, adjustments are based upon the proportion ($C_{qm}/$

$Q_M)/(C_{qb}/Q_B)$.

(iv) The contract must not have used phrase "fixed rate item" in relation to the item in the bill. This criterion is specified in order to allow for some bill items having very provisional quantities. Alternatively, if the bill includes items which have no quantity Q_B (the phrase "rate only item" might have been used, for example), there will be no "quantity of this item in the Bill of Quantities or other Schedule" and the rates for these items would not be adjusted under sub-clause 12.3.

If the four criteria in sub-paragraph (a) are satisfied, the bill rate would typically be changed in proportion to such change in cost per unit quantity which was the direct result of the change in quantity. In other words, the rate per unit quantity would be adjusted pro rata to the proportion $(C_{qm}/Q_M)/(C_{qb}/Q_B)$. Having passed the three criteria which specify percentages, the new rate should not take account of the criterion percentages in (a) (i), (ii) and (iii). Although these percentages preclude adjustment of some items, changes to the rates should not be based upon differences between actual and criterion percentages.

Typically, C_{qm}/Q_M would be expected to be equal to C_{qb}/Q_B and criterion (a)(iii) would not be satisfied. For example, mass excavation costs are typically proportional to the quantity excavated, so $C_{qm}/Q_M = C_{qb}/Q_B$. For some other construction operations, the costs (C_{qm} and C_{qb}) may include an element which is fixed and does not depend upon the quantity executed; so if $Q_M > Q_B$, then $C_{qm}/Q_M < C_{qb}/Q_B$. The fixed element might be the provision of temporary works or the assembly of contractor's equipment, unless the cost of such operations is included in other Bill items and not in C_{qm} and C_{qb}.

Sub-paragraph (b) specifies criteria relating to work instructed under clause 13, which includes variations, work under provisional sums, and (possibly) some types of work valued under the provisions in the daywork schedule. In these cases, a new rate or price will be considered appropriate if there is no bill rate or price for work of similar character and executed under similar conditions.

If a new rate or price is to be assessed, it may be derived from relevant rates and/or prices in the bill of quantities or other appropriate Schedules, and/or from reasonable costs.

The idea which the statements made above attempt to get across to the readers can be best formulated as a mathematic model is referred to as 'break-even analysis'.

Break-even analysis is a method for presenting costs and profits in a form designed to aid interpretation and analysis. The presentation can be graphical or tabular. Typically, this information presents profits as a function of a percentage of 'fixed capital' (for example, contractor's equipment, temporary works, head office overheads, etc) utilization, reveals point where operations showing no profit and are just breaking even, and determines the point at which it will pay to suspend operations.

As a simple example, consider a contract for canal excavation having fixed costs C_F for the time for completion which does not change with the volume of excavation q (quantity).

Other costs will be variable and will change with the quantity q, such as materials and labor. Suppose the cost per cubic excavation is v as estimated by the contractor. Then, if a linear relationship exists, the variable cost V itself will be $v \times q$. Note that v is the cost per cubic meter of excavation and may be constant and that $V = v \times q$ is the actual variable cost. If C_T is the total cost for the time for completion, then

$$C_T = V + C_F = v \times q + C_F \tag{7.1}$$

Equation representing C_T (7.1) can be plotted against the quantity q as shown in Figure 7.2.
Equation (7.1) is very simple because the following assumptions are made for the purpose of this explanation:

(1) The cost per cubic excavation v is constant; hence the variable costs $V = v \times q$ are linearly dependent on the quantity q.
(2) Fixed costs C_F are independent of quantity q.
(3) There are no financial costs.
(4) There is no income other than that from excavation.
(5) All excavation is accepted by the employer at the same rate.

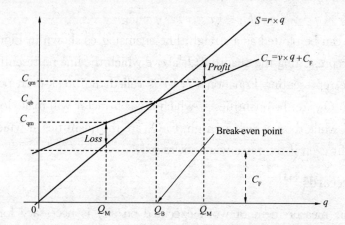

Figure 7.2 Break-even Analysis in Graphical Form

Now let us consider the tender price T which the contractor considers represents his total cost C_T, that is:

$$T = C_T = v \times Q_B + C_F \tag{7.2}$$

Then the rate which the contractor entered in the bill of quantities for excavation is:

$$r = T/Q_B = (v \times Q_B + C_F)/Q_B = v + C_F/Q_B \tag{7.3}$$

If the measured quantity Q_M is exactly the quantity stated in the bill of quantities, that is Q_B, then the payment due to the contractor, denoted by S_B, is:

$$S_B = r \times Q_B = T \tag{7.4}$$

Then the contractor is at 'break-even', meaning that he will recover all the costs, that is the fixed cost C_F and the variable costs $v \times Q_B$, but no profit.

If the measured quantity $Q_M < Q_B$, then the payment due to the contractor, denoted by S_L, is:

$$S_L = r \times Q_M < r \times Q_B = T, \tag{7.5}$$

which means that the contractor will not recover all the costs and make a loss. In order to make up the contractor's loss, the A/E and the contractor shall agree a new rate, denoted by r_N, so that $r_N \times Q_M = T > r \times Q_M$, that is $r_N > r$.

On the other hand, however, if the measured quantity $Q_M > Q_B$, then the payment due to the contractor, denoted by S_H, is:

$$S_H = r \times Q_M > r \times Q_B = T, \tag{7.6}$$

which means that the contractor will not only recover all the costs but also make a profit. In order to keep the contractor's profit in check, the A/E and the contractor shall agree a new rate, denoted by r_N, so that $r_N \times Q_M = T < r \times Q_M$, that is $r_N < r$.

Let S be the payment due to the contractor when he has excavated q m^3, then

$$S = r \times q \tag{7.7}$$

Further, let P be the gross profit, then

$$P = S - C_T = r \times q - v \times q - C_F = (r - v) \times q - C_F \tag{7.8}$$

Equation (7.7) can be plotted as a straight line against q as shown in Figure 7.2.

The point representing the quantity Q_B above which the line representing Equation (7.7) cross the line representing Equation (7.1) is called the break-even point. The points which are left to Q_B are the quantities at which the contractor will make loss if the rate has not been raised, while those which are right to Q_B are the quantities at which the contractor will make a profit even if the rates were not raised.

7.2.5 Cost records [15]

The periodic measurement of work executed on site is necessary for the purpose of comparison of progress with programme and of assessment of the payments due to the contractor. The measurements can be taken from record drawings or physically on the site. The quantities of materials stored on the site will be assessed by direct measurement and observation.

In measuring the amount of work executed for purpose of interim payments, it is sometimes convenient to agree average cross-sections or volumes when computing the volume of earthwork executed, concrete placed and other commonly used operations. To simplify the interim measurement calculations an alternative method is to compute accurately the final volume and to allow a percentage of that figure for each valuation. Whatever method is used, care must be taken to ensure that the cumulative approximate valuations do not exceed the value of the accurate final total.

Graphs such as bar chart or networks, are often used in comparing the actual progress of the works against the programmed requirements for the sections or major parts of the works both in quantitative and monetary terms. The records shall be normally made and kept of labour (type, hours and rate), contractor's equipment (type, hours and rate) and materials (type, quantity and rate).

The records help assess the cost of new or varied works at current or contract rates, and check the value of any claims or estimates submitted by the contractor, including the verification of items of work done on a daywork basis.

Words, Phrases and Expressions 7.2

[1] reasonable advance notice in writing 提前足够时间的书面通知
[2] break-even analysis 盈亏平衡分析
[3] financial cost 财务费用(包括资金筹集费用和资金使用费用)
[4] break-even point 盈亏平衡点
[5] gross profit 毛利润,利润总额(未缴纳所得税之前的利润)

7.3 Daywork

7.3.1 Necessity of daywork

Daywork is the method of valuing work on the basis of time spent by operatives, materials and plant incorporated in the works, and contractor's equipment employed, with an allowance to cover oncosts and profit. Under sub-clause 13.6 of the *Red Book*, for work of a minor or incidental nature, the engineer may instruct that a variation shall be executed on a daywork basis. The work shall then be valued in accordance with the daywork schedule included in the contract.

Before ordering goods for the work, the contractor shall submit quotations to the engineer. When applying for payment, the contractor shall submit invoices, vouchers and accounts or receipts for any goods.

Except for any items for which the daywork schedule specifies that payment is not due, the contractor shall deliver each day to the engineer accurate statements in duplicate which shall include the following details of the resources used in executing the previous day's work:
(a) the names, occupations and time of contractor's personnel,
(b) the identification, type and time of contractor's equipment and temporary works, and
(c) the quantities and types of plant and materials used.

One copy of each statement will, if correct, or when agreed, be signed by the engineer and returned to the contractor. The contractor shall then submit priced statements of these resources to the engineer, prior to their inclusion in the next statement.

The daywork as a method of valuing contractor's work should apply only where the normal process of measurement and valuation at billed rates or rates analogous thereto is not practicable. This method can be convenient to both the resident engineer and the contractor's representative, as a contractor who is satisfied with the operative method of payment is likely to cooperate more readily in the execution of a complex variation or one involving unpredictable factors. Lengthy disputes over valuations can be reduced by the payment of dayworks based upon a relatively straightforward analysis of site records.

Sub-clause 13.6 of the *Red Book* sets the procedure for a work to be executed on a daywork basis and for the contractor to apply for payment therefor, the contractor's representative and his staff shall follow it to prevent any subsequent forfeiture of rights to payment.

7.3.2 Daywork schedule

A daywork schedule should be included if the probability of unforeseen work, outside the items included in the bill of quantities, is relatively high. To facilitate checking by the employer of the realism of rates quoted by the tenderers, the daywork schedule should normally comprise:

(a) a list of the various classes of labor, materials, and contractor's equipment for which basic daywork rates or prices are to be inserted by the tenderer, together with a statement of the conditions under which the contractor will be paid for work executed on a daywork basis; and

(b) a percentage to be entered by the tenderer against each basic daywork subtotal amount for labor, materials, and plant representing the contractor's profit, overheads, supervision, and other charges.

A typical daywork schedule illustrating this approach is shown in Figure 3.12~3.15 (see 3.2.8).

7.4 Certificates and payments

7.4.1 Monthly statements

Under sub-clause 14.3 of the *Red Book*, the contractor, when applying for payment due to him, shall submit a statement in six copies to the engineer after the end of each month, in a form approved by the engineer, showing in detail the amounts to which the contractor considers himself to be entitled, together with supporting documents which shall include the report on the progress during this month.

The statement shall include the following items, as applicable, which shall be expressed in the various currencies in which the contract price is payable, in the sequence listed:

(a) the estimated contract value of the works executed and the contractor's documents produced up to the end of the month (including variations but excluding items described in sub-paragraphs (b) to (g) below);

(b) any amount to be added and deducted for changes in legislation and changes in cost;

(c) any amount to be deducted for retention, calculated by applying the percentage of retention to the total of the above amounts, until the amount so retained by the employer reaches the limit of retention money (if any) stated in the contract;

(d) any amount to be added and deducted for the advance payment and repayments in accordance with the contract;

(e) any amount to be added and deducted for plant and materials in accordance with the contract;

(f) any other addition or deduction which may have become due under the contract or otherwise; and

(g) the deduction of amounts certified in all previous payment certificates.

When the employer agrees to pay the contractor for the plant and materials (should be those of which substantial quantities are involved) specified, as agreed between the parties, in the Appendix to Tender and delivered by the contractor on the site for incorporation in the permanent works but not incorporated in such works and a percentage of the invoice value of specified materials is stated in the Appendix to Tender, the interim payment certificates shall include (i) an amount for plant and materials which have been sent to the site for incorporation in the permanent works, and (ii) a reduction when the contract value of such plant and materials is included as part of the permanent works.

The engineer shall determine and certify each addition if the following conditions are satisfied:

(a) the contractor has: (i) kept satisfactory records (including the orders, receipts, costs and use of plant and materials) which are available for inspection, and (ii) submitted a statement of the cost of acquiring and delivering the plant and materials to the site, supported by satisfactory evidence; and

(b) the relevant plant and materials: (i) are those listed in the schedules for payment when shipped, (ii) have been shipped to the country, en route to the site, in accordance with the contract, and (iii) are described in a clean shipped bill of lading or other evidence of shipment, which has been submitted to the engineer together with evidence of payment of freight and insurance, any other documents reasonably required, and a bank guarantee in a form and issued by an entity approved by the employer in amounts and currencies equal to the amount due hereunder: this guarantee shall be valid until the plant and materials are properly stored on site and protected against loss, damage or deterioration; or

(c) the relevant plant and materials: (i) are those listed in the Appendix to Tender for payment when delivered to the site, and (ii) have been delivered to and are properly stored on the site, are protected against loss, damage or deterioration, and appear to be in accordance with the contract.

The additional amount to be certified shall be the equivalent of eighty percent of the engineer's determination of the cost of the plant and materials (including delivery to site), taking account of the documents above mentioned and of the contract value of the plant and materials.

7.4.2 Payments to contractor

On condition that the employer has received and approved the performance security, the engineer shall, within 28 days after receiving a statement and supporting documents from the contractor, issue to the employer an interim payment certificate (see Figure 7.3 for a sample) which shall state the amount which the engineer fairly determines to be due, with supporting particulars.

However, prior to issuing the taking-over certificate for the works, the engineer shall not be bound to issue an interim payment certificate in an amount which would (after retention and other deductions) be less than the minimum amount of interim payment certificates (if any) stated in the Appendix to Tender. In this event, the engineer shall give notice to the contractor accordingly.

INTERIM PAYMENT CERTIFICATE

A/E's name and address: Serial No.:
Employer's name and address: Valuation date:
                               Installment date:
Main Contractor's name and address: Job reference:
                                     Issue date:

I/We certify that in accordance with Clause 60.4 of Conditions of Contract
under the Contract
dated:
for the Works:
situated at:

Interim payment as detailed below is due from
The Employer to the Main Contractor   **HK $**
Total value: 57,257,348.00
includes the value of works by Nominated Subcontractors as detailed on
direction form no.   dated
Less retention   2,812,030.00
after deducting any retention sums released previously or herewith
Balance (cumulative total amount certified for payment)   54,445,318.00
Less cumulative total amount previously certified for payment   47,694,618.00

Amount due for payment on this certificate   6,750,700.00

(in words) SIX MILLION, SEVEN HUNDRED AND FIFTY THOUSAND, AND SEVEN HUN DOLLARS ONLY

                                             Signed   A/E

Figure 7.3   A Sample Interim Payment Certificate[33]

An interim payment certificate shall not be withheld for any other reason, although:

(a) if any thing supplied or work done by the contractor is not in accordance with the contract, the cost of rectification or replacement may be withheld until rectification or replacement has been completed; and/or

(b) if the contractor was or is failing to perform any work or obligation in accordance with the contract, and had been so notified by the engineer, the value of this work or obligation may be withheld until the work or obligation has been performed.

The engineer may in any payment certificate make any correction or modification that should properly be made to any previous payment certificate. A payment certificate shall not be deemed to indicate the engineer's acceptance, approval, consent or satisfaction.

The employer shall pay to the contractor:

(a) the first installnent of the advance payment within 42 days after issuing the Letter of Acceptance or within 21 days after receiving the documents in accordance with sub-clause 4.2 and sub-clause 14.2 of the *Red Book*, whichever is later;

(b) the amount certified in each interim payment certificate within 56 days after the Engineer receives the statement and supporting documents; and

(c) the amount certified in the final payment certificate within 56 days after the employer receives this payment certificate.

Payment of the amount due in each currency shall be made into the bank account, nominated by the contractor, in the payment country (for this currency) specified in the contract.

If the contractor does not receive payment as provided in the contract, the contractor shall be entitled to receive financing charges compounded monthly on the amount unpaid during the period of delay. This period shall be deemed to commence on the date for payment specified in the contract, irrespective of the date on which any interim payment certificate is issued.

Unless otherwise stated in the contract, these financing charges shall be calculated at the annual rate of three percentage points above the discount rate of the central bank in the country of the currency of payment, and shall be paid in such currency.

The contractor shall be entitled to this payment without formal notice or certification, and without prejudice to any other right or remedy.

Where the contract is subject to a price fluctuation clause, allowance will be needed to cover changes in the cost of labour and materials, usually calculated on the basis of the price adjustment formula. The adjustments are generally made on the basis of CPI (consumer price index) indices for which the base date was 1978 (100) in China and the indices covering the principal labour and material items are published monthly by the China State Statistic Bureau in China. Furthermore, variations, extras and contractor's claims all need considering and including in the valuation where appropriate.

7.4.3 Interim valuations

A typical example of an interim valuation for a road contract is illustrated in Table 7.1 to show the general format and approach, starting with the adjustment of preliminaries and general items and followed by the various work sections, nominated subcontractors, retention and materials, less the amount previously certified to give the amount due to the contractor under this valuation. The contract period is 20 months.

Table 7.2 is an example of summarized interim valuation for a building contract.

Table 7.1 Interim Valuation

Valuation for Certificate nr 8:14 October 2003	HK $	HK $
Bill nr 1 Preliminaries and General Items		
Performance bond	288 000	
Insurances	1464 000	
Offices for Engineer's staff-establishment	72 000	
Ditto-maintenance, $ 144 000 × 8/20	57 600	
Attendance upon Engineer's staff, $ 180 000 × 8/20	72 000	
Testing of materials	60 000	
Traffic regulation-establishment	84 000	
Ditto-operation and maintenance, $ 264 000 × 8/20	105 600	
Pumping plant-establishment	64 800	
Ditto-operation and maintenance	80 400	
Ditto-standing by	39 600	
Site accommodation-establishment	180 000	
Ditto-maintenance, $ 192 000 × 8/20	76 800	
Concrete mixing plant-establishment	192 000	
Ditto-operation and maintenance, $ 240 000 × 8/20	96 000	
Hardstandings	114 000	3 046 800
Bill nr 2 Site Clearance		
Bill total	612 000	
Removal of 15 nr tree stumps, 0.5~1.0 m diam., @ $ 336	5 040	617 040

续表

Valuation for Certificate nr 8:14 October 2003	HK $	HK $
Bill nr 3 Earthworks		
General excavation of material for disposal,		
maximum depth 0.5~1 m, 19 500m³@ $ 72	1 404 000	
Ditto-maximum depth 1~2 m, 13 600 m³@ $ 78	1 060 800	
General excavation of material for reuse,		
maximum depth 0.5~1 m, 15 600m³@ $ 48	748 800	
General excavation of topsoil for reuse,		
maximum depth not exceeding 0.25 m, 7 200 m³@ $ 36	259 200	
Trimming of slopes, 6 600m³@ $ 6	39 600	
Filling and compacting 150 mm of excavated		
topsoil to slopes, 6 600m²@ $ 6	39 600	
Imported hardcore, 15 000 m³@ $ 96	1 440 000	4 992 000
Bill nr 4 Carriageway		
Granular base, 150mm deep, 50 000m²@ $ 42	2 100 000	
Concrete carriageway slab, 225mm deep,		
45 000m²@ $ 180	8 100 000	
Steel fabric reinforcement, 45 000m²@ $ 30	1 350 000	
Expansion joints, 20 000 m @ $ 180	3 600 000	
Precast concrete kerb, 14 000m @ $ 120	1 680 000	16 830 000
Bill nr 5 Footpaths		
Granular base, 75 mm deep, 1600m²@ $ 24	38 400	
Bituminous macadam base course, 50mm		
deep, 1200m²@ $ 48	57 600	
Bituminous macadam wearing course,		
10 mm deep, 1200m²@ $ 18	21 600	117 600
Bill nr 6 Bridges		
Bridge 1		
Bill total $ 2 760 000-20 per cent complete		552 000
Bill nr 7 Culverts		
Bill total $ 8 040 000-25 per cent complete		2 010 000
Bill nr 8 Surface Water Drainage		
Bill total $ 2 880 000-40 per cent complete		1 152 000
Bill nr 9 Retaining Walls		
Bill total $ 1 128 000-20 per cent complete		225 600

续表

Valuation for Certificate nr 8:14 October 2003	HK $	HK $
Bill nr 10 Fencing		
Chain link fencing, 900 m @ $ 180	162 000	
Wood post and rail fencing, 480 m @ $ 144	69 120	
Chestnut pale fencing, 660m @ $ 96	63 360	294 480
Bill nr 11 Dayworks		172 800
Variation nrs 1-13		152 040
Nominated subcontractor:		
Electrics Ltd-street lighting	936 000	
Add for profit, 5 per cent	46 800	
Attendance	74 400	1 057 200
		31 219 560
		936 588
Less retention (3 per cent)		30 282 972
		929 028
Materials on site		31 212 000
Less total of certificates 1~7		26 352 000
Total amount due	HK $ 4 860 000	

Note: This is an adaptation based on [15].

Table 7.2 Interim Valuation [33]

Contract information	Date	Interim valuation No.14
Architect:	**Valuation of work in progress**	
Employer:		$
Main contractor:	Main contractor's work	43,602,000.00
Contractor No.:	Materials on site	4,381,000.00
Contract Sum: $ 102,541,379.91	Profit & attendance on P.C. items	192,000.00
Contingencies: $ 4,000,000.00	Sub-total	48,175,000.00
Limit of Retention $ N/A	Nominated Sub-contractors as Statement	9,082,348.00
Date of possession:	Nominated Suppliers as statement	9,082,348.00
	Sub-total	9,082,348.00
Notes	Gross Valuation	57,257,348.00
1. It is assumed that the architect	Total retention	2,812,030.00
/Engineer will make any adjustment		
in respect of work which he considers	Net valuation	54,445,318.00

improperly executed or of unsuitable standard.

2. The nominated Suppliers and Subcontractors have been informed of amounts recommended to them.

3. This valuation includes the sum of $9,082,348.00 gross in respect of building in accordance with statements received Consultants.

4. Unless specifically stated no deduction has been made for Liquidated and Ascertained Damages

Note: N.S.C.s stands for Nominated subcontractors.

Certificates Nos 1~13	47,694,618.00
Amount due	**6,750,700.00**
Statement of Retention	
Gross retention held (5%)	2,408,750.00
Partial release of retention on Practical completion.	
Net retention	2,408,750.00
Retention held from N.S.C.s	403,280.00
Total retention as shown in valuation above	2,812,030.00

7.4.4 Payment of retention money

Retention money means an accumulated sum of moneys which the employer sometimes retains when payments shall be made to the contractor to cover costs of repair of unforeseen defects and other costs that the contractor may have been responsible for. The purpose of the employer is to secure the contractor's obligations to complete the works and remedy the defects. He holds retention money as a trustee for the contractor but without obligation to invest. The employer shall pay the retention money to the contractor in accordance with the provisions of conditions of contract. Sometimes, the contractor can receive the retention money immediately by furnishing the employer a bond or bank guarantee for such sum.

Sub-clause 14.9 of the *Red Book* provides that when the taking-over certificate has been issued for the works, the first half of the retention money shall be certified by the engineer for payment to the contractor.

If a taking-over certificate is issued for a section or part of the works, a proportion of the retention money shall be certified and paid. This proportion shall be two-fifths (40%) of the proportion calculated by dividing the estimated contract value of the section or part, by the estimated final contract price. Promptly after the latest of the expiry dates of the defects notification periods, the outstanding balance of the retention money shall be certified by the engineer for payment to the contractor. If a taking-over certificate was issued for a section, a proportion of the second half of the retention money shall be certified and paid promptly after the expiry date of the defects notification period for the section. This proportion shall be two-fifths (40%) of the proportion calculated by dividing the estimated contract value of the section by the estimated final contract price.

However, if any work remains to be executed as required in the contract, the engineer shall be entitled to withhold certification of the estimated cost of this work until it has been executed.

When calculating these proportions, no account shall be taken of any adjustments for changes in legislation and/or cost.

The practice that the employer retains a sum when payments shall be made to the contractor to cover costs of likely repair has imposed a significant financial burden on many contractors, such as the cost of financing the delayed repayment of the second half of retention of money under sub-clause 14.9 of the *Red Book*. The situation is worse if the repayment of even the first half of the retention money is also delayed as required by the employer. The cost of financing in most cases is not negligible having regard to the fact that the business of contracting has become less profitable than some thirty years ago.

As a result there has emerged a tendency that when the taking-over certificate has been issued for the works and the first half of the retention money has been certified for payment by the engineer, the contractor shall be entitled to substitute a guarantee in a form and provided by an entity both approved by the employer, for the second half of the retention money. On receipt by the employer of the required guarantee, the engineer shall certify and the employer shall pay the second half of the retention money.

7.4.5 Advance payment

Generally, advance payment is a sum of money paid by one party to another in anticipation of receiving goods, services, or other assets, or in order to finance specific ensuing operations, which money must be accounted for in connection with such operations or returned. Particularly it is a payment by the employer to a contractor in anticipation of, and for the purpose of, performance under a contract or contracts.

When the employer receives a guarantee as specified in the contract and a total advance payment is stated in the Appendix to Tender he shall make an advance payment, as an interest-free loan for mobilisation. The total advance payment, the number and timing of installments (if more than one), and the applicable currencies and proportions, shall be as stated in the Appendix to Tender.

The engineer shall issue an interim payment certificate for the first instalment after receiving a statement as specified in the contract and after the employer receives (i) the performance security as specified in the contract and (ii) a guarantee in amounts and currencies equal to the advance payment. This guarantee shall be issued by an entity and from within a country (or other jurisdiction) approved by the employer, and shall be in the form annexed to the particular conditions or in another form approved by the employer.

The contractor shall ensure that the guarantee is valid and enforceable until the advance payment has been repaid, but its amount may be progressively reduced by the amount repaid by the contractor as indicated in the payment certificates. If the terms of the guarantee specify its expiry date, and the advance payment has not been repaid by the date 28 days prior to the expiry date, the contractor shall extend the validity of the guarantee until the advance payment has been repaid.

Unless stated otherwise in the Appendix to Tender, the advance payment shall be repaid through percentage deductions from the interim payments determined by the engineer in accordance with contract, as follows:

(a) deductions shall commence in the next interim payment certificate following that in which the total of all certified interim payments (excluding the advance payment and deductions and repayments of retention) exceeds 10 percent of the accepted contract amount less provisional sums; and

(b) deductions shall be made at the amortisation rate of one quarter (25%) of the amount of each payment certificate (excluding the advance payment and deductions and repayments of retention) in the currencies and proportions of the advance payment, until such time as the advance payment has been repaid.

If the advance payment has not been repaid prior to the issue of the taking-over certificate for the works or prior to termination under clause 15, clause 16 or clause 19 (as the case may be), the whole of the balance then outstanding shall immediately become due and payable by the contractor to the employer.

Words, Phrases and Expressions 7.4

[1] certify *vt.* 签发证书,以证明……(declare (usually by giving a certificate) that one is certain of *sth.*, that *sth.* is true, correct, in order)
[2] final payment certificate 结算证书
[3] monthly statements 月报表
[4] applying for payment due to him 申请到期应支付给他的款项
[5] Interim valuations 期中估价
[6] outstanding balance of the Retention Money 保留金未支付的剩余部分
[7] amortisation rate 摊提比例

Notes 7.4

疑难词语解释

[1] showing in detail the amounts to which the contractor considers himself to be entitled
【译文】 详细列明承包商认为自己有权得到的各款数额。
【解释】 which 前面的 to 是 be entitled 要求的。
[2] The statement shall include the following items, as applicable, which shall be expressed in the various currencies in which the contract price is payable, in the sequence listed:
(f) any other addition or deduction which may have become due under the contract or otherwise; and
(g) the deduction of amounts certified in all previous payment certificates.
【译文】 该表应写明下列符合情况的款项,并分别按用以支付合同价的各种不同货币填报，其顺序如下：

(f) 根据本合同或其他理由所有其他可能到期应予增加或扣除的数额，以及

(g) 从所有以前签发了付款证书的数额中扣除的总额。

【解释】 the contract price is payable in various currencies = the contract price is to be paid for in various currencies. 至于 amounts certified in all previous certificate 若是直接翻译，应当是"用以前各种证书证明了的数额"，其中的 in 应理解为"用"，而不是"在"。当然，翻译成"列在以前各付款证书中的金额"也行。

[3] When the employer agrees to pay the contractor for the plant and materials (should be those of which substantial quantities are involved) specified, as agreed between the parties, in the Appendix to Tender and delivered by the Contractor on the Site for incorporation in the permanent works but not incorporated in such works and a percentage of the invoice value of specified materials is stated in the Appendix to Tender, the interim payment certificates shall include (i) an amount for plant and materials which have been sent to the site for incorporation in the permanent works, and (ii) a reduction when the contract value of such plant and materials is included as part of the permanent works.

【译文】 在业主同意为承包商运至现场准备用于工程但尚未使用的由双方商定列入标书附录中的设备与材料（应当是需要量大的材料）向承包商支付款项，并在标书附录中列入了上述设备与材料发票价值的百分比时，期中付款证书应当：(i) 加入已经运至现场准备用于永久工程的设备与材料款项，并 (ii) 在上述设备与材料的合同价值已经构成永久工程价值一部分时将其扣除。

【解释】 should be those of which substantial quantities are involved 中的 involve 是"需要"的意思。incorporation 是 incorporate 的名词，incorporate 在这一句里的意思是"将……变成……的组成部分"。

[4] on condition that the employer has received and approved the performance security, the engineer shall, within 28 days after receiving a statement and supporting documents from the contractor, issue to the employer an interim payment certificate which shall state the amount which the engineer fairly determines to be due, with supporting particulars.

【译文】 只要业主已收到并批准了履约保证，工程师就应在收到承包商一份报表及其证明文件后的28天内，向业主签发期中付款证书，说明工程师合理确定为到期应支付的数额，并以具体详情证明之。

【解释】 On condition that 的意思是：(只有) 在……条件下，条件是……，设若，如果。当 due 放在名词后面做后置定语时（后面常有 to 跟随），意思是：应付（给）的，到期的；例如，assess the payments due to the contractor（估计应当到期支付给承包商的款项）。注意，不要将其中的 due to 同短语（be) due to 混淆。当 due 用做表语时，意思是：(预定) 应到的，预定的，预期的；例如，any other addition or deduction which may have become due under the contract or otherwise

（根据本合同或其他理由预定要增加或扣除的任何其他数额）。当due放在名词前面做定语时，意思是：适当的，适宜的，相当的，正当的，应有的，应该的，当然的。例如，in due course (of time) 在适当的时候，及时（地），经相当时候，顺次；in due form 正式（地），照例，以适当［规定］的形式，in due time 在适当的时候（时机），时机一到。

[5] This period shall be deemed to commence on the date for payment specified in the contract, irrespective of the date on which any interim payment certificate is issued.

【译文】 应当认为，这一时期于合同规定的付款日期开始，与签发任何期中付款证书的日期无关。

【解释】 此句是被动语式，但在译成汉语时，最好译成无人称句。

[6] Unless otherwise stated in the contract, these financing charges shall be calculated at the annual rate of three percentage points above the discount rate of the central bank in the country of the currency of payment, and shall be paid in such currency.

【译文】 除非合同另有说明，否则，计算这些筹资费用的利率应当是工程所在国中央银行为支付货币颁布的贴现率再加三个百分点，并应用这种支付货币支付。

【解释】 discount rate 应译成贴现率，不应译成折现率。而项目财务评价在计算项目现金流的财务净现值时，应译成折现率，不应译成贴现率。

[7] The contractor shall be entitled to this payment without formal notice or certification, and without prejudice to any other right or remedy.

【译文】 承包商有权取得这笔款项，不需正式通知或付款证书，也不损害任何其他权利或补救权利。

【解释】 注意 prejudice 的含义和短语 without prejudice to 的用法。

[8] Where the contract is subject to a price fluctuation clause, allowance will be needed to cover changes in the cost of labour and materials, usually calculated on the basis of the price adjustment formula.

【译文】 当本合同有价格调整条款时，要将劳动力和材料费用的增减考虑在内，一般是根据价格调整公式计算。

【解释】 若将 is subject to 直译，则是：当本合同受某个价格调整条款制约时。若将 allowance will be needed to cover changes in the cost 直译，则是：需要留出预备费用，以便应付费用的增减。

[9] the indices covering the principal labour and material items

【译文】 适用于基本人工和材料事项的指数。

【解释】 cover 在这里的意思是：适用于，有关，对付，处理（deal with）。例如，Do the rules cover all possible cases? （这些规则适用于所有可能的情况吗？）/A new section introduced in CESMM3 to cover small items of building work, such as pumping stations and other ancillary buildings. （在CESMM3中新增了一节，用于小型建筑工程，例如泵站和其他辅助建筑物）。

[10] The engineer shall determine and certify each addition if the following conditions are

satisfied:

【译文】 如果满足下列条件，则工程师应确定每一笔增加额并为其签发证书。

7.5 Adjustments of contract price

7.5.1 Adjustments for changes in legislation

Majority of the standard forms of contract provide clauses protecting the parties from the consequences of changes in legislation made after a date (referred to as Base Date in the *Red Book*) specified in the contract in the country where the works is carried out. No protection is provided in respect of changes in the laws of other countries.

It may be necessary to obtain legal advice as to whether a particular event constitutes a "change:
(a) in the laws of the country (including the introduction of new laws and the repeal or modification of existing laws) or
(b) in the judicial or official governmental interpretation of such laws".

The contract price, under the protective clauses, is to be adjusted to take account of any such increase or decrease in cost resulting from a change in the laws of the country (including the introduction of new laws and the repeal or modification of existing laws) or in the judicial or official governmental interpretation of such laws, made after the Base Date, which affect the contractor in the performance of obligations under the contract.

If the change delays or will delay the contractor and/or increases or will increase his cost, he is required to give notice under the clauses or sub-clauses. The notice should be given within a specified time after he became aware of the circumstances, in accordance with the contract. If the change decreases the contractor's cost and the employer considers himself to be entitled to a reduction in the contract price, the employer is required to give notice under the contract. Notices should be given "as soon as practicable".

The *Red Book* requires the contractor to send a copy of his notices to the employer.

The protective sub-clauses describe the contractor's entitlements in terms which are used elsewhere in the general conditions.

Under sub-clause 13.7(b), the final contract price is to include the "additional cost as a result of these changes in the laws or in such interpretations". As with other provisions which describe the contractor's entitlements, such additional cost must only be included once, namely without duplication. In this case, it would be inappropriate for the final contract price to include a specific element of additional cost under sub-clause 13.7(b), if and to the extent that such additional cost has already been included in the adjustments for changes in cost.

After receiving this notice, the engineer shall proceed to agree or fairly determine these matters with due consultation with the parties, having regard to all the circumstances.

7.5.2 Adjustments for changes in cost

With regard to a building or civil engineering contract, fluctuations may be defined as adjusting the contract price for subsequent changes in cost during the currency of the contract.

The contract price fluctuations clause is optional and is normally included in contracts of over two years duration. It consists of a cost indices system of variation of price which superseded the former laborious procedure of calculating the price fluctuations from wage sheets and invoices.

The amount to be added to, or deducted from, the contract price is the net amount of the increase or decrease in cost to the contractor in carrying out the works.

Sub-clause 13.8 of the *Red Book* provides formulae to adjust the contract values to reflect escalation of costs due to inflation, which is intended to place greater risk on the contractor.

The formulae require data which is to be specified in a table of adjustment data for each payment currency, the tables being included in the Appendix to Tender. Table 7.3 is a sample table for this purpose and it illustrates the format of each table of adjustment data. In sub-clause 13.8, "table of adjustment data" means the completed table of adjustment data for local and foreign currencies included in the contract. If there is no such table of adjustment data, this sub-clause shall not apply.

Table 7.3 Table of Adjustment Data when sub-clause 13.8 applies

Coefficient: scope of index	Country of origin; currency of index	Source of index; Title/definition	Value on stated date(s)*	
			Value	Date
a=0.10 (fixed)				
b= labour				
c=				
d=				
e=				

* These values and dates confirm the definition of each index, but not define Base Date indices.

It must define the coefficients (proportions) and cost indices (reference prices) which are to be used to adjust the other amounts included in each currency in each payment certificate. Typically, the employer will have defined the fixed (non-adjustable) coefficient "a" before the tender documents are issued to tenderers, but may prefer each tenderer to define the other coefficients and all the sources of the cost indices in the table for each currency, so that they can fairly reflect: the proportions of cost (for example, different tenderers may anticipate different percentages for labour and equipment), and the sources of the cost indices (each of which should relate to the currency of cost, which may also differ between tenderers).

The total of the proportions ($b + c + d + \cdots etc.$) in each table of adjustment data must be checked mathematically to ensure that it does not exceed $(1 - a)$.

For each index, the source and title/definition should be stated in each table. Typically, tenderers will not know the value as at the base date, unless (unusually) its value is published immediately before the tender submission date. Therefore, the fourth column of the table may be used to define the value of the index at another recent date, which is then inserted in the fifth column. This "recent date" does not become a substitute Base Date, and is only used as a reference linked to the index's value which is stated in the fourth column. Knowing that the index had a certain value on a certain date, the engineer should be able to examine the published "source of index" and determine the index which has the published "title/definition" stated in the table and had such value on such date. This purpose is clarified in the fifth from last paragraph of the sub-clause 13.8, which follows immediately after the definitions of the expressions used in the formula.

For each index, the "country of origin" and "currency of index" should be stated in each table. Typically, this country will be that of the "currency of index" but may not be the same as the particular currency of payment to which the table relates.

For example, the contractor may incur costs in many currencies but only be paid in one foreign currency. Bearing in mind that each index reflects costs in a particular currency, it is necessary to convert indices from the "currency of index" to the "currency of payment" (unless these currencies are the same) in accordance with the fourth from last paragraph of sub-clause 13.8.

The adjustment multiplier P_n is "to be applied to the estimated contract value". Note that P_n will usually exceed 1 and is a multiplier. In order to calculate an amount to be added for changes in legislation and changes in cost in accordance with sub-clause 13.7 and sub-clause 13.8, then the mathematical expression $(P_n - 1)$ must be multiplied by the estimated "contract value" of the work carried out in period "n". This "contract value" is the value that the engineer, after due consultation with each party in an endeavour to reach agreement, agrees or fairly determines in accordance with the contract by evaluating each item of work, applying the measurement agreed or determined in accordance with the sub-clauses 12.1 and 12.2 and the appropriate rate or price for the item, taking due regard of all relevant circumstances and is subject to adjustments in accordance with the contract.

When there occurs a change in cost, the amounts payable to the contractor shall be adjusted for rises or falls in the cost of labour, goods and other inputs to the works, by the addition or deduction of the amounts determined by the formulae prescribed in sub-clause 13.8. To the extent that full compensation for any rise or fall in costs is not covered by the provisions of this or other clauses, the accepted contract amount shall be deemed to have included amounts to cover the contingency of other rises and falls in costs.

The adjustment to be applied to the amount otherwise payable to the contractor, as valued in accordance with the appropriate schedule included in the contract and certified in payment certificates, shall be determined from formulae for each of the currencies in which the contract price is payable. The formulae shall be of the following general type:

$$P_n = a + b L_n/L_0 + c E_n/E_0 + d M_n/M_0 + etc.$$

where P_n is the adjustment multiplier to be applied to the estimated contract value in the relevant currency of the work carried out in period "n", this period being a month unless otherwise stated in the Appendix to Tender.

"a" is a fixed coefficient, as stated in Table 7.3, representing the non-adjustable portion in contractual payments; "b", "c", "d", etc. are coefficients representing the estimated proportion of each cost element related to the execution of the works, as stated in Table 7.3; such tabulated cost elements may be indicative of resources such as labour, equipment and materials.

"L_n", "E_n", "M_n", ... are the current cost indices or reference prices for period "n", expressed in the relevant currency of payment, each of which is applicable to the relevant tabulated cost element on the date 49 days prior to the last day of the period (to which the particular payment certificate relates); and "L_0", "E_0", "M_0", ... are the base cost indices or reference prices, expressed in the relevant currency of payment, each of which is applicable to the relevant tabulated cost element on the Base Date.

The index figures are available from various sources. For example, those compiled by the Department of the Environment in the U.K. comprise:
(1) the index of the cost of labour in civil engineering construction;
(2) the index of the cost of providing and maintaining constructional plant and equipment; and
(3) the indices of constructional materials prices in respect of aggregates, bricks and clay products generally, cements, cast iron products, coated roadstone for road pavements and bituminous products generally, fuel for plant to which the DERV index will be applied, fuel for plant to which the gas oil index will be applied, timber generally, reinforcement (cut, bent, and delivered), other metal sections, fabricated structural steel, and labour and supervision in fabricating and erecting steelwork.

If their source is in doubt, it shall be determined by the engineer. For this purpose, reference shall be made to the values of the indices at stated dates (quoted in the fourth and fifth columns respectively of the table) for the purposes of clarification of the source; although these dates (and thus these values) may not correspond to the base cost indices.

The set of coefficients as set out in Table 7.4 and the cost indices as stated in Table 7.5 were used in a hydropower station works in China during 1990s.

Table 7.4 Used in a Hydropower Station Works in China[34]

Coefficient: scope of index	Country of origin; currency of index	Source of index; Title/definition	Value on stated date(s)*	
			Value	Date
a=0.15				
b=0.25 foreign labour	Germany	Germany		
c=0.29 Contractor's Equipment	Germany; Mark	Germany		
d=0.05 Contractor's Equipment	United States; Mark	United States		
e=0.04 Marine transportation	United States; Mark	United States		
f=0.02 Section steel	Germany; Mark	Germany		
g=0.16 Miscellaneous expenses	Germany; Mark	Germany		
h=0.04 Miscellaneous expenses	United States; Mark	United States		

Based on the adjustment data stated in the two tables, the formulae shall be as follows:

$P_n = 0.15 + 0.25 EL/EL_0 + 0.29 EQ_1/EQ_{10} + 0.05 EQ_2/EQ_{20} + 0.04 MT/MT_0 + 0.02 ST/ST_0 + 0.16 MI_1/MI_{10} + 0.04 MI_2/MI_{20}$

In cases where the "currency of index" (stated in Table 7.3) is not the relevant currency of payment, each index shall be converted into the relevant currency of payment at the selling rate, established by the central bank of the country, of this relevant currency on the above date for which the index is required to be applicable.

Table 7.5 Cost Indices Used in a Hydropower Station Works in China[34]

Title of index	Current cost indices	Base cost indices
b=0.25/foreign labour	EL = 175	EL_0 = 147.5
c=0.29/Contractor's Equipment	EQ_1 = 132	EQ_{10} = 125.9
d=0.05/Contractor's Equipment	EQ_2 = 140.5	EQ_{20} = 151.3
e=0.04/Marine transportation	MT = 107.5	MT_0 = 116.0
f=0.02/Section steel	ST = 74.5	ST_0 = 69.3
g=0.16/Miscelaneous expenses	MI_1 = 141.5	MI_{10} = 120.2
h=0.04/Miscelaneous expenses	MI_2 = 149.6	MI_{20} = 144.4

Until such time as each current cost index is available, the A/E shall determine a provisional index for the issue of interim payment certificates. When a current cost index is available, the adjustment shall be recalculated accordingly.

If the contractor fails to complete the works within the time for completion, adjustment of prices thereafter shall be made using either (i) each index or price applicable on the date 49 days prior to the expiry of the time for completion of the works, or (ii) the current index or price: whichever is more favourable to the employer.

The weightings (coefficients) for each of the factors of cost stated in the table(s) of adjustment data shall only be adjusted if they have been rendered unreasonable, unbalanced or inapplicable, as a result of variations.

Words, Phrases and Expressions 7.5

currency of index 编制指数用的货币

7.6 Delay damages

7.6.1 Definition of delay damages

The liquidated damages are the damages that had been specified in most standard forms of contract before the *Red Book* was published in 1999 and is an amount of money payable as compensation should there be a breach.

The word *liquidated* means that the principle to pay money as compensation for a breach of contract has been established, as opposed to '*unliquidated*'. For example, in the majority of building or civil engineering contracts, there is a provision to the effect that should the contractor fail to complete the works during the specified contract period the employer is entitled to be paid an agreed sum of money by the contractor as compensation. The employer need only prove that there has been a breach in order to automatically claim the sum. Liquidated damages are pre-determined by the parties themselves and should be a genuine pre-estimate of possible loss. The court normally awards the amount of liquidated damages so agreed. Sometimes, however, the amount of damages payable on breach is not merely an agreed and reasonable compensation, but is more in the nature of a *penalty*. Where a minor breach occurs and a heavy payment is made by way of compensation, there is obviously injustice. Certain principles have been laid down in the giving of relief, where the sum specified has been inserted in the contract as a frightener to ensure performance. The court hearing the appeal of the party in breach will base its decision on the following principles:

(1) the sum agreed must treated as a penalty, if it is extravagant and unreasonable in amount, by comparison with the greatest loss that can follow from a breach of the contract;

(2) where the payment of a smaller sum is secured by a larger sum, the latter is a penalty; and/or

(3) when a single lump sum is payable by way of compensation, on the occurrence of one or more, or all of several events, some of which may cause serious or others not so serious damage, there is a presumption that the sum inserted is a penalty.

The relief afforded by the court where a penalty is found to exist, is to excuse payment of that amount. The court has the power to substitute its own award of damages, computed on the basis of compensation for the loss sustained. On the other hand, where an agreed sum is in the nature of liquidated damages, no greater sum will be awarded by the court, even if it is proved that the consequences of breach have been more serious than the parties had foreseen.

If the damages are not specified in the contract, then they are referred to as unliquidated damages and an arbitrator or the court would be responsible for assessing their value as compensation for actual loss suffered subject to the principle of remoteness. They may be in the form of:

(a) substantial damages which put the claimant in the same situation as he would have been if the contract had been performed; or

(b) nominal damages which apply if no loss is suffered and represent only a token sum awarded for the infringement of a contractual right.

Apart from the remedies for breach of contract, it is worth mentioning exemplary or punitive damages which are awarded in certain tort cases. The purpose of awarding this type of damages is not to compensate the plaintiff, nor even to strip the defendant of his profit, but to express the court's disapproval of the defendant's conduct, for example, where he has deliberately committed a wrong (such as defamation) with a view to profit.

Damages may also be classified as general damages which are payable as compensation for the loss presumed to follow from a breach of contract or from a tort; and special damages which are payable as compensation for particular losses not presumed, but which in fact have followed in a particular case.

The third and fourth editions of the *Red Book* used a more definite term 'liquidated damages for delay' rather than simply a 'liquidated damages'.

The term 'liquidated damages for delay' was substituted with 'delay damages' in the *Red Book* when it was published in 1999, maybe for easy understanding by the users who are not familiar with it.

Delay damages is defined in the *Red Book* as the sum to be paid by the contractor for every day which shall elapse between the relevant time for completion and the date stated in the taking-over certificate if he fails to complete the whole of the works, and each section (if any), within the time for completion for the works or section (as the case may be), including other items specified in the contract.

The sum is usually inserted in Appendix to Tender and will be the amount payable for each day, week or other prescribed period for which completion is delayed. A sum inserted for delay damages to be enforceable must be a genuine pre-estimate of damages, and should represent the likely financial loss or cost incurred by the employer if delay occurs. If the amount is not a genuine pre-estimate of damage it could be held by the courts to be a penalty. In these circumstances, the employer can only recover his actual loss and not the amount of the penalty. Whether the sum is a penalty or is delay damages will be largely influenced by the terms and inherent circumstances. In many building or civil engineering contracts the delay damages are related to those included in previous contracts of a similar nature.

The precedent set by the courts for a valid assessment of damages was summarized as follows:[13]

(1) If the parties made a genuine attempt to pre-estimate the loss likely to be suffered, the sum stated will be liquidated damages and not a penalty, irrespective of actual loss.

(2) The sum will be a penalty if the amount is extravagant having regard to the greatest possible loss that could be caused by the breach.

Why so many words have been said about the distinction between the terms 'penalty' and 'liquidated damages' can be explained by the fact that under English law, a penalty is subject to the rules of equity.

Since equity regards penalties as inequitable, it will not therefore be bound by a description, such as 'liquidated damages', if in truth the sum specified is a penalty.

The following approach provides a reasonable basis for the assessment of delay damages:

(1) Loss of interest on the cost of the contract works, based on the assumption that 80 per cent of the money is paid at theoretical completion and that annual interest payable is 16 per cent. This amounts to 0.2 per cent of the contract price per week but will need monitoring to take account of significant fluctuations in interest rates. [15]

(2) Professional fees of the A/E and site staff, calculated as a percentage of the contract sum.

(3) Further costs, such as costs of a temporary nature awaiting the completion of the project, such as temporary housing, and additional costs of the Employer which are best assessed on an ad hoc basis.

(4) Fluctuations, where applicable, normally calculated by reference to the index numbers relating to the valuation period.

It will be more difficult to assess the rate of delay damages for work that on completion will not be a source of direct gain to the employer, such as the construction of a motorway. The delay in completion of a sea wall by three months may only result in loss of amenity, but the situation would be entirely different if the contract extended into the winter and an exceptionally high tide occurred during the period of delay and caused extensive damage to adjoining property.

Although the estimate of delay damages is not expected to be precise, it must be reasonable. Thus the inclusion of damages of $ 30 000 per week in respect of a pumping station costing $ 912 000 to construct would be considered grossly excessive.

Substantial completion of the contract will normally release the contractor from his obligation to pay delay damages, particularly where the employer has taken possession of the whole or part of the structure. An employer could not claim delay damages in respect of non-completion of a power station, merely because an ancillary building, which would not prevent the power station being operated, was incomplete. On the other hand, where statutory consents are required, the project is incomplete and cannot be used, and the damages for delay are payable until the necessary consents have been obtained.

The advantages of delay damages can be listed as follows: [15]
(1) they do not have to be proved;
(2) they are agreed between the parties in that they are known to the contractor at the time of tendering and he can allow for them in his tender total; and
(3) the employer can simply deduct them without having to issue a writ through the courts.

The total amount due, under sub-clause 8.7 of the *Red Book*, shall not exceed the maximum amount of delay damages (if any) stated in the contract.

These delay damages shall be the only damages due from the contractor for such default, other than in the event of termination under sub-clause 15.2 prior to completion of the works. These damages shall not relieve the contractor from his obligation to complete the works, or from any other duties, obligations or responsibilities which he may have under the contract.

7.6.2 Correspondence regarding delay damages

The A/E may request the employer's instructions regarding liquidated damages by writing him a letter like the one shown in Figure 7.4.

When the employer has decided to recover the delay damages as deductions from the payments due to the contractor, the A/E may write the employer a letter like the one shown in Figure 7.5 if the contract uses the form of *Red Book*.

A/E's address
29 October 2005

Employer's address
Dear Mr. Wang,

Re: Delay damages due from the contractor

I enclose my certificate in accordance with sub-Clause 8.7 of the General Conditions of Contract.

You may now take steps to recover delay damages at the rate stated in the Appendix to Tender or I can include the appropriate amount as a deduction in the Contract Price and from the next interim payment certificate under sub-Clause 2.5, due to be issued on or about 15 November 2005.

If you decide that the latter course is most convenient, I should be pleased to receive your written instructions as soon as possible.

Yours sincerely,

Fang Jianguo
Architect's Representative

Figure 7.4   Letter Requesting Employer's Instructions
Regarding Delay Damages[35]

Under sub-clause 2.5, the employer shall give notice to the contractor as soon as practicable after he became aware, or should have become aware, of the event or circumstances that entitle him to make any deduction from an amount certified in a payment certificate. Figure 7.6 is a letter from the employer to the contractor when he deducts delay damages in a payment certificate.

3 November 2005

Employer's address
Dear Mr. Wang,

Re: Deduction from payment due to the contractor as delay damages

Thank you for your letter of 1 November 2005 regarding the deduction of delay damages.

You are entitled to deduct delay damages at the rate of 1.2% of the final Contract Amount per day for full day, in Hong Kong dollar and 30% in which the Contract Price is payable between 25 September 2005 at which the contractor should have completed and 20 October 2005 on which the works were completed and the site cleared, provided that you have notified the contractor that, in your opinion, he is not entitled to any or any further extension of time.

Yours sincerely,

Fang Jianguo
Architect's Representative

Figure 7.5   Letter Regarding Deduction from Payment
due to the Contractor as Delay Damages[35]

Employer's address
9 November 2005

The contractor's address
RECORDED DELIVERY

Dear Mr. Liu,

Re: Deduction from payment certificate as delay damages

I enclose my cheque in the sum of $ 568,000 as payment due under certificate number 050817 of 30 October 2005. Delay damages have been deducted in accordance with sub-Clause 8.7 and Sub-Clause 2.5 of the contract. They have been calculated as follows:

Commencement Date: 12 October 2004
Time for Completion: 360 days

Practical completion date stated in the Taking-Over Certificate for the Works: 12 October 2005

The delay damages due to us = 7 days @ $ 100,000 per day = $ 700,000

Yours sincerely,

Wang Dagui

Employer's Representative

Figure 7.6　Letter Regarding Deduction from Payment Certificate[35]

Words, Phrases and Expressions 7.6

［1］　liquidated damages 先约赔偿费
［2］　liquidated damages for delay 误期赔偿费
［3］　delay damages 误期赔偿费
［4］　as opposed to 'unliquidated' 与'unliquidated'不同,与'unliquidated'相反
［5］　to the effect 意思是……（with the meaning＜issued a statement to the effect that he would resign 发表了声明,意思是他将辞职＞）
［6］　relief n. 法律上的补救、补偿（legal remedy or redress）
［7］　excuse vt. 使（某人）免除（责任、规定、惩罚等）（set（sb.）free（from a duty, requirement, punishment, etc.））

Notes 7.6

疑难词语解释

［1］　The relief afforded by the court where a penalty is found to exist, is to excuse payment of that amount. The court has the power to substitute its own award of damages, computed on the basis of compensation for the loss sustained. On the other hand, where an agreed sum is in the nature of liquidated damages, no greater sum will be awarded by the court, even if it is proved that the consequences of breach have been more serious than the parties had foreseen.

【译文】　发现有罚金时,法庭给予的救济是不支付这一款项的。法庭有权以自己裁定的损失赔偿费代替之,数额根据对所受损失的补偿额计算。另一方面,当双方商定的数额性质属于损失赔偿费之时,即使事实表明违反合同的后果已经超过合同双方的预先估计时,法庭裁决的数额也不会超过这种损失赔偿费。

【解释】computed 限定逗号前面的 damages,属于非限定性后置定语。in the nature 是状语,修饰系表结构 an agreed sum is of liquidated damages。it is proved that 属于主语从句。

［2］　Delay damages is defined in the *Red Book* as the sum to be paid by the contractor for every day which shall elapse between the relevant time for completion and the date stated in the taking-over certificate if he fails to complete the whole of the works, and each section (if any), within the time for completion for the works or section (as the case may be), including other items specified in the contract.

【译文】 误期赔偿费在红皮书中的定义是:如果承包商未能在整个工程或某单项工程(具体视情况而定)的竣工时间内完成整个工程,以及每一可能的单项工程,包括合同事先规定的其他工作,则应由承包商为有关竣工时间与接收证书载明日期之间的每一天支付的数额。

【解释】 elapse 不必译出。"每一可能的单项工程"是从 each section (if any) 中的 if any 译出的。

[3] The sum is usually inserted in Appendix to Tender and will be the amount payable for each day, week or other prescribed period for which completion is delayed. A sum inserted for delay damages to be enforceable must be a genuine pre-estimate of damages, and should represent the likely financial loss or cost incurred by the employer if delay occurs. If the amount is not a genuine pre-estimate of damage it could be held by the courts to be a penalty. In these circumstances, the employer can only recover his actual loss and not the amount of the penalty. Whether the sum is a penalty or is delay damages will be largely influenced by the terms and inherent circumstances. In many building or civil engineering contracts the delay damages are related to those included in previous contracts of a similar nature.

【译文】 这一数额通常填写在标书附录之中,且应是按天、周或其他规定的竣工延误时间单位支付的数额。填写的误期赔偿费数额要能够实现,就必须事先认真估算,体现业主在实际发生延误时的可能损失或开销的大小。如果这一数额未经过事先认真估算,则法庭会将其视为一种罚金。无论该数额是误期赔偿费还是罚金,都在很大程度上取决于合同条款和具体情况。许多建筑或土木工程合同的误期赔偿费都同以前类似合同的误期赔偿费有关。

[4] The precedent set by the courts for a valid assessment of damages was summarized as follows:

(1) If the parties made a genuine attempt to pre-estimate the loss likely to be suffered, the sum stated will be liquidated damages and not a penalty, irrespective of actual loss.

(2) The sum will be a penalty if the amount is extravagant having regard to the greatest possible loss that could be caused by the breach.

【译文】 有人曾将法庭为正确判断误期赔偿费而确立的先例归纳如下:(1) 如果合同双方的确事先认真努力估计可能蒙受的损失,则不管实际损失有多大,合同载明的这一数额应属于误期赔偿费而非罚金。(2) 如果这一数额同违约所造成的最大可能损失相比大得过分,则该数额应视为非罚金。

[5] The following approach provides a reasonable basis for the assessment of delay damages:

(1) Loss of interest on the cost of the contract works, based on the assumption that

80 per cent of the monies is paid at theoretical completion and that annual interest payable is 16 per cent. This amounts to 0.2 per cent of the contract price per week but will need monitoring to take account of significant fluctuations in interest rates.

(2) Professional fees of the A/E and site staff, calculated as a percentage of the contract sum.

(3) Further costs, such as costs of a temporary nature awaiting the completion of the project, such as temporary housing, and additional costs of the Employer which are best assessed on an *ad hoc* basis.

(4) Fluctuations, where applicable, normally calculated by reference to the index numbers relating to the valuation period.

【译文】下面的办法是估算误期赔偿费的合理出发点：(1) 合同工程费用额的利息损失，计算时假设80%的金额是在理论上竣工时支付，且每年应支付的利息是16%，这样一来，利息损失额就是每周合同价的0.2%，但是需要随时监视利率是否有显著的变动，以便将其考虑在内。(2) 建筑师/工程师及其现场人员的酬金，按合同额的某一百分比计算。(3) 其他费用，例如临时住房等在项目竣工之前的临时性费用，以及业主增加的费用。后者最好根据具体情况专门计算。(4) 当物价改变时，一般根据估价时期的指数计算物价的变动数额。

[6] It will be more difficult to assess the rate of delay damages for work that on completion will not be a source of direct gain to the employer, such as the construction of a motorway. The delay in completion of a sea wall by three months may only result in loss of amenity, but the situation would be entirely different if the contract extended into the winter and an exceptionally high tide occurred during the period of delay and caused extensive damage to adjoining property.

Although the estimate of delay damages is not expected to be precise, it must be reasonable. Thus the inclusion of damages of $30 000 per week in respect of a pumping station costing $912 000 to construct would be considered grossly excessive.

【译文】对于竣工时不会直接为业主带来收益的工程，例如修筑高速公路，延误的单位时间损失赔偿费更难估算。某防波堤延误三个月可能仅仅损失些许便利，然而，如果工程拖延到冬季，并且在延误期间遇上罕见的涨潮，给附近的财产造成巨大的损失，则情况就会迥然不同。虽然估算误时损失赔偿费很难精确，但必须合理。如此说来，费用为91万2千元的泵站若在合同中规定延误一个星期赔偿3万元就太过分了。

[7] Substantial completion of the contract will normally release the contractor from his obligation to pay delay damages, particularly where the employer has taken possession of the whole or part of the structure. An employer could not claim delay damages in respect of non-completion of a power station, merely because an ancillary building, which would not prevent the power station being operated, was incomplete. On the

other hand, where statutory consents are required, the project is incomplete and cannot be used, and the damages for delay are payable until the necessary consents have been obtained.

【译文】 工程的基本竣工常常会解除承包商支付误期赔偿费的义务，特别是在业主已经接收了整个或部分工程的时候。发电站的业主不能仅仅因为某个不妨碍发电站运行的辅助建筑物未完成而声称发电站未完成，并就此要求误期赔偿费。另一方面，在必须要由当局同意时，若项目因未完成而不能使用，就必须取得当局的同意，否则就必须支付误期赔偿费。

[8] there is a provision to the effect that should the contractor fail to complete the works during the specified contract period the employer is entitled to be paid an agreed sum of money by the contractor as compensation.

【译文】 有一条规定，意思是：如果承包商未能在规定的合同期内完成工程，业主就可以由承包商支付一笔事先商定的款项，给予补偿。

Chapter 8　Contractors' Claims and Their Settlement

8.1　Introduction

'Claim' is a term frequently talked about in construction industry. However, there is still some misunderstanding of it in China, due to the lack of experience with them on both sides of the employers and the contractors. For example, some contractors still believe, of course incorrectly, that their tenders can be priced low deliberately so as to win a contract hoping to make good their losses by way of subsequent claims. The belief has rarely proved true. Their experience has not shown that it is a reliable way of doing business. As a matter of fact the chance is unpredictable unless the employers have proved consistently blind to the incompetence of architects or engineers appointed by them. On the contrary, it has been found that the employers in China financed by the international financial institutions, such as the World Bank and Asian Development Bank, and the private investors from developed parts of the world and their architects or engineers are much more competent in contract administration than their Chinese contractors, which has made it difficult for the local Chinese contractors to succeed in making claims. On the other hand, many of the Chinese contractors working on the public works, even some property developments have been made humble by the sharp competition and tried their best to keep on good terms with the employers. As a result, the Chinese contractors usually do not claim for their entitlements under the contract.

Most of those involved in the construction industry know from their own experience that claims occur and divert considerable resources in terms of human resources at the expense of ongoing construction. The claims cause budgetary difficulties and financial embarrassment to the employers and financial difficulties, restriction of cash flow, losing of financial liquidity or worse to the contractors and hence the subcontractors.

It has been realized that the problem of claims is probably the most difficult and controversial matter affecting relations in the construction industry and it would not only be in the interests of both the employers and the contractors but also of the society as a whole if the impact of claims were substantially reduced.

There were sayings in the United Kingdom in 1980s that claims are unavoidable and the cause cannot be diagnosed, much less treated and that "more than a quarter of the money paid by employers to contractors is based on negotiations which take place after the contract is signed." If claims represent a situation which is unacceptable and avoidable, then something should be done about it. If they are unavoidable, then they should be at least dealt with efficiently and expeditiously and the cost in terms of both money and human resources minimized.

Contracting is a high risk business and therefore the contractor at the time of tendering has the right to be able to plan and expect to proceed with his work in an orderly manner. The employer, on the other hand, has the right to expect experience and competence from the contractor. The contractor should not attempt to make the employer pay for his mistakes and the latter must not expect the former to bear the cost of errors or changes made by him or the design team. A contractor, when pricing a tender, is asked to take risks. When the contract is let and underway, and the employer or his agents prevent the contractor from carrying out his contractual obligations properly and effectively, then he will probably find it necessary to make a claim.

The only way in which a contract can be successful is that, in addition to proper documentation and tendering, administration of the contract during construction must be properly observed by the parties to it.

Types, initiation, preparation and assessment of any procedures for contractors' claims, the relevant conditions of contract and their application, additional cost for loss or expense, disruption of work resulting from variations and the operation of delay damages are briefly described in this chapter.

Words, Phrases and Expressions 8.1

[1] claim *n*. 索赔，索补
[2] initiate *vt*. 开始，着手，发起，（主动）提出（begin, set (a scheme, etc.) working）

8.2 Definition and types of claims

8.2.1 Definition of a claim

8.2.1.1 Definition

Many contractors in China, even today, consider it is an "*infra dignitatem*" or even sharp practice to submit a claim, but this should not be, because the word "claim" is defined as:

"A demand for something due or believed to be due" (*Merriam Webster's Collegiate Dictionary, Tenth Edition*);

"Demand for something due, an assertion of a right to something" (*Oxford English Dictionary*).

The main point in the definitions is the words 'right' and 'due'. The word 'claim' therefore should be used to denote a request by a contractor for some additional payment arising out of carrying out new work, remedying defects in the works or providing a service under contract, whether the value of the work is large or small. In fact any work carried out by means of a contract can give rise to claims.

The construction activities are complex involving numerous trades and the conditions in which they take place vary considerably from site to site. Climate and market conditions, characteristics of the works and availability of resources are a few of the variables, each of which can have a significant effect on the operation of the contract.

Provision is made in most construction contracts for these complexities and uncertainties by means of clauses permitting the contractor to claim for loss or expense resulting from specific contingencies. Attempts have been in the *Red Book* to clarify the contractual requirements and remove any ambiguities as far as possible. In the absence of these provisions, contractors would have to include in their tenders for many more uncertainties than they do now, which would cause tender prices to increase significantly. However, under the standard form of contract, the employer will only have to meet the cost of such contingencies if they arise and have been duly verified.

The term "claim" as used in this context is a request by the contractor for recompense for some loss or expense that he has suffered, or an attempt to avoid the requirement to pay delay damages. It is in this light that claims should be viewed seriously by both the employers and the contractors. Unscrupulous claims by contractors to redress the effects of inefficiency or profit shortfall are unlikely to receive sympathetic consideration by the A/E or the employer. It has been justifiably argued that the term "claim" should be used only in respect of fundamental breaches of the contract and that the remainder is contractual entitlements.[15]

A claim to be successful must be well prepared, based on the appropriate contract clauses and founded on facts that are clearly recorded, presented and provable. It follows, therefore, that the conditions of contract and the other contract documents used will need to be examined when deciding whether or not a claim is permissible. Any ambiguity or conflict within the contract documents will make it more difficult to decide if a claim is permissible.

Under the sub-clause 17.5 of the *Red Book* "claim" is assigned a particular meaning: a claim (or proceedings pursuing a claim) alleging an infringement which in this sub-clause means an infringement (or alleged infringement) of any patent, registered design, copyright, trade mark, trade name, trade secret or other intellectual or industrial property right relating to the works.

Discussion of intellectual and industrial property rights is obviously beyond the scope of this book.

8.2.1.2 Contractor's action

The contractor must apply in writing for the issue of instructions, details, drawings, levels, or for the nomination of sub contractors appropriate. He must also give written notice to the A/E of any cause of delay in the progress of the work and written notification in respect of any claims that he is contemplating making in respect of variations or loss and/or expense. He must take positive steps to ensure that the A/E's instructions are issued in writing or oral instructions confirmed. Alternatively, the contractor may confirm in writing any oral instruction of A/E. If the latter has not contradicted confirmation in writing forthwith, it shall be deemed to be his written instruction as provided under sub-clause 3.3 of the *Red Book*. The contractor should ensure that the various certificates required under the contract be issued by the Engineer, particularly in respect of completion, defects liability and extensions of time, to prevent any unnecessary problems arising in the future. It is always a better policy to avoid disputes rather than being involved in their settlement. [15]

8.2.2 Types of claims

A point of view has been held by some that claims can be only in respect of matters for which specific provision is made in the conditions of contract or other contract documents. But there is no dispute over the fact that matters can and indeed arise in the course of many contracts for which no provision is made. For example, some employers offer or undertake to supply materials for use by the contractors and the materials have been found defective when delivered on the site. The employer will certainly be deemed to be liable. However, most standard conditions of contract do not stipulate what action shall be taken in such a case, but there are legal decisions in the common law countries in respect of such a matter in the other cases which may indicate the position.

On the other hand, there are those who argue that claims can be only in respect of matters not provided for in the contract. For example, a contractor is dissatisfied with an architect's valuation of varied work and gives notice to the architect requesting to agree and/or determine higher rates and/or prices. The architect, without having regard to the validity of the contractor's claim under the applicable law of the contract and attempting to brush it off, responds with a letter stating that the conditions of contract make no provision for such claims and therefore the matter should be left until the works has been completed. The contractor, however, followed with a letter as shown in Figure 8.1.

Contractor's address
21 October 2006

Architect's address

Dear Mr. Wang,

<u>Re: Your wrong decision of 19 October 2006 on our request of 12 October 2006</u>

We inform you that we cannot accept the decision as stated in your letter of 19 October 2006. We agree that there is no express provision in the contract covering our request. However, sub-clause 20.1 does provide that we can give you a notice when we consider ourselves to be entitled to additional payments under any clause of the conditions of contract or *otherwise* in connection with the contract. Attention shall be drawn to the word 'otherwise'. We mean by 'otherwise' the applicable law of our contract.

We are not prepared to postpone the matter until the works has been completed; the amount of money involved is large and the cost of financing is high, and we could not be able to keep the progress abreast with the programme if our claim has not been settled within three weeks.

We must formally notify you of our intention to claim the additional cost of financing in respect of the sum due to us until the new rates are agreed or determined under sub-clause 12.3 and payments are received.

Yours sincerely,

Fang Jianguo
Contractor's Representative

Figure 8.1   Letter to the Architect Asking to Agree New Rates of Varied work

This example shows that even though there is no express provision as to what will happen in some events the contractors still have rights to expressing his opinions and notifying his intentions.

A conclusion can be drawn that it does not matter what the nature of the claim may be or whether it is justified or not.

There are three main types of claim:[15]

(1) *Contractual claims*

These are claims that are founded on specific clauses within the terms of the contract. This type of claim will be considered in more detail later in this chapter, with particular reference to the *Red Book*.

(2) *Ex-contractual claims*

These claims are not based on clauses within the terms of a contract, although the basis of the claim may be circumstances that have arisen out of the contract and have resulted in loss or expense to the contractor. On occasions a sympathetic employer has settled an ex-contractual claim, because the contract was on time although the contractor suffered exceptional misfortune.

However, in general these claims are unlikely to succeed as there is no contractual obligation for payment and any payments made are in the nature of *ex gratia* payments (act of grace or out of kindness). Typical examples are where late deliveries of materials by a supplier on a firm price contract resulted in substantial price increases on the materials, or where difficulty was experienced by the contractor in recruiting adequate labour and he was obliged to pay high additional costs to attract them.

(3) *Common Law claims*

Many of the clauses in the standard forms of contract are stated to be without prejudice to any other rights and remedies. Such rights are established by taking action through the courts for damages for breach of contract, tort, repudiation, implied terms and other related matters.

The claims can be generated at any stage during the running of the contract, from the formation of contract throughout the cessation of the employer's liability towards the contractor as specified in the contract, on one hand. On the other hand, the claims may arise out of every matter covered by the provisions of the contract, such as contract documentation; commencement and progress; certificates and payments, delays and disruptions; completion and defects liability; determination (termination); subcontractors and suppliers, settlement of dispute, etc..

8.2.3　A/E's instructions and their possible consequences

A/E is authorized to issue instructions to the contractor in the majority of standard forms of contract which are prepared based upon his appointment by the employer as the contract administer. For example, sub-clause 8.1 of the *Red Book of Fourth Edition* provides that the engineer shall have authority to issue to the contractor, from time to time, such supplementary drawings and instructions as shall be necessary for the purpose of the proper and adequate execution and completion of the works and the remedying of any defects therein. The contractor shall carry out and be bound by the same.

As far as the *Red Book* is concerned the engineer has the similar authority to do so.

The instructions in most cases will cause consequences in terms of money or time, or both. The consequences must be taken into account when initiating and preparing a claim. For example, the following consequences may result from an A/E's instruction:[15]

(1) work in several areas will be disrupted;

(2) labour and contractor's equipment have to be transferred to other areas;

(3) new material has to be ordered, and replaced materials either scrapped or transported to another site;

(4) other related work is more costly;

(5) delay is caused and extension of time is necessary; and

(6) reimbursement of loss and expense is sought.

Figure 8.2 illustrates the contractor's notice to the architect his intention to claim for both extension of time and additional payment based on the item 3 '*A/E's instructions, variation orders and site instructions*' in Table 8.1 hereinafter(see 8.7.3).

Contractor's address
22 September 2006

Architect's address

Dear Mr. Luo,

Re: Additional cost of compliance with AI No. 32 and extension of time

We hereby give you notice under sub-clause 20.1 of the General Conditions of Contract that compliance with your AI No. 32 of 18 September 2006 received by us on 20 September 2006 is likely to cause delay to the progress of the Works. Compliance with this instruction forms an event giving rise to our claim as stated under sub-clause 20.1.

The particulars of the expected effects of compliance are that the additional work will take a further three weeks to complete and no other work will be possible while this is being executed. As you will see from the enclosed copy of our programme, on which we have indicated in red the work required by your instructions, we are at a critical stage. The intervention of this work will also affect the work of subsequent trades which will lead to still further delay in the execution of that work as we have indicated in blue on the enclosed programme.

In these circumstances we estimate that the likely delay to completion of the Works beyond the current Time for Completion will be five weeks. Please confirm that the particulars and estimate given above are sufficient to enable you to grant the appropriate extension of time as required by sub-clause 8.4 of the General Conditions of Contract. If you require any further particulars please specify their nature within the next seven days, failing which we will assume that the information we have given is sufficient.

Clearly, the work of all nominated subcontractors will also be affected and we are therefore sending a copy of this notice to each of them.

Yours sincerely,

Fang Jianguo
Contractor's Representative

Enclosure
Note: AI stands for Architect's Instruction

Figure 8.2 Notification of delay caused by an architect's instruction

Case study 8.1-Notice as a condition precedent to entitlement[36]

Some contracts, including the *Red Book*, require the contractor to give notice to the engineer as a "condition precedent" to be entitled to an extension of time. This means that if the contractor does not give a notice, he is not entitled to this. But, in the United Kingdom, because of an old rule in the common law system, this may in fact work against the employer; so that he is not able to claim delay damages.

As generally accepted in the common law system, if an employer causes delay to the works, then the contract completion date will no longer be binding on the contractor and the employer will lose his right to delay damages. The provisions for extension of time are needed in the contract to enable the employer (or his A/E) to set a new contract completion date and revive the right to delay damages. The underlying legal rationale is that the employer cannot insist on a contractual obligation being performed by the contractor, when the employer has himself prevented the contractor from completing on time. This is known as "the prevention principle". Therefore, any well drafted contract contains provisions for time to be extended if the employer causes delay.

Most clauses dealing with extension of time place an obligation on the contractor to give notice within a certain time and/or in a certain manner. This is good practice and causes no difficulty, except where the service of the notice is expressed to be a "condition precedent" i.e. if the contractor does not give notice correctly, then he will not be entitled to an extension of time.

Strictly speaking, however, that will not stop the prevention principle applying to make the contract completion date no longer binding (and the delay damages provisions unenforceable). So, the contractor is theoretically better on by not giving a notice. Thus apparently toughening up the extension of time clause in favour of the employer (by making notice a condition precedent) in fact works against the employer.

This problem is ameliorated when the A/E or the employer himself is able to extend the completion date irrespective of whether the contractor has given notice.

There have been two recent cases in Australia which have illustrated this issue.

The first is *Gaymark Investments v Walter Construction* (1999). Gaymark employed Walter Construction to build a hotel. Gaymark caused a delay to the work. Under the contract, Walter Construction had an obligation to give notice to Gaymark in order to obtain an extension to time, but failed to do so.

While a standard form of contract had been used various clauses had been amended. The result of the amendments was that Walter Construction's obligation to give notice was a condition precedent and no other party had the power to grant an extension of time if notice had not been given. The arbitrator held that Walter Construction was able to rely on the prevention principle. Gaymark appealed, but the Court dismissed the appeal.

Consequently, Gaymark lost the right to delay damages and Walter Construction was only obliged to complete the works within a reasonable time.

The second case is *Abigroup Contractors v Peninsula Bamain* (2002), where Peninsula employed Abigroup to reconstruct and refurbish factory buildings. Peninsula delayed the works. Under the contract, it was a condition precedent that Abigroup give notice, which it failed to do. The court referred the matter to a referee, who found that under the contract the superintendent had the power to grant an extension of time, whether or not Abigroup had given notice.

The referee also found that while the superintendent was the agent of Peninsula, the wording of the contract meant that the superintendent's power to extend time was unilateral (i. e. not for the sole benefit of the Peninsula, but of both parties) and so the superintendent could and should have exercised that power in Abigroup's favour. The court agreed.

The position is the same under English law. Although we are not sure that an English court would allow a contractor to benefit from not complying with its contractual obligations (like Walter Construction did).

Sub-clause 20.1 of the *Red Book* says "If the contractor fails to give notice of a claim within such period of 28 days, the Time for Completion shall not be extended······". This wording is potentially subject to the same problem as was illustrated in the Gaymark case, as the employer does not appear to have the right to grant an extension of time without having received notice from the contractor.

It is, therefore, advisable for the employer to carefully consider the perceived benefits of making giving notice a condition precedent to an extension of time simply to be tough with the contractor. The employer shall be satisfied with his clear ability to waive compliance with the condition precedent. It is also vital to authorize the A/E to grant extensions of time even if the contractor has failed to give notice.

Words, Phrases and Expressions 8.2

[1] *infra dignitatem* 有失身份，有失尊严
[2] trade *n.* 工种，职业，（手工业内）的行业（occupation; way of making a living, especially a handicraft, such as mason, carpenter, bricklayer, tailor, etc.）
[3] contradict *vt.* 反驳，驳斥（say (that *sth.. said* or *written*) is not true, deny (the words of a person)）
[4] contractual claims 合同索赔，根据合同提出的补偿要求
[5] ex-contractual claims 非合同索赔，根据合同以外的理由提出的补偿要求
[6] ex gratia payments（act of grace or out of kindness）道义索赔
[7] common law claims （根据）普通法（提出的）索赔
[8] breach of contract 违反合同，违约
[9] running of the contract 执行合同，履行合同
[10] cessation of the employer's liability towards the contractor 业主停止向承包商支付款项

Notes 8.2

疑难词语解释

[1] The term 'claim' as used in this context is a request by the contractor for recompense for some loss or expense that he has suffered, or an attempt to avoid the requirement to pay delay damages.

【译文】 这里用"索赔"表达承包商因受损失或开销而提出的补偿要求，或不想服从支付误期赔偿费这一要求的意图。

【解释】 将 claim 译成"索赔"可能属于"历史的误会"。"赔"字在汉语（至少现代汉语）中的意思是：因给他人造成损失而给予补偿。而在工程业主与承包商之间，即使损失不是业主给承包商造成的，业主也会根据合同满足承包商的合理要求。承包商在根据合同向业主提出合理要求时，就用 claim 这个词。显然，"索赔"不能正确地反映这种情况。不过，我国建筑业，以及许多其他行业人士已经习惯这种说法，让他们改过来，比撼山还难。

[2] It is in this light that claims should be viewed seriously by both the employers and the contractors. Unscrupulous claims by contractors to redress the effects of inefficiency or profit shortfall are unlikely to receive sympathetic consideration by the A/E or the employer.

【译文】 正是在这一方面，业主与承包商都应严肃地看待索赔。承包商为弥补自己效率低造成的后果或利润下降而别有用心的索赔，建筑师/工程师或业主不会出于同情而给予考虑。

[3] It is always a better policy to avoid disputes rather than being involved in their settlement.

【译文】 避免争议或陷入之后再力图解决，两种方针（策略）总是前者好。

【解释】 policy 最好别译成"政策"，政策总与政府有关，承包商是企业，无"政"可言。

8.3 Procedure for claims

8.3.1 Contractor's action

The contractor under a construction contract submits claims for consideration by the A/E. The A/E has a contractual duty to consider them, whether they encompass time or money or both, in an independent and impartial manner without showing bias towards either party, and make a decision on its validity and quantification.

The contractor must accept the A/E's decision on most matters until the completion of the works, subject to the provisions of sub-clause 20.4 regarding the decision of DAB and those regarding the arbitration of sub-clause 20.6, if necessary, in the case of the *Red Book*.

The majority of claims on construction contracts relate to adverse physical conditions and artificial obstructions (sub-clause 4.12); extension of time (sub-clause 8.4), delay claims arising out of sub-clauses 1.9, 2.1, 4.24, 8.4, 8.9, 10.2, 10.3, 16.1, 18.4 and 19.4, instructed variations (clause 13) and valuation of instructed variations and rate fixing (clause 12).

The employer's failure to give possession of the site on the date as usually specified in the Appendix to Tender is quite a common occurrence and the contractor may write to the employer for his entitlements under the contract as shown in Figure 8.3.

Contractor's address
21 September 2006

Employer's address
Copy to Architect

Dear Mr. Wang,

Re: Delay in giving possession

Sub-clause 2.1 of the General Conditions of Contract between us provides for you to give us possession of the site on 25 September 2006 as specified in Appendix to Tender.

The Architect has today informed us that you are not in a position to give us possession of the site on that date because of your difficulties in acquisition of the lot of the land.

As you have no doubt been advised, this amounts to a breach of the contract by you and we must therefore inform you with regret that we reserve our common law rights in this matter and in particular our right to claim compensation for loss or damage which we suffer as a result.

Yours sincerely,

Fang Jianguo
Contractor's Representative

Figure 8.3 Letter to the Employer about Delayed Giving of Possession

Under sub-clause 20.1 of the *Red Book* the contractor is also required, when giving notice in writing of his intention to make a claim to the engineer, to submit any other notices which are required by the contract, and supporting particulars for the claim, all as relevant to such event or circumstance, and shall thereafter under the same sub-clause keep such contemporary records as may be necessary to substantiate any claim, either on the site or at another location acceptable to the engineer. Figure 8.4 illustrates such a notice to the engineer of the subsidence of the site.

Within 42 days after the contractor became aware (or should have become aware) of the event or circumstance giving rise to the claim or within such other period as may be proposed by the contractor and approved by the engineer, the contractor shall send to the engineer a fully detailed claim which includes full supporting particulars of the basis of the claim for an extension of time and/or additional payment.

Contractor's address
29 September 2006

Architect's address

Dear Mr. Luo,

Re: Disruption caused by subsidence of the site

Sub-clause 4.12 of the General Conditions of Contract requires us to give notice to you whenever we encounter adverse physical conditions we consider to have been unforeseeable.

We hereby inform you that the progress of the Works is being and likely to continue to be delayed by a subsidence of the site seriously preventing our work from continuing on site. As you know, this work currently consists of excavation and concreting for the foundations to the hotel section and it is not possible for us, at this stage of the Works, to find alternative work on the site.

These circumstances amount to an adverse unforeseeable physical obstruction which is an event qualifying as a ground for extension of time under sub-clause 8.4. Although we programmed to allow for the subsurface and hydrological conditions determined by us based on the Site Data made available by the Employer to us in accordance with sub-clause 4.10. It is of our opinion that the subsidence has caused exceptionally serious consequences in terms of time and cost.

Our investigation and analysis have convinced us that the subsidence is likely to have continuing effect. We will provide you with the particulars as you may require. In the meantime we will continue to executing the Works, using such proper and reasonable measures as we can take to prevent delay, but as we have already indicated there is very little that we can do at the present time.

Will you please inspect and investigate these physical conditions as soon as possible so that (1) whether or not our encounter were unforeseeable and if so to what extent it were, and (2) to what extent our entitlement were to extension of time under sub-clause 8.4 can be determined.

Yours sincerely,

Fang Jianguo
Contractor's Representative

Figure 8.4 Notification of Delay Caused by Unforeseeable Physical Conditions

If the event or circumstance giving rise to the claim has a continuing effect:
(a) this fully detailed claim shall be considered as interim;
(b) the contractor shall send further interim claims at monthly intervals, giving the accumulated delay and/or amount claimed, and such further particulars as the engineer may reasonably require; and
(c) the contractor shall send a final claim within 28 days after the end of the effects resulting from the event or circumstance, or within such other period as may be proposed by the

contractor and approved by the engineer.

8.3.2 A/E's action

A/E should not take a letter from a contractor asking for details or information as a prelude to a claim. The chance for the contractor to do so subsequently should be reduced if he receives the information promptly.

On the other hand, however, the A/E should realize that most contractors will be looking for opportunities of submitting claims, whether fully justified or otherwise, especially in periods of economic depression, and should not take offence when adequate notice is given. It is in the interests of all parties to a contract to deal with extensions of time and evaluation of monetary entitlement at the time when the relevant events occur and the pertinent facts are still fresh in everyone's minds. Furthermore, their satisfactory resolution has the effect of keeping the contractor on target both for money and for time. Another advantage is that a contractor tends to keep his best management and employees upon those contracts that he knows have effective completion dates and cost targets and where claims will be dealt with expeditiously.[15]

It is wise for the A/E to encourage the contractor to keep him informed of anything that is happening or has happened involving the possibility of additional expense. In some cases the work in hand may be concerned or affected and the A/E may be able to take remedial action, where the fault rests with him or the employer, or avoiding or mitigating action, where the cause is one for which the employer has accepted the risk.

It is the main actions for the A/E, when having received a claim from the contractor, to identify:[15]

(1) the clauses of the contract which apply and the way to interpret them;
(2) the cost items involved; and
(3) the contractor's monetary entitlement.

The architects/engineers shall get rid of the dislike as to dealing with claims and become ready to do so. The dislike arises from the facts:[15]

(1) it is time and energy consuming and could be devoted to other work that would be more profitable to them;
(2) it would certainly increase costs to the employer and displease him;
(3) it seems to the employer some lack of care on the part of the A/E, which could be embarrassing and result in action by the employer against the A/E;
(4) they sometimes experience difficulty in handling claims in an effective manner; and
(5) should the A/E's decision be disputed, the matter in dispute may be referred to arbitration or courts of law.

In most cases assessment of a claim by the A/E requires the master programme often in a bar chart, a method statement showing the contractor's general arrangements intended to carry out his work, and a detailed breakdown of the costs of preliminaries and general items, and if any of the data appears to the A/E to be unrealistic he can and should require the further details and information that he considers necessary.

8.3.3 Engineer's action under the *Red Book*

Where a contractor notifies of his intention to claim under sub-clause 20.1 of the *Red Book* for additional cost incurred in dealing with unforeseeable physical conditions and/or obstructions, which he could not reasonably have foreseen, the engineer under sub-clause 4.12 may if he thinks fit and without admitting the employer's liability:

(1) require the contractor to provide an estimate of the cost and timing of the measures he is taking or proposes to take;

(2) approve in writing such measures with or without modification;

(3) issue instructions in writing as to the method of dealing with the physical conditions or artificial obstructions;

(4) instruct a suspension under sub-clause 8.8 or a variation under clause 13.1; or

(5) monitor the record-keeping and/or instruct the contractor to keep further contemporary records.

The contractor shall permit the engineer to inspect all these records, and shall (if instructed) submit copies to the engineer.

Within 42 days after receiving a claim or any further particulars supporting a previous claim, or within such other period as may be proposed by the engineer and approved by the contractor, the engineer shall respond with approval, or with disapproval and detailed comments. He may also request any necessary further particulars, but shall nevertheless give his response on the principles of the claim within such time.

8.3.4 Assessment of claims

Should the engineer decide that the claim is acceptable in whole or in part, he should under the *Red Book* proceed to:

(1) determine any extension of time for completion under sub-clause 8.4;

(2) certify payment of reasonable costs for additional work done;

(3) certify payment of reasonable costs for additional constructional plant used;

(4) certify payment of reasonable costs incurred by the contractor by reason of any unavoidable delay or disruption of working suffered by him; and

(5) certify payment of a reasonable percentage addition in respect of costs of additional work done and plant used, but not on delay and disruption costs.

Each payment certificate shall include such amounts for any claim as have been reasonably substantiated as due under the relevant provision of the contract. Unless and until the particulars supplied are sufficient to substantiate the whole of the claim, the contractor shall only be entitled to payment for such part of the claim as he has been able to substantiate.

The requirements of sub-clause 20.1 of the *Red Book* are in addition to those of any other sub-clause which may apply to a claim. If the contractor fails to comply with this or another sub-clause in relation to any claim, any extension of time and/or additional payment shall take account of the extent (if any) to which the failure has prevented or prejudiced proper investigation of the claim, unless the claim is excluded under the second paragraph of sub-clause 20.1.

The A/E often has to carry out extensive investigations so that a claim can be assessed. For instance, when the contractor makes a claim concerning the additional cost of executing work due to delay in the supply of certain drawings, he would need to:[15]

(1) ascertain the period between the latest date when drawings were required and the date at which they were actually supplied to the contractor;

(2) ascertain on which work the men were employed who would otherwise have been engaged on the work detailed on the missing drawings;

(3) determine whether they were transferred to this latter work immediately the drawings were supplied; and

(4) ascertain the overall programme and labour position at the time the arrival of the drawings was being awaited.

The claim may be based on additional cost due to operatives standing idle or to indirect problems such as inconvenience and loss of efficiency caused by reprogramming of the works.

Another possible source of claims is where several contractors are working at the same time on the same site where building and civil engineering, mechanical and electrical contracts are in operation, with a distinct possibility of interference between the contractors, despite the careful wording of the contracts and excellent co-ordination arrangements on site. The superstructure contractor may be delayed while certain plant is being installed, then he has to leave access gaps in the structure and subsequently to close them at a much later date, possibly beyond the end of his contract period.

When the A/E ascertains, as his duty, the amount of the direct loss and/or expense, he must require the contractor to supply all necessary factual information. Sub-clause 13.3 of the *Red Book* provides that the contractor must submit on request such details of the loss and/or expense as are reasonably required by the Engineer. This may not necessarily mean that the contractor shall submit the full cost items and their amounts he claims, although it is usual for the contractor to supply them in accordance with sub-clause 13.3. On a building contract the details might include such aspects as comparative programme/progress charts pin-pointing the effect upon progress, together with the relevant extracts from wage sheets, invoices for hiring of the contractor's equipment and other related documents.

When evaluating loss and expense, the A/E has always to compare what has happened with what would have happened if the delay or disruption had not occurred. Conjecture is inevitable but it is a professional duty of the A/E, to assess the matter, possibly with assistance from a consultant quantity surveyor, in a fair and impartial manner and to exercise due professional skill and judgment.

Words, Phrases and Expressions 8.3

[1] procedure for claims 索赔程序
[2] adverse physical conditions 恶劣的外在条件，不利的物质条件
[3] give possession of the site（业主向承包商）交付现场
[4] acquisition of the lot of the land 征用土地
[5] contemporary records 当时记录（不宜译成"同期记录"）
[6] substantiate vt. 列举事实以支持，证实，证明（give facts to support（a claim, statement, charge, etc.）
[7] assess vt. 估定（税款、罚金等）的数额，估计（decide of fix the amount of（a tax, fine, or other payment））
[8] assessment of claims 估计索赔的数额
[9] operatives standing idle 窝工的作业人员
[10] instructed (ordered) variations 指示的变更

Notes 8.3

[1] 疑难词语解释
1) The employer's failure to give possession of the site on the date as usually specified in the Appendix to Tender is quite a common occurrence and the contractor may write to the employer for his entitlements under the contract as shown in Figure 8.1.
【译文】 业主不能在通常于标书附录中注明之日交付施工现场乃常见之事，承包商可致函业主要求合同赋予自己的权利，函件之例如图 8.1 所示。
【解释】 give possession of the site 是全球工程承包业约定俗成的用语，意思是：将施工现场的土地使用权暂时交给承包商，交付之后，他人能否进入，除合同规定之外，全由承包商决定。
2) Under sub-clause 20.1 of the *Red Book* the contractor is also required, when giving notice in writing of his intention to make a claim to the engineer, to submit any other notices which are required by the contract, and supporting particulars for the claim, all as relevant to such event or circumstance, and shall thereafter under the same sub-clause keep such contemporary records as may be necessary to substantiate any claim, either on the site or at another location acceptable to the Engineer.
【译文】 红皮书还要求承包商在书面通知工程师自己的索赔意图之时，将合同要求的其他通知，以及凡是与引起该索赔的事件与情况有关，可证明索赔的细节交给工程

师，然后按照同一款的要求保留可用于证明任何索赔的现场或工程师认可的其他地点的当时记录。

3) The engineer under sub-clause 4.12 may if he thinks fit and without admitting the employer's liability:

(1) require the contractor to provide an estimate of the cost and timing of the measures he is taking or proposes to take;

【译文】工程师可根据4.12款在其认为适当且不造成业主会承担责任的误解条件下：

(1)要求承包商提出费用估算书，以及承包商已经或准备采取的各项措施时间安排；

【解释】注意，if he thinks fit 是插入语，而 without admitting the employer's liability 是状语，修饰 require。

[2] 同义词辨析

estimate/value, assess/evaluate 都有"估计"、"估算"、"估价"的意思。一般说来，estimate 和 value 侧重于估算数量的过程。assess 和 evaluate 则要从分析对象多种变量之间的因果关系开始，最后估算量。另一方面，value 仅估算价值量，estimate 则估算更多的量，例如人数，材料数量，施工机具台班量等。与此类似，evaluate 主要考虑价值方面，而 assess 则考虑更多的性质，可以是人，材料，施工机具等的能力、品质、性能等。上述解释不很准确，但有助于读者选用。

8.4 Justifying claims by contract

Application to a claim of appropriate clauses or sub-clauses of the conditions of contract is one of the essential elements for the claim to become successful and therefore the contractor's representative must get familiar with them and correctly relate the events or matters which have caused him to make a claim to the clauses or sub-clauses.

8.4.1 Unforeseeable physical conditions

Majority of construction projects involve a substantial proportion of work below ground. Under sub-clause 4.12 of the *Red Book*, for example, a claim may be submitted as a result of the natural physical conditions, man-made and other obstructions and pollutants, including sub-surface and hydrological conditions but excluding climatic conditions, which the contractor encounters at the site during the execution of the works and he could not reasonably have been foreseen at the time of tendering, even if he is experienced. This sub-clause has a wide range of implications and the final decision on the validity of any claim rests with the engineer who, on occasions, seems to adopt a rather harsh and uncompromising stance. There is provision for reference of disputes to arbitration in most conditions of contract but in practice the contractor rarely uses this option for a variety of reasons, including the high cost and the possible adverse effect on future invitations to tender.

A claim submitted under the provisions of sub-clause 4.12 of the *Red Book*, may be for additional payment to cover the cost of dealing with the natural physical conditions, man-made obstruction(s) or pollutants. The sub-clause prescribes that this shall include the reasonable cost incurred by the contractor by reason of any unavoidable delay or disruption of working suffered as a consequence of encountering the said conditions, obstructions or pollutants.

The type of unforeseeable physical conditions that can occur may range from running sand, unpredictable water tables and subsidence to geological faults, while obstructions could encompass old tips and foundations to unrecorded mine workings and underground services.

8.4.2 Extension of time for completion

The contractor can request and the engineer can grant extensions of time for completion of the works with reference to sub-clause 8.4 if *Red Book* is used. It should be noted that an extension granted under this sub-clause does not automatically entitle the contractor to any additional costs incurred by him during the period of extension.

When determining each extension of time under sub-clause 20.1, the Engineer shall review previous determinations and may increase, but shall not decrease, the total extension of time.

The Engineer is bound to notify the contractor of any extensions that are granted and of any claims for extension of time to which he considers the contractor is not entitled under sub-clause 8.4. He should also keep the employer fully informed of the decisions made and they may affect the payment of delay damages.

8.4.3 Claims for delay

Where the contractor can demonstrate that the actual delays constitute genuine delays for which payment is specifically provided or implied in other clauses of conditions of contract, then he is entitled to an additional payment. This category of delay arises when the contractor is unable to deploy his labour and equipment in such a way that the intended output can be achieved as related to the critical path on the approved programme. The principal clauses in the *Red Book* which specifically refer to payment for delays are:

☐ Sub-clause 1.9 Delayed Drawings or Instructions;
☐ Sub-clause 2.1 Right of Access to the Site;
☐ Sub-clause 4.24 Fossils (delay due to carrying out the Engineer's instruction for dealing with the same);
☐ Sub-clause 8.4 Testing;
☐ Sub-clause 8.9 Consequences of Suspension;
☐ Sub-clause 10.2 Taking Over of Parts of the Works;
☐ Sub-clause 10.3 Interference with Tests on Completion;
☐ Sub-clause 16.1 Contractor's Entitlement to Suspend Work;
☐ Sub-clause 18.4 Consequences of Employer's Risks (Failure to give possession of site); and
☐ Sub-clause 19.4 Consequences of *Force Majeure*.

A request from the contractor to the Engineer for further drawings is illustrated in Figure 8.5.

Contractor's address
15 September 2006

Architect's address

Dear Mr. Luo,

Re: Reinforcing bar schedules for foundations to the hotel section

We hereby apply for issuing the bar schedules for erecting the reinforcing cages for the foundations to the hotel section of the Works. You will see from our Master Programme, as last revised, that we intend to commence erection of the steel bars for the foundations on 29 September 2006 and in order to allow for ordering, delivery and fabrication of the steel bars we shall require the schedules no later than 22 September 2006.

Yours sincerely,

Fang Jianguo
Contractor's Representative

Figure 8.5 A Request from the Contractor for Further Drawings

Assume the contractor did not receive the required bar schedules until 27 September 2006 and the works is likely to be delayed as a result of the engineer's failure to issue the reasonably notified bar schedules, the contractor shall give a further notice to the engineer and shall be entitled subject to sub-clause 1.9 to:

(a) an extension of time for such a delay, if completion is or will be delayed, under sub-clause 8.4, and

(b) payment of any such cost plus reasonable profit, which shall be included in the contract price.

Such a further notice is illustrated in Figure 8.6.

Contractor's address
28 September 2006

Architect's address

Dear Mr. Luo,

Re: Late bar schedules for foundations to the hotel section

We have now received from you the bar schedules which we specifically requested you to issue in our application of 15 September 2006. The date on which it was necessary for us to receive them, having regard to the Completion Date currently determined, was 22 September 2006. The delay in receipt of these bar schedules was therefore 5 working days.

As you know, the foundation work is on the critical path of our programme as approved by you. It has become evident that the Works is certainly to be delayed 4 days as we estimate, having regard to the likely measures that we can take at the present time. On the other hand, the delay has caused the subcontractor to become idle for 5 days and, as a result, he has incurred additional cost. We have already asked him to give us the particulars and an estimate of the consequences in three days.

No doubt, you will consolidate these into a single grant of extension of time and an addition to the contract price after taking all factors into account.

We are looking forward to your grant of an extension of time updating the Completion Date and an addition to the contract price within the period specified in sub-clause 20.1.

Yours sincerely,

Fang Jianguo
Contractor's Representative

Figure 8.6 A Request from the Contractor for Further Drawings

8.4.4 Claims under other clauses of conditions of contract

There are various provisions in the *Red Book*, under which the contractor may be entitled to claim for additional payment. These are now listed to assist the reader in locating them.

☐ Sub-clause 4.6 sets out: if any instruction is issued and to the extent that it causes the contractor to incur unforeseeable cost then it shall constitute a variation.

☐ Sub-clause 8.4 entitles the contractor to recover the cost of tests not provided for in the contract.

☐ Sub-clause 10.2 states: if the contractor incurs cost as a result of the employer taking over and/or using a part of the works, other than such use as is specified in the contract or agreed by the contractor, the contractor shall (i) give notice to the engineer and (ii) be entitled to payment of any such cost plus reasonable profit, which shall be included in the contract price.

☐ Sub-clause 11.2 under which the contractor shall be notified immediately by (or on behalf of) the employer if and to the extent that such work is attributable to any other cause. In this case the work shall be treated as a variation and deal with undr sub-clause 13.3.

☐ Sub-clause 11.8 states that the contractor makes search for the cause of any defect, under the direction of the engineer. Unless the defect is to be remedied at the cost of the contractor under sub-clause 11.2, the cost of the search plus profit shall be agreed or fairly determined by the engineer with due consultation with the employer and contractor, having regard to all relevant circumstances, and shall be included in the contract price.

☐ Sub-clause 12.3 under which if the measured quantity of a work item is changed by more than 10% from the quantity of this item in the bill of quantities or other schedule, a new rate or price shall be appropriate for the item.

☐ Sub-clause 13.5 entitles the contractor to adjust provisional sums on condition that he produces quotations, invoices, vouchers and accounts or receipts in substantiation as required by the engineer.

☐ Sub-clause 14.8 under which the contractor shall be entitled to receive financing charges compounded monthly on the amount unpaid during the period of delay if he does not receive payment as provided in the contract. Payment shall be made into the bank account, nominated by the contractor, in the payment country (for the relevant currency) specified in the contract.

☐ Sub-clause 18.1 requires the employer to indemnify and hold harmless the contractor, the contractor's personnel, and their respective agents, against and from all claims, damages, losses and expenses (including legal fees and expenses) in respect of (1) bodily injury, sickness, disease or death, which is attributable to any negligence, willful act or breach of the contract by the employer, the employer's personnel, or any of their respective agents, and (2) the matters for which liability may be excluded from insurance cover, such as (i) the employer's right to have the permanent works executed on, over, under, in or through any land, and to occupy this land for the permanent works, (ii) damage unavoidably resulting from the contractor's obligations to execute the works and remedy any defects, and (iii) a cause attributable to the employer's risks, except to the extent that cover is available at commercially reasonable terms.

8.4.5 Variations

8.4.5.1 Definition of variation

The terms 'variation' and 'change' are synonymous, although 'change' is used instead of 'variation' in the United States.

There is no industry-wide accepted definition of what is a variation. Each standard form of contract has its own definition but clearly 'variation', in the generic sense, refers to any alteration to the basis upon which the contract was entered into. This means the 'variation' embraces not only changes to the work or matters appertaining to the work, but also those to the contract documents themselves. A variation can be taken to be any, a combination of any or all of the following:

(1) the alteration or modification of the design, quality or quantity of the works, as shown

upon the contract drawings and described by or referred to in the contract bills of quantities, and includes the addition, omission or substitution of any work, the alteration of the kind or standard of any of the materials and goods to be used in the works, and the removal from the site of any work, materials or goods executed or brought thereon by the contractor for the purposes of the works other than work or material or goods which are not in accordance with the contract.

(2) the instructions regarding the expenditure of provisional sums, prime cost sums and those concerning the nature of the work which are not specifically termed a variation in the contract documents.

(3) the alteration of contract document by agreement between the parties after execution of the original contract this is a variation of the contract terms or conditions.

(4) variation of price clause which enables the contract price to be adjusted for changes in the cost of labor or materials. It is clear that such variations are directly attributable to matters not being as stated or as required in the contract documents. This occurs either because circumstances actually change or because circumstances upon which the contract documents were based were misinterpreted. The former is a matter which we can easily appreciate and comprehend; nevertheless, it still has two distinct aspects. Firstly, circumstances may change in such a way, over which the parties can have no control, that the documentation can now be seen to be defective. Alternatively, circumstances may require that the employer determine a choice of action, with the resulting choice creating a variation. That documentation is erroneous because of misinterpretation, lack of skill or whatever, is more difficult to accept. Nevertheless, this problem exists and is often a direct result of how and when the design is developed. Any instructions from the employer concerning amendments to the design or to the work already done, on account of altered requirements, should invariably be given to the A/E. The A/E should then instruct the contractor by means of a variation. He should also advise the employer on variations found necessary or desirable and inform the employer of the effect of all variations on the programme and the cost of the works.

8.4.5.2 A/E's right to vary

Usually the A/E has power to vary any part of the works, including temporary works, and instructed variations may include changes in the specified sequence method or timing of construction. The A/E's powers do not however, extend to variations of the terms and conditions of the contract.

As stated in sub-clause 13.1 of the *Red Book*, variations may be initiated by the engineer at any time prior to issuing the taking-over certificate for the works, either by an instruction or by a request for the contractor to submit a proposal. The contractor shall execute and be bound by each variation, unless the contractor promptly gives notice to the engineer stating (with supporting particulars) that the contractor cannot readily obtain the goods required for the variation. Upon receiving this notice, the engineer shall cancel, confirm or vary the instruction. Each variation may include:

(a) changes to the quantities of any item of work included in the contract (however, such changes do not necessarily constitute a variation),

(b) changes to the quality and other characteristics of any item of work,

(c) changes to the levels, positions and/or dimensions of any part of the works,

(d) omission of any work unless it is to be carried out by others,

(e) any additional work, plant, materials or services necessary for the permanent works, including any associated tests on completion, boreholes and other testing and exploratory work, or

(f) changes to the sequence or timing of the execution of the works.

The contractor shall not make any alteration and/or modification of the permanent works, unless and until the engineer instructs or approves a variation.

A variation instruction should contain the following details:

(1) a concise but adequate description identifying the variation and including the relevant item number from the bill of quantities where applicable;

(2) the location in the works, including a cross-reference to any drawing on which it is shown;

(3) a description of the nature of the change, indicating whether it is an addition, omission or modification;

(4) sufficient details to quantify the extent of the change;

(5) details of timing where this is a significant factor;

(6) any other information which the contractor may reasonably require to carry out the work; and

(7) the basis of valuation, whether by billed rates or on a daywork basis.

8.4.5.3 Encouraging contractor's proposals

Sub-clause 51.2 of the *Red Book of Fourth Edition* prescribes that the contractor should not make any variation unless he receives a written instruction or written confirmation of a oral instruction from the engineer.

As opposed to the *Red Book of Fourth Edition*, sub-clause 13.3 of the *Red Book* encourages the contractor, at any time, to submit to the engineer a written proposal which (in the contractor's opinion) will, if adopted, (i) accelerate completion, (ii) reduce the cost to the employer of executing, maintaining or operating the works, (iii) improve the efficiency or value to the employer of the completed works, or (iv) otherwise be of benefit to the employer.

The proposal shall be prepared at the cost of the contractor and shall include the items listed in sub-clause 13.3. If a proposal, which is approved by the engineer, includes a change in the design of part of the permanent works, then unless otherwise agreed by both parties:

(a) the contractor shall design this part,

(b) the contractor shall submit to the engineer the contractor's documents for this part in accordance with the procedures specified in the contract. These contractor's documents shall be in accordance with the specification and drawings, written in the language for communications defined in the contract, including additional information required by the engineer to add to the drawings for coordination of each party's designs. The contractor shall be responsible for this part and it shall, when the works are completed, be fit for such purposes for which the part is intended as are specified in the contract, and, prior to the commencement of the tests on completion, submit to the engineer the "as-built" documents and operation and maintenance manuals in accordance with the specification and in sufficient detail for the employer to operate, maintain, dismantle, reassemble, adjust and repair this part of the works, and

(c) if this change results in a reduction in the contract value of this part, the engineer shall proceed to agree or determine a fee, which shall be included in the contract price. This fee shall be half (50 %) of the difference between the following amounts:

(i) such reduction in contract value, resulting from the change, excluding adjustments for changes in legislation and cost, and

(ii) the reduction (if any) in the value to the employer of the varied works, taking account of any reductions in quality, anticipated life or operational efficiencies.

However, if amount (i) is less than amount (ii), there shall not be a fee.

8.4.5.4 Variation procedure

Sub-clause 13.3 of the *Red Book* states that if the engineer requests a proposal, prior to instructing a variation, the contractor shall respond in writing as soon as practicable, either by giving reasons why he cannot comply (if this is the case) or by submitting:

(a) a description of the proposed work to be performed and a programme for its execution,

(b) the contractor's proposal for any necessary modifications to the programme as stated in the contract and to the time for completion, and

(c) the contractor's proposal for evaluation of the variation.

The engineer shall, as soon as practicable after receiving such proposal, respond with approval, disapproval or comments. The contractor shall not delay any work whilst awaiting a response.

Each instruction to execute a variation, with any requirements for the recording of costs, shall be issued by the Engineer to the contractor, who shall acknowledge receipt.

8.4.5.5 Evaluation of variations

Each variation shall be evaluated in accordance with the contract, unless the Engineer instructs or approves otherwise in accordance with clause 13.

The engineer is required, under sub-clause 3.5 of the *Red Book*, to consult with the contractor prior to ascertaining the value of variations instructed under sub-clause 13.1.

Normally evaluation will be secured by agreement between the contractor and the engineer; failing this the engineer is empowered to determine a provisional rate or price for the purposes of interim payment certificates in accordance with the principles outlined and to notify the contractor.

Variations may render some rates or prices entered by the contractor in the bill of the quantities unreasonable or inapplicable and the contractor may give notice to the A/E that rates or prices should be varied. Such notice shall be given before commencement of the varied work or as soon thereafter as practicable. The A/E shall agree or fairly determine such rates or prices as he considers reasonable and proper, presumably having regard to the component elements of the original rates or prices.

Where the variation results in an increase in the quantity of an item in the bill, the bill rate is likely to become unreasonable or inapplicable. For example, the contractor to a contract for a waterway channel of 1.6 m in depth, 2 m in width and 1800 m long is instructed to excavate an extra length of 300 m, the rate in the bill may be used to value the additional earthwork as it is of similar nature and executed under similar conditions. However, when the additional work is not directly comparable to that in the bill, for example, if the engineer instructs to excavate down further 1 m the sections of the channel which have already been formed as specified in the contract, then the original rate entered by the contractor at the time of tendering against the excavation in the bill will certainly become inapplicable, although it may be used as the basis for valuation as far as may be reasonable. This process requires consultation between the engineer's site staff and the contractor, for example, by a comparison of the billed work with the varied work and probably by analysis of the billed rate to enable a new rate to be determined.

Figure 8.7 is the notice from the contractor to the engineer informing that the original rate of the earthwork in the bill will certainly become inapplicable.

Contractor's address
18 October 2006

Architect's address

Dear Mr. Luo,

Re: Inapplicable rate for extra earthwork

We have today received your EI No. 15 of 16 October 2006 which purports to have been issued under sub-Clause 13.1 of the General Conditions of Contract. Compliance with this instruction would mean that the scope and character of the work, as described in the contract documents, would be so substantially changed that it would no longer be the work which we contracted to carry out.

Our work, as specified in the contract documents, is for formation of a waterway channel of 1.6 m in depth, 2 m in width and 1800 m long. We have already completed a section of 350 m long. You instruct us to excavate further 1 m down. This is clearly not an additional work of the design, quality or quantity of the work as shown upon the Drawings and as described by or referred to in the Bill of Quantities. Your instruction will involve us in excavating below the water table and we will certainly incur the cost of extra dewatering as a result of the excavation below the water table as shown in the borehole logs which you made available to us at the time of tendering.

The rates as entered by us in the Bill of Quantities at the time of tendering, therefore, will be no longer applicable to the work to be varied. Please refer to the enclosed particulars for our suggested new rates.

Will you please agree with us as to the new rates for the work as you instruct so that our work can be evaluated in accordance with sub-Clause 12.3 based on the new rates.

We are looking forward to hearing from you within the period of time as specified in sub-Clause 201.

Yours sincerely,

Fang Jianguo
Contractor's Representative

Note: EI stands for Engineer's Instruction.

Figure 8.7 A notice from the contractor for a new rate

It should be possible to reach agreement provided the principle of analogous rates is accepted and the mutually agreed returns covering the varied work are available.

It may be that the nature and scope of the instructed work is such that no appropriate billed rate exists. In such cases the *Red book* requires the resident engineer and the contractor's representative to consult in order to agree a fair valuation, which generally takes the form of the reasonable cost of the work, plus a percentage for profit. Where the contractor's representative considers the billed rate to be inappropriate, the resident engineer must give him the opportunity to put forward his proposals for assessing a new rate or price, supported by all necessary particulars, including the build up of his contract rates. The resident engineer will check their accuracy and assess their relevance to the valuation.

Where the parties are unable to agree on the method of valuation, the resident engineer must resolve it himself. If the contractor has failed to submit full particulars in substantiation of his rate and/or price, the resident engineer is not obliged to present his own detailed calculations to the contractor's representative. There is no requirement to show how the value has been calculated which could, if provided, encourage the contractor to start a protracted negotiation process. The resident engineer must resist the temptation to arrive at a valuation unduly in favour of the employer, as he must be fair and impartial in all his dealings with the contractor.

For example, a contractor erecting a pumping station was instructed to change the base level of the pumping chamber after the excavation had been completed and trimmed. The only suitable machine available had to travel from the far end of the large site across an area that had become badly rutted by construction traffic in adverse weather conditions. Although the excavation and re-trimming work took only a few hours, the machine and operator were not available for any other work for a full day.

The valuing of variations may not be straightforward, and the resident engineer should give full consideration to all possible side effects, indirect losses, delay and disruption, and consequential costs. No alterations should be made to site records to make provision for these factors. If, for example, the excavator could operate at only 40 per cent efficiency then this should be stated and the resident engineer's confirmation requested. If operations elsewhere are delayed or deprived of supplies, the contractor's representative's estimate of the resultant loss should be submitted for the engineer's inspection and assessment.

Special attention shall be given to the following two situations:[31]
(1) the valuation of omissions, and
(2) the assessment of variations in which the payment of the full billed rate for additional work would provide the contractor with excessive profit.

As to the first situation, sub-clause 12.4 provides that whenever the omission of any work forms part (or all) of a variation, the value of which has not been agreed, if:
(a) the contractor will incur (or has incurred) cost which, if the work had not been omitted, would have been deemed to be covered by a sum forming part of the Accepted Contract Amount;
(b) the omission of the work will result (or has resulted) in this sum not forming part of the contract price; and
(c) this cost is not deemed to be included in the evaluation of any substituted work; then the contractor shall give notice to the engineer accordingly, with supporting particulars. Upon receiving this notice, the engineer shall proceed to agree or fairly determine this cost, with due consultation with the parties and having regard to all the circumstances, which shall be included in the contract price.

On the other hand, since sub-clause 13.3 concludes by referring to variations being valued under clause 12, and the quantity (for the purposes of evaluation) of an omitted item is zero, clause 12 concludes by entitling the contractor to compensation for the costs reasonably incurred in the expectation of carrying out work subsequently omitted under the variation.

It is important to bear in mind that the word "cost" referred to above is in its usual usage, including profit.

The significance of this point is best illustrated by a typical example. If the contractor had ordered formwork for a work which was subsequently omitted by variation, the Accepted Contract Amount would typically have included direct cost plus reasonable profit in respect of this formwork. The contractor would then be entitled to recover cost and profit. However, if the employer was thus required to pay for the full cost of an item, he may also be entitled to recover it as his property.

The second situation arises when part of the billed rate relates to fixed overheads which do not increase pro rata with the quantity of work which has been described in 6.2.4.

On issuing a variation instruction, the resident engineer has a duty to evaluate its likely effect, as it could result in no cost difference, a cost saving or extra cost. It could be that a modification results in no delay, disruption or additional work, and should therefore have no financial consequences.

When the contractor is instructed to omit work, there is sometimes a presumption that the contract price can be reduced by deducting the value of the quantity deleted at the billed rates. It could, however, adversely affect the contractor as he may already have placed orders for materials which are subject to cancellation charges or the goods may have been delivered and payments made for them. Contractor's equipment and other resources may have been assigned to the site, involving the contractor in direct expenditure. Preparatory or temporary works may have been executed in readiness for the omitted work and other operations may have been programmed around it, which could have been carried out more economically had the contractor's representative received earlier notice. In these circumstances the contractor is entitled to payment to reimburse him for the direct loss and expense incurred.

8.4.5.6 Evaluation by DAB of variations

In respect of the pricing of variations, the engineer under sub-clauses 12.3 is authorized to determine certain rates or prices. If the contractor intends to claim a higher price than that determined he may under sub-clause 20.4 of the *Red Book* refer his request in writing to the DAB for decision, with copies to the other party and the engineer. Such reference shall state that it is given under sub-clause 20.4.

Both parties shall promptly make available to the DAB all such additional information, further access to the site, and appropriate facilities, as the DAB may require for the purposes of making a decision on such request. The DAB shall be deemed to be not acting as arbitrator(s).

Within 84 days after receiving such reference or other reasonable period the DAB shall give its decision, which shall be reasoned and state that it is given under this sub-clause. The decision shall be binding on both parties, who shall promptly give effect to it unless and until it shall be revised in an amicable settlement or an arbitral award. Unless the contract has already been abandoned, repudiated or terminated, the contractor shall continue to proceed with the works in accordance with the contract.

If either party is dissatisfied with the DAB's decision or the DAB fails to give its decision within the specified period, then either party may, within 28 days after receiving the decision or DAB's failure, give notice to the other party of its dissatisfaction and intention to commence arbitration.

After giving notice that he intends to claim under this clause, the contractor must send to the engineer a first interim account giving full and detailed particulars of the amount claimed to date and the grounds upon which the claim is based. The contractor shall update this information when the full facts and extent of costs are known.

Case study 8.2 Settlement of disputes by DAB

A project is now can be used to illustrate the procedure for settlement of contractor's claims by a DAB.

Background of the project

A hydropower project in a Central America country comprises a double-curvature concrete arch dam, an underground powerhouse (built by a separate contractor) and a series of underground tunnels.

Contractor: A joint venture formed of three international contractors

Employer: DEP (Department of Energy and Power)

Accepted Contract Amount: Equivalent to US $ 220 million

Financiers: World Bank, Inter-American Development Bank, the governments of the three countries where the three contractors are from.

DAB:

Carlos Wang, from country C (chairman)

J. E. Chen, from country A, replaced on his death by an engineer from country D

I. V. Leto, from country I

Claim No. 1 Application of Formula for adjustments for changes in cost

The contract price was adjusted for inflation-by means of the adjustment formulae as described under sub-clause 13.8 of the *Red Book*, with "table of adjustment data" entered by the contractor at the time of tendering having been applied to the value of the work executed by the contractor.

The dispute centered on what constituted the "value of the work executed by the contractor".

The contractor argued that it should include all items (excluding only "extra work") making up the tender amount, thereby including the cost of the performance bond/bank guarantee and insurances.

The employer's insisted that the latter were excluded because: 1) they did not involve physical work; 2) they were covered by lump-sum (rather than unit-price) bid prices; and 3) because the fixed 20% of the escalation formulae included possible increases of the above costs.

DAB's recommendation for Claim No.1

After hearing the represented employer and contractor, DAB recommended unanimously as follows:

"…the contract provides that bank guarantee and insurances, both quoted in individual lump sums, are no doubt part of the "Schedule of rates and/or prices"; that all items which are part of the "Schedule of rates and/or prices" are subject to adjustments for changes in cost; and that the value of the works…amounting to…includes the cost of the bank guarantee and insurances.

We find that the contract, in establishing the value of the work subject to adjustments for changes in cost, definitely does not consider the character of work or classification of expenditures. Quite obviously, for one thing, the cost of overhead incurred by the contractor was to have been prorated over the items of work, and this element necessarily includes in the tender price for very many matters intangible in character compared to the physical effort of executing the work. Nonetheless, since they contribute to forming the physical items of work and are part of the "Schedule of rates and/or prices", adjustments are not to be applied to their value."

Claim No. 2 Compensation for increased local prices of gasoline and diesel

The adjustment formulae dealt with increases in the cost of the above fuels as follows:

Foreign currency: through the cost of fuel in its country of origin

Local currency: no direct provision, some indirect provision through the CPI.

The claim arises from the fact that there have been several increases in the local official prices of diesel and gasoline, as a result of an executive order issued by the secretary of the Department of Economy of the host government. The contractor claimed compensation based on sub-clause 13.8 of the *Red Book*-Adjustments for Changes in Cost.

The employer argued that 1) sub-clause 13.7 does not apply to the above-mentioned "executive order"; 2) the contractor was compensated (directly or indirectly) through the adjustment formulae; 3) the contractor had knowingly taken said risk during the tender negotiations.

Sub-clause 13.7[Adjustments for Changes in Legislation] provides as follows:

The contract price shall be adjusted to take account of any increase or decrease in cost resulting from a change in the laws of the country (including the introduction of new laws and the repeal or modification of existing Laws) or in the judicial or official governmental interpretation of such laws, made after the 28 days prior to the latest date for submission of the tender, which affect the contractor in the performance of obligations under the contract.

DAB's recommendation for Claim No. 2

Having reviewed the contractor's formal submission and heard both parties represented, recommended unanimously that:

1. Sub-clause 13.7 [Adjustments for Changes in Legislation], does not include "executive orders".

2. Prior to tender opening, the contractor had suggested the inclusion of a fuel element in the local currency adjustment formula. As this suggestion was not taken up, the contractor consciously assumed the risk of not having fuel increases reimbursed.

3. The contractor might be afforded some measure of recovery of the increased local fuel costs through the application of the foreign currency adjustment formula, but not necessarily.

4. The contractor is afforded a measure of recovery of the increased local fuel costs through the application of the local currency adjustment formula.

The contractor's claim for additional cost due to the increase in the local price of fuel which was not reimbursed accordingly dismissed.

Claim No. 3 Compensation for cement transported in excess of tender quantity

The claim stems from the fact that the cost of transporting cement is not part of the rates for cement/concrete and the quantity transported (480,000 tons) considerably overran the billed quantity (400,000 tons).

The tender bills of quantities and schedules of rates and/or prices set out the lump sums to be paid for providing all the major items of equipment, including the transport of cement.

The contractor submitted that these lump sums were intended to reimburse him for the depreciation and wear and tear of the trucks to transport the 400,000 tons of cement specified in the contract.

Clearly, if the quantity of cement to be transported is increased as ordered by the engineer, then the depreciation and wear and tear is similarly increased and the salvage value of the equipment is decreased.

The employer argued that the transportation equipment was abused, was not properly maintained and was subjected to excessive accidents.

DAB's recommendation for Claim No. 3

Having examined the contractor's formal submission and a heard both parties as represented, recommended unanimously that:

"The board is not convinced that the employer's criticism is valid: 1) in regard to the accident record, it considers the lost vehicle days due to accidents to be quite acceptable; 2) the alleged abuse and lack of maintenance would not reduce the additional depreciation, etc. caused by transporting 80,000 tons of additional cement.

The board orders that the contractor be paid the additional depreciation, etc. of its cement transportation equipment caused by every ton of cement transported in excess of 400,000 tons (exclusive of cement intended for its own use)…"

Claim No. 4 Compensation for interruption in the issuance of local labor index

The 'table of adjustment data' for local currency adjustment formula entered by the contractor in the Appendix to Tender contains a local labor index, published on July 1 of every year.

The claim arises from the fact that no such decree was issued after 1 July 2001, and the obsolete index cannot obviously enable the contractor to recover the cost.

The contractor contemplated prior to submission of tenders the possibility of such an event occurring…in which case there shall be formal, and mutual agreement stipulating the conditions of such change…

The dispute arose on the extent or method of making such a change.

The contractor proposed an index based on 85% of the actual increases awarded to labor after a certain date.

The employer produced a revised adjustment formula based on a totally academic statistical approach.

DAB's recommendation for Claim No. 4

DAB, after having reviewed the contractor's formal submission and a heard both parties as represented, recommended unanimously that:

The contractor's submissions imply that the adjustment formula should reimburse it for 100% of the actual increase in local wages. The board cannot agree with that concept. It is the bidder's responsibility to estimate how much of the actual increases will be recovered by way of the formula and to allow for the balance in his tender.

There is no contractual rationale for the contractor's proposal.

The Board finds itself unable to adopt the employer's formula. First, the rate of increase in the local minimum labor salary is appreciably greater than the rate of increase of the Honduras consumer price index proposed by the employer as a Replacement for the labor index. Second, it does not correspond in any way with the formula included in the contract.

In summary, this board sees no reason to abandon the adjustment formula, upon which the contractor's tender is based. The suspension of publication of one of the indices used in that formula was contemplated in the contract and a procedure for dealing with such an eventuality was specified. It is the Board's opinion that the terms of the contract must be observed to the fullest extent possible.

The board therefore conceives its duty to establish labor indices with effect from 1 July 2001.

During the period 1July 1997 through 20 June 2001, a trend in the growth of the published indices for labour cost was established, If that trend is extrapolated through 30 June 2001. The Board opines that the labor indices for intervening years can be determined thereupon.

Words, Phrases and Expressions 8.4

[1]　valuation of variations 确定变更的价值，估算变更的价值，变更估价

[2] variation procedure 变更程序

[3] justifying claims by contract 利用合同说明索赔的根据与理由

[4] Unforeseeable physical conditions 不可预见的物质条件

[5] Force Majeure 不可抗力

[6] appreciate and comprehend 认识与理解

[7] repudiate vt. 拒绝偿付债务，拒绝履行义务（refuse to pay a debt or to perform an obligation）

[8] evaluation of variations 确定变更价值，变更估价

Notes 8.4

疑难词语解释

[1] but clearly 'variation', in the generic sense, refers to any alteration to the basis upon which the contract was let.

【译文】 但是，显然，"变更"一般指改变本合同发包时的依据。

[2] Engineer shall proceed in accordance with sub-clause 3.5 to agree or determine the contract price by evaluating each item of work, applying the measurement agreed or determined in accordance with the above sub-clauses 12.1 and 12.2 and the appropriate rate or price for the item.

【译文】 工程师应根据第3.5款着手确定每一分项工程的价值，将按上面第12.1和12.2款商定或确定的这一分项工程计量结果乘上适合于这一分项工程的单价或价钱，进而商定或确定合同价。

【解释】 apply 在这里的意思是四则运算中的"乘"。

[3] Majority of construction projects involve a substantial proportion of work below ground.

【译文】 大多数建设项目要有很大一部分工作在地表以下。

【解释】 involve 在此句中是"免不了"，"需要"的意思。

[4] There is provision for reference of disputes to arbitration in most conditions of contract but in practice the contractor rarely uses this option for a variety of reasons, including the high cost and the possible adverse effect on future invitations to tender.

【译文】 大多数合同条件有将争议提交仲裁的条文，但承包商由于种种原因实际上很少这样做，例如费用高、业主将来可能不再邀请自己投标。

[5] A claim submitted under the provisions of sub-clause 4.12 of the *Red Book*, may be for additional payment to cover the cost of dealing with the natural physical conditions, man-made obstruction (s) or pollutants.

【译文】 根据红皮书第4.12款提出的索赔可能要求增加支付的数额，用于处理自然物质条件、人为障碍或污染物。

【解释】 cover 在此是"（钱）用于……"的意思，且是所需款项的下限，即"够用"，而非"绰绰有余"。

[6] Sub-clause 13.5 entitles the contractor to adjust provisional sums on condition that he produces quotations, invoices, vouchers and accounts or receipts in substantiation as required by the engineer.

【译文】 第13.5款赋予了承包商在按照工程师要求出示用于证明的报价（单）、发票、票据与账目或收据之后调整暂定金额的权利。

【解释】 on condition that 前提是

[7] The contractor shall be entitled to receive financing charges compounded monthly on the amount unpaid during the period of delay if he does not receive payment as provided in the contract.

【译文】 如果承包商未收到合同规定的款项，则有权获得支付延误期间的财务费用，数额等于以拖延未付数额为本金按月计算的复利。

【解释】 compound on an amount 计算某数额的复利

[8] Sub-clause 18.1 requires the employer to indemnify and hold harmless the contractor, ……, against and from all claims, damages, …… in respect of（1）bodily injury, sickness, disease or death, which is attributable to any negligence, willful act or breach of the contract by the employer, ……

【译文】 第18.1款要求业主保障承包商不因业主、……任何疏忽、恣意妄为或违约造成的人身伤害、小恙、疾病或死亡方面承担所有索赔、赔偿费、……的责任，使之免受其害。

[9] the alteration of contract document by agreement between the parties after execution of the original contract this is a variation of the contract terms or conditions.

【译文】 在原来合同生效之后，合同双方经协商对合同文件的改变，就是合同条款或条件的变更。

【解释】 注意，此处的execution不能译成"执行"。

[10] No alterations should be made to site records to make provision for these factors.

【译文】 不得为将来把这些因素计算在内而修改现场记录。

【解释】 注意，此处provision for 是"为……做准备"的意思。

[11] sub-clause 13.3 concludes by referring to variations being valued under clause 12, and the quantity (for the purposes of evaluation) of an omitted item is zero, ……

【译文】 第13.3款最后一句提到按照第12条估算变更的价值，在估价时假设删减事项的数量是零，……

【解释】 注意，此处concludes是"结束"的意思。

[12] deploy his labour and equipment in such a way that the intended output can be achieved as related to the critical path on the approved programme.

【译文】 部署劳动力和设备的方式要能够实现批准的施工计划关键路线上的原定成果。

【解释】 注意，此处as related to the critical path on the approved programme 是"intended output"的后置定语。

8.5 Additional loss or expense

Disturbance of the regular progress of the works often results in direct loss and/or expense and may arise out of lack of instructions, drawings, details or levels; the opening up or uncovering for inspection or testing if found unnecessary; discrepancy among contract drawings, specification and bill of quantities; delay on the part of other contractors employed by the employer and A/E in connection with instructions covering postponement of any work under the contract.

If the contractor follows the procedure provided for in the contract with regard to notification and the A/E is of the opinion that the contractor has suffered direct loss and/or incurred direct expense then the A/E will determine the amount of such loss or expense. This loss as determined is added to the contract price and included in an interim certificate when it is to be issued.

The probable causes of interruption can also entitle the contractor to extension of time, as they are under the control of the employer or A/E. There are also matters for which an extension of time may be granted which would not necessarily provide grounds for a monetary claim, such as *force majeure*; exceptionally adverse weather; strikes and lock-outs; and delays in securing labour or materials.

The Engineer may issue to the contractor at any time instructions and additional or modified drawings which may be necessary for the execution of the works and the remedying of any defects, for example, under sub-clause 3.3 in the *Red Book*. If an instruction constitutes a variation the additional cost or expense shall be evaluated in accordance with the contract.

The timing of requests by the contractor to the A/E for instructions is very important and must be realistic. For example, it would be unreasonable for a contractor to request a bar bending schedule in the second week of commencement for reinforced concrete work to be executed in the thirty-third week and then to regard the immediate non-provision as a basis for claim.

Accurate daily or weekly reports to the A/E, together with an agreed programme of work preferably supported by a network in the form as approved by the A/E and a work method statement, should provide a realistic forecast of when information is reasonably required. This can be further refined by the supply by the contractor of key dates when specific details and information will be required.

The contractors' claims arising from additional loss and/or expense embrace a wide range of matters as follow:[15]

(1) bringing to the site, contractor's equipment, plant and workforce whose work was finished, and their subsequent transfer to other sites;

(2) extra hire periods of contractor's equipment;

(3) delay in the non-use of contractor's equipment;

(4) moving unwanted or unused contractor's equipment;

(5) waste of materials through deterioration caused by long delays. The contractor must, however, provide adequate storage facilities and use the materials in the proper sequence;

(6) uneconomic use of contractor's equipment and labour when diverted to other work, to avoid waiting or standing time;

(7) extra water, power, lighting and watching;

(8) extra dewatering, as a result of the extension of underground working;

(9) ready mixed concrete instead of site mixed concrete, after the removal of the batching plant; and

(10) labour operations as against mechanised operations, after the removal of contractor's equipment.

Changed mode, manner and/or methods of construction may cause the works not to make regular progress but extension of the time for completion is not necessarily needed. It is the case when a varied work is a non-critical item and can be undertaken in float time. It is disturbance that costs money although the work in its entirety is not delayed. Such claims may revolve around acceleration with the need for increased productivity of labour or contractor's equipment, the necessity of providing additional supervision and other matters not time-related.

The A/E's late supply of drawings or other information could delay delivery of materials and disrupt the work. The contractor may then have to arrange overtime work to keep the work on programme and so incur additional cost recoverable from the employer.

The contractor's tender is based upon work being carried out in an orderly sequence with the most efficient use of labour and contractor's equipment. A disturbance to the sequence due to the lack of details could significantly increase the cost of the work, such as by 150, 200, or 450 per cent. Productivity can be reduced by other factors, such as exceptionally adverse weather, go-slow tactics on the part of the workforce, poor supervision, inefficient subcontractors and poor or insufficient labour.

Words, Phrases and Expressions 8.5

standing time 窝工时间

8.6 Disruption due to variations[15]

The certain variation instructions can significantly affect the efficiency of a construction project, productivity and profitability. For example, an extension, as a variation, to a sewage treatment works under construction may impinge upon the contractor's main access to the site and disrupt the movement of operatives, plant and materials on and off the site.

Any claim shall be treated on its merits, although there are some important fundamental principles for its preparation and assessment.

8.6.1 Programme

A properly prepared and detailed programme is a vital aid to proving disruption of the programme, particularly the one with critical activities having been identified.

8.6.2 Resource allocation

To aid in proving the loss due to disruption the contractor had better to show, in a simple form on the programme, the resources he has planned to use throughout its various stages, for example, the number of operatives to be employed per day or per week, and the amount and type of contractor's equipment expected to be used. These can then be compared with the records of the actual resources used on the works and the cost differential established. The resident engineer and his staff normally keep their own records as an independent check. The contractor may also incorporate the information described in a method statement.

8.6.3 Acceleration

There is no provision in most standard forms of contract of a right to the employer or contract administrator to require accelerating the progress in order to achieve a completion date earlier than as provided for in the contract. Where the employer's objectives change and such earlier completion becomes worth paying a premium, then such a proposal may be put to the contractor as request for an agreed modification of the contract terms.

Distinction must be made between the terms 'accelerate (acceleration) or speed up (speeding up)' and 'expedite (expediting)'. The term 'accelerate (acceleration) or speed up (speeding up)' refers to the employer's initiative after the signing the agreement (contract), while the term 'expedite (expediting)' refers to the employer's response when in his opinion actual progress is too slow to complete within the time for completion, and/or progress has fallen (or will fall) behind the current programme.

Although the *Red Book* does not provide an authority to accelerate completion, it does implicitly allow the Engineer to instruct to do so by requiring from time to time the contractor to revise the programme.

Sub-clause 8.3 requires the contractor to submit a revised programme to the engineer if, at any time, the engineer gives notice to the contractor that a programme fails (to the extent stated) to comply with the contract or to be consistent with actual progress and the contractor's stated intentions.

Sub-clause 8.6 of the *Red Book* states that if, at any time, (a) actual progress is too slow to complete within the time for completion, and/or (b) progress has fallen (or will fall) behind the current programme approved by the engineer in accordance with the contract, other than the extension of time granted by the engineer as authorized by the contract, then the engineer may instruct the contractor to submit a revised programme and supporting report describing the revised methods which the contractor proposes to adopt in order to *expedite* progress and complete within the time for completion. Unless the engineer notifies otherwise, the contractor shall adopt these revised methods, which may require increases in the working hours and/or in the numbers of contractor's personnel and/or goods, at the risk and cost of the contractor. If these revised methods cause the employer to incur additional costs, the contractor shall subject to the contract pay these costs to the employer, in addition to delay damages (if any).

If an instruction issued by the A/E has caused delay it may be in the best interests of all concerned to make up the time lost by acceleration. It would, however, be unwise for the contractor to expend money in this way without first obtaining the A/E's agreement and his confirmation that he is prepared to meet the cost.

The contractor may alternatively request an extension of time, with an associated request for compensation under sub-clause 8.4 of the *Red Book*, the engineer will usually decide to accept the cost of acceleration, as this is likely to cost less than the other available alternatives and will ensure that the works still be completed on time. The cost of such an acceleration claim is normally obtained by comparing the revised programmes with the original one, but incorporating the same details, should be supplied by the contractor to the engineer.

8.6.4 Prolongation claims[15]

Prolongation claims are those made by the contractors based on the loss and/or expense which they incur as result of the extension of time for completion.

These claims may be presented as a result of extensive variations causing it necessary to extend time, or as a result of interference with the regular progress by the employer and/or legally constituted authorities.

The various elements likely to appear in this type of claim are now separately considered.

8.6.4.1 Preliminaries and general items

Contractors often claim an adjustment of preliminaries and general items because an extension of time has been granted. These claims are only admissible where the delay is attributable to an action of the employer or a legally constituted authority, or has resulted from the effect of unforeseeable physical conditions. If the claim is admissible, only reasonable additional costs verifiable by the A/E should be included in the contract price and paid.

The value of preliminaries and general items may be increased because of a number of the components being directly related to the contract period. A claim under this head should preferably be based on actual expense, and not merely to pro rata the prices inserted against preliminaries and general items. The contractor should record the expenses actually incurred during the period of extension. Another approach is to sub-divide the preliminaries items into the three categories of lump sums related to specific events, time based and value based. The main expenses involved are likely to be incurred on the following matters:[15]

(1) salaries of site staff (time-related);
(2) contractor's equipment and vehicles retained on site (time-related);
(3) temporary lighting (time based but adjusted for the time of the year);
(4) offices and stores retained on the site and their upkeep (time-related but with allowance made for erection, dismantling and transport);
(5) safety measures (time-related);
(6) protection of the works (time-related);
(7) insurance premiums (value-based);
(8) telephone (specific event);
(9) electricity (specific event); and
(10) rates on site buildings (specific event).

8.6.4.2 Season when work is carried out

The shift of a work from summer into winter may lead to extra costs as a result of reduced working hours, stoppages through cold weather and other associated additional costs. Carefully kept site records will clearly identify these additional costs comparable with the programme and resource allocation.

8.6.4.3 Extended attendance[15]

When the contractor prices attendance on nominated subcontractors these may, in certain circumstances, relate to the period of time that the nominated subcontractor is on the site. When as a result of variations or extensions of time these attendances are extended, the contractor should record the actual costs incurred.

8.6.4.4 Overheads

When a contractor makes a prolongation claim he will also include site overhead and in some cases general overheads in it. Additional payment would be claimable when the prolongation causes the site and head offices to incur additional cost and/or expense on the contract. The additions should certainly be proved the direct result of the prolongation.

Overheads or "overhead costs", refers to the expenses of a contractor as an enterprise, such as office staff salaries, as distinct from the wages and other emoluments of labourers and tradesmen; and office rents and rates, as distinct from the rents for equipment hire. The amount of the expenses is independent of the quantities of work executed on the site by the contractor.

Overheads are usually calculated as a budgeted amount on the anticipated year's expenditure on the items as mentioned above, and are related to an annual average turnover budgeted based on the turnovers in the previous years. A minimum percentage on the average must be achieved to recover costs. If the annual average turnover is significantly reduced then the percentage of overheads needs increasing to produce the required sum. If the contract's time for completion is extended, work of the same value is spread over a longer period and, as a result, turnover in a given year is reduced. To recover his overheads the contractor may need to increase his overheads percentage.

The term is used loosely to embrace all the foregoing costs and is sometimes qualified by 'head office or general' or 'site' to distinguish those output-unrelated costs which are incurred in respect of the contractor's organization generally as distinct from those of a specific contract. The term does not, however, have a precise meaning, and whether or not a particular expense is recoverable as part of a specific claim is, in the absence of agreement between the parties, often a matter for adjudication by DAB, award by arbitrators, or decision by the court.

Overheads are sometimes also referred to as establishment charges and mobilization charges although these terms more accurately refer to those expenses which whilst contract-related are independent of the quantity of work to be executed.

It is complex and difficult to deal with addition for overheads in a claim, both to prove and to value, and it usually involves significant amounts of money.

There are, however, a number of methods nowadays available for this purpose in practice.
(1) Eichleay Formula[37]

"Eichleay" method got its name from a successful court case involving the Eichleay Corporation. In the case, a contractor suffered a suspension while waiting for the Engineer's instruction on a variation. The work could not be abandoned and replacement work was not feasible. To recover the cost of unabsorbed overhead (general and administrative expenses) applicable to the delay, Eichleay proposed a daily rate method which won an award in its favour.

Eichleay method of arriving at a 'daily overhead rate' is shown as follows:

$$O_A = \frac{Amounts\ claimed\ but\ disputed}{Total\ value\ of\ contract} \times O_T$$

$$O_R = \frac{O_A}{Days\ of\ contract\ performance}$$

$$O_C = O_R \times number\ of\ days\ of\ delay,$$

where the 'Total value of contract' refers to the value of contract generated during period of dispute; O_A, O_T, O_R, and O_C is 'Overhead applicable to disputed contract', 'Total overhead', 'Daily contract overhead rate', and 'Claimed overhead', respectively.

Case study 8.3 An example illustrative of how Eichleay method works

The amounts which the contractor claims and those which are verified by the engineer are listed below.

	Disputed	Others	Total
Payments due to the contractor	¥120,000	¥240,000	¥360,000
Cost verified	144,000	192,000	336,000
Gross profit (loss)	¥(24,000)	¥48,000	¥24,000

Expenses on general and site overheads	¥36,000
Days of delay	60
Days of contract performance	300

The required calculations to arrive at the addition for overheads are shown below.

$$O_A = \frac{¥120,000}{¥360,000} \times ¥36,000 = ¥12,000$$

$$O_R = ¥12,000/300 \text{ days} = ¥40.00/\text{day}$$

Addition for overheads as claimed by the contractor is,

$$O_C = ¥40.00/\text{day} \times 60 \text{ days} = ¥2,400.00$$

There are a few obvious weaknesses in the Eichleay method:
(a) all fixed overheads are allocated to the delayed contract based on the value of the contract. In the case 8.3 above, this relationship may not be meaningful because the overheads shall not be based on the 'amounts claimed but disputed' but on costs. In this example, the 'amounts claimed but disputed' accounts for one third of the overheads since it is one third of 'total value of contract'. But if the costs are used in the Eichleay formula, then the 'amounts claimed but disputed' would get approximately 43 percent of overheads (¥144,000 divided by ¥336,000), as opposed to 33 percent under the Eichleay method using 'amounts claimed but disputed'.
(b) the "days of contract performance" used to compute the "daily contract overhead rate" includes the "days of contract delay," which understates the rate and the resulting unabsorbed fixed overheads applicable to the delay.
(c) the method is less adaptable in a manufacturing environment than in a construction environment.

The advantages of the Eichleay method were its simplicity and its acceptance in previous court cases. The method also makes it easy for the experienced financial personnel to determine unabsorbed fixed overheads.

In addition, the Eichleay method can be easily explained to the people not with sufficient financial knowledge.

(2) Total direct cost allocation [37]

Another approach widely used to deal with the overheads in contract claims is based on total direct costs. It is a method easy to use although it is not suitable in certain circumstances.

This method generally assigns overhead costs to an individual contract in proportion which the direct cost of the contract bears to the total direct costs of all contracts in the contractor's hand. The example below illustrates an application of this method.

Case study 8.4

The overheads determined based on total direct costs

	Disputed contract	All contracts in hand	Total
Direct costs	¥3,000,000	¥12,000,000	¥15,000,000
Total expenses on general overheads			¥1,500,000
Overheads/direct costs	1,500,000/15,000,000 = 0.10		
Overheads assigned to disputed contract	(0.10) ×(¥3,000,000) = ¥300,000		

In theory, this method may not be suitable for many contractors because it does not allow for differences among various types of contracts. For example, some accountants may argue that contracts that consist mostly of subcontractor fees or material costs should not be burdened with the same percentage of overhead costs as contracts that consist primarily of labor costs from the contractor's staff.

In practice, this method is most applicable to contractors that use approximately the same mix of material costs, labor costs, and subcontract costs for most contracts. If the cost of major components changes significantly from job to job, direct cost allocation probably is not a good method to use.

Other allocation methods may be more useful when a significant time delay is experienced on a long-term contract. If a delay results in significantly reducing the amount of work performed in a year (thereby reducing actual direct costs), direct cost allocation will not provide a contractor with a reasonable allocation of indirect overhead costs for contract claims purposes. In other words, if no direct job costs are incurred during the delay period, overhead will not be allocated to the delayed job. This problem also happens when overhead allocation is based upon direct labor hours or direct labor cost.

If replacement work is not available to use the excess capacity created by a delay, the total allocation base used to distribute indirect overhead costs will be reduced. This, in turn, will increase the overhead rate used on active contracts. Overhead that should be allocated to the delayed contract will then be allocated to other active jobs, thus reducing profits on those jobs.

In summary, the total direct cost allocation method of assigning overhead costs to individual contracts is best suited to companies which:
(a) typically maintain a relatively constant mix of inputs (e.g., material costs, labor costs, subcontract costs) in relation to total contract costs; and

(b) have not experienced significant shutdown periods on their contracts. The primary benefit of using this method is its simplicity.

8.6.4.5 Loss of profit[12][15]

Most contractors' claims will include some element of profit. However, a claim of this kind is very difficult to prove, since profit can be lost by inefficiency or insufficient tender before any extension of time occurs. There are various schools of thought:

(1) It can be argued that an overall shortfall of $ 4 800 000 will result in a profit shortfall in twelve months of say 3 per cent of $ 4 800 000 = $ 144 000.

(2) Another argument is that since the total value of the work was not reduced, although it was spread over a longer period, neither was the profit reduced; thus there was no actual loss.

(3) Yet another approach could be that keeping key operatives on the site for a longer period than anticipated reduces their profit earning capacity elsewhere.

The definition that loss is something the contractor should have received but did not is untenable. It could possibly be argued that in case the contractor as a company will not be able to pay the desired return to its investors on their invested capital as a direct result of the anticipated shortfall arising from the extension of the contract. The most common result of a loss of profit claim is a compromise.

Since the contractor has been able to determine his additional costs as the result of claim entitlement, including recovery of additional general overheads, it would seem that the contractor should be entitled to a fair profit. However, there seems always doubt about equity, profit percentages, and the base from which profit percentages should be derived. There have been cases in arbitration in which the contractor has clearly substantiated his entitlement (damages), but in attempting to please both parties the arbitrators have not given the contractor overheads, profit, and interest.

However complex the situation may be, the contractor should include an addition for profit in his claim, and it should be supported in the best possible manner.

To support the percentage for arriving at the amount of profit, the best approach is usually to present a schedule of profitable contracts completed in the previous years based on which an average profit percentage can be achieved. The contracts on which the contractor made a loss should not be considered. The schedule of profitable contracts can either be a list of contracts or a summary of the contractor's income statements for several previous years, adjusted for the losing contracts. The table below illustrates the approach.

There is also doubt as to what base should be used for arriving at the profit percentage. A schedule should be presented showing the contractor's actual final costs and overheads in the same format in which the contract was tendered. The profit percentage should be applied in the same way that it would have been at the tender stage had the contractor known what the total scope of costs was going to be.

Case study 8.5 The profit percentage determined based on previous profits[37]

HISTORICAL PROFIT ANALYSIS

Based on Audited Income Statement Data

	Years				Total
	1987	1988	1989	1990	
Total cost (all profitable contracts)	$9,600,000	$8,400,000	$10,800,000	$12,000,000	$40,800,000
Gross profit (all profitable contracts)	1,020,000	1,056,000	1,068,000	1,140,000	4,284,000
Percentage	10.6%	12.6%	9.9%	9.5%	10.5%

Words, Phrases and Expressions 8.6

[1]　critical activities　关键活动,关键工序

[2]　cost differential　费用差额

[3]　expedite *vt*. 催促,督促(help progress, speed up(business, etc))

[4]　budget *n*. 预算 (A planned program for a fiscal period in terms of estimated costs. 一财政期在估算费用方面制定的计划。/A budget is a plan that deals with future allocation and utilization of various resources to different organization activities over a given period of time. Typically expressed in numerical terms, budgets are perhaps most frequently thought of in financial (monetary) terms. However, they are also used to plan allocation and utilization of labor, raw material, floor space, machine hours, and material usage. In this broader sense, a budget is simply a tool which managers use to translate future plans into quantitative terms. In addition to being planning tools, budgets serve as a control device for evaluating organization operations. 预算是把各种资源分配和应用于组织将来一定时期不同活动上去的一项计划。预算一般表示成数字,人们在提及"预算"时常常只知道它的财务含义。但是,预算也用于计划人力、原材料、建筑面积、机时和材料用途的分配和使用。广义而言,预算就是管理人员把将来的计划换算成数字的一种工具。除了计划工具,预算还是评价组织经营管理状况的控制手段。)

[5]　budget *vt*. 预算编制,编制预算(The process of translating approved resource requirements into time-phased financial requirements. 将批准的资源要求编成按时间阶段划分的财务要求的过程。)

[6]　budgeted *adj*. 计划的,预计的

[7]　extended attendance　对分包商的管理延长了时间,管理分包商的时间延长

[8]　schools of thought　思想学派,思想流派

[9]　spread over　将……分摊到……

Notes 8.6

疑难词语解释

Prolongation claims are those made by the contractors based on the loss and/or expense they incur as result of the extension of time for completion as provided in the contract. These claims may be presented as a result of extensive variations and be associated with extension of time, or as a result of interference with the regular progress by the employer and/or legally constituted authorities. The various elements likely to appear in this type of claim are now separately considered.

【译文】 受阻延期索赔乃承包商因竣工延期蒙受损失或开销而根据合同所提出。此类索赔可能因变更太多且同延长时间有关,也可能因为业主或有关当局干扰了工程正常进展。下面逐一考虑这类索赔中可能出现的各种不同因素。

8.7 Preparation of claims

It is the responsibility of the contractor to formulate his claim in detail and to furnish the evidence on which the claim is based. A properly supported claim shall be carefully examined to establish whether the facts are properly based, whether the matters submitted match the circumstances provided for in the contract and whether the amount claimed can be justified.

8.7.1 Initiation of claims

8.7.1.1 Being on-site during course of works[15]

It is essential that the event causing the contractor to suffer loss or expense should be identified and the supporting information and facts collected as soon as possible if the claim is to be successful.

The contractor's site staff is likely to experience a variety of embarrassing situations where it is difficult for a claim to succeed entirely, which can be illustrated by the following examples:

(1) the site surveyor observes an operation proceeding on site but cannot determine clearly how it is going to be reimbursed;

(2) the contracts manager experiences difficulty in obtaining information from the engineer despite a number of approaches; and

(3) the managing director is unhappy about the profitability of the contract.

In each of these circumstances immediate investigation is warranted to identify the cause and to determine whether it provides a realistic basis for a claim and, if so, to give notice of the claim.

The data and information regarding the actual cost of a contract compared with its estimated cost or value is often produced far too late with many contractors. A detailed cost study should be carried out on the more important elements of work. One approach is to relate actual cost to budgeted cost from the estimator's records on carefully selected elements for which this form of comparison can be done readily, albeit approximately, and regularly on site.

Full records of all pertinent facts should be kept and notice given to the A/E when it has become obvious that the contractor would sustain a loss likely to form the subject of a claim. Contractors' claims may be generated by a variety of factors, including:
(1) discrepancies between drawings, specification, schedules and the bill of quantities;
(2) late or hurried preparation of detailed information;
(3) multiplicity of A/E's instructions;
(4) constant revisions to drawings;
(5) changes of mind on the part of the employer; and
(6) poor coordination between two or more parties involved in the contract.

8.7.1.2 Periodic checks of contract particulars[15]

It should be the duty of one of the contractor's site staff members, for example, the quantity surveyor, to check if any major change has occurred between the contract documentation and the works as such. Some examples may illustrate such major changes:
(1) the site area and accesses, which may be more restricted than was indicated in the contract documents;
(2) the substructure, the details may be incomplete at the design stage;
(3) the proportions and quantities of each item of work in the bills of quantities-it is just possible that 700 000 m^3 of excavation has been incorrectly billed as 70 000 m^3; and
(4) the nature and character of the works, to determine whether any significant change has occurred between the works as executed as indicated in the contract documents and that subsequently required on the site.

A check should also be made of the effect of variations on:
(1) the programme and time for completion; and
(2) the smooth running of the contract.

When checking any such changes or on-site problems, it is important to be aware of all the relevant information on which the estimator's assumptions are based at the time of tendering and make them available.

The contractor must be alerted on any discrepancies or divergences between the contract documents, such as substantial differences between scaled and figured dimensions on drawings, incorrect concrete mixes and areas of fabric reinforcement.

The varied work may be so different in character and/or rate from that in the original bill. A substantial variation in the conditions under which the work is to be carried out will also necessitate varied rates such as the substitution of restricted access to the site for free access, curved work in lieu of straight work or the need for manual work in place of mechanical work.

Loss caused by changes on the site does not necessarily imply that the contractor has grounds for making a claim. When profits are lower than expected, the contractor should examine and compare his progress against the programme as approved by the A/E. Some of the problems may be attributable to his own making or responsibility, whereas others could emanate from actions of the employer or A/E. Responsibility can normally be identifiable from the contract, although there are likely to be some borderline cases.

A comparison of progress against the programme often reveals delays, out of sequence working, disruption and out of order. The cause digging may identify a justifiable claim to be made. Similarly, a comparison of expense on labour with productivity may show excess cost, which may happen with the contractor's equipment and preliminaries.

If the contract is running late the contractor tends to claim for an extension of time for completion, which shall be scrutinized by the A/E.

8.7.2 Invoking appropriate clauses of contract and giving of notice[15]

Invoking relevant clause or sub-clause in the conditions of contract to justify a claim is essential for it to become successful. The contractor can use the past judgements and/or appropriate literature as guide in selecting appropriate clause or sub-clause.

The contractor must give notice within the time as required in the contract or his entitlements may be lost or reduced. Alternatively, the contractor may have to use another clause as the basis for his claim. The A/E may have authority to make an award based on the information subsequently brought to his attention when the notice has been given late or not at all. In these circumstances the award may be assessed on the low side.

Failure to notify a claim often forces a contractor to claim his entitlements under a less favourable clause. Insufficient records may restrict his choice or force him to change later. Good records must be retained after completion of the contract until all matters have been settled.

Once the A/E has accepted the claim in principle the contractor should present a fully quantified claim. If the A/E does not accept the claim even in principle it nevertheless still is a manifestation of the contractor's the intention to seek a formal decision of the A/E. Should the contractor be dissatisfied with the decision, the dispute can be referred to the DAB or Dispute Board (DB), and even arbitration, when the fully prepared quantified claim forms part of the points of claim prepared for the decision of the DAB or DB or the arbitration as the case may be.

8.7.3 Supporting particulars

Claims are submitted principally to cover the extra cost and/or expense resulting from disruption of the work or prolongation of the contract. In either case a substantial amount of supporting information is needed in order to prepare a sound and logical claim. One useful approach is for the contractor to require site staff to insert daily comments against a numbered list of topics to avoid significant omissions. It is also wise to keep two sets of important records in different offices to guard against possible loss in the event of fire.

In preparing a claim, the contractor may need to refer to any of the documents as shown in Table 8.1.

Table 8.1 Documents to be Referred to in Preparation of a Claim[15]

	Description	Likely use or role
1	Correspondence between the parties to a contract	It seems straightforward at first sight but can have more important implications.
2	Approved minutes of head office and site meetings concerning the contract	It may contain instructions, variations and additional requirements.
3	A/E's instructions, variation orders and site instructions	They could be the most important single item to support a claim.
4	Contract and working drawings, including all revisions and bar bending schedules, and other contract documents	They can be used to identify divergences and inconsistencies between them.
5	Labour allocation sheets	Location, tasks and standing time are shown.
6	Correspondence with and claims from subcontractors and suppliers	They may indicate additional requirements from main contractor or causes of delay.
7	Site diary	It must contain accurate and comprehensive entries and will often highlight problems.
8	Daily weather reports	They are certainly, for example, the evidence of adverse weather conditions.
9	Register of received drawings	They can be at least used to identify revisions to drawings as compared with those on which tender was based.
10	Progress photographs, dated by the photographer	They can show lack of progress and identify disruptions as basis for prolongation claims.
11	Site survey and level details	They form basis for earthwork quantities.
12	Site records showing delays and disturbances	They could be an important element in substantiating claims for loss and expense.
13	Photographs and report detailing condition of site at date of possession	They could show obstructions and part only of site available.

	Description	Likely use or role
14	Records showing time period between date of tender and date of possession, or order to start work	They may show delay at outset.
15	Build up of tender	Allocations for preliminaries and general items, site and general overheads and profit are particularly useful.
16	Extension of time claims and allowances certified by A/E	They may be produced at request of DAB, arbitrator, etc..
17	Measurement files and interim certificate measurements	They are obviously useful.
18	Materials schedule (quantities received and delivery dates)	They may show additional costs.
19	Invoice lists	They may show additional costs under fluctuations clause, where appropriate.
20	Records of contractor's equipment	They indicate standing time and number of times brought to and from site.
21	Statutory undertakers' records	They may substantiate the delays caused by authorities.
22	Site costs and finance	They show how money has been spent and extent of reimbursement.
23	Authorised daywork schedule	They cover varied work which cannot be valued at billed rates.
24	Programmes and progress charts	They show Contractor's anticipated programme and actual performance.
25	Borehole logs	They show if actual soil conditions are different from those which the Contractor could reasonably have anticipated.
26	Work method study	It identifies extent to which disruption has occurred and its effect.
27	Variation data sheets	Nature and effect of variations are shown.
28	Interim applications, certificates and payments	They show amounts and pattern of payments - cash flow aspects.
29	Schedules of defects	They may support additional costs.

8.7.4 Basis of a successful claim[15]

Any contractor who is well organized and experienced in the specific type of contract could have provided for and be able to recognize the occurrence of events likely to cause him to suffer loss or incur expense. Notice and contemporary records of such events should be sent to the A/E in accordance with the appropriate time limits.

Failure in or late giving notice and submission of claims by a contractor would inevitably cause difficulties, since neither the employer nor the A/E would have had the opportunity to check the factors giving rise to the additional expense, at, or about the time of their occurrence. A contractor can so easily fail to obtain reimbursement of monies to which he is entitled simply because of his late submission of claim or notification of his intention to submit one.

The potential loss and expense must be clearly identified, quantified and valued. In addition, other parties to the contract must be convinced that they are valid claims and that the integral parts are claimable and correctly valued.

The quality of the records is an important factor for a claim to be settled in favour of the contractor. Only if his management appreciates the likelihood of loss and expense situations with contracts, and then has established an appropriate, efficient and effective system (which can be referred to as record management system) to identify and record all relevant background information and data the claims can be prepared well so that an accurate and well-founded evaluation can be made promptly.

A records management system is a specific set of procedures, related control functions, and automation tools that are consolidated and combined into a whole, as part of the project management information system. A records management system is used by the project manager to manage contract documentation and records. The system is used to maintain an index of contract documents and correspondence, and assist with retrieving and archiving that documentation.

The procedures detail all the activities, interactions, and related roles and responsibilities of the contractor's team members for the claim. The procedures also include integrated activities needed to collect, verify and analyze the records, information and data and archive them for future use by the contractor.

The system should continually look ahead to diagnose on-site and off-site operations and all the relevant activities and then identify the problems which are likely to arise.

There have been quite a number of claims are based on the hope that the A/E's sympathy will suffice. There may be grounds for an *ex gratia* payment but these are rarely made, particularly in public sector contracts. An entitlement within the contract must be shown, stated, and proved. The clauses or sub-clauses shall be referred to so that a claim can be well justified. Most clauses in the conditions of contract shall have set out the employer's responsibility clearly, including action or inaction of the A/E and others.

Entitlement to an extension of time shall be claimed for whenever appropriate. It may be needed to reduce delay damages as, for example, in the case of unforeseeable physical conditions where the contractor is normally to incur additional cost and expense. An extension of time may further entitle a contractor to increase preliminaries and overheads, or extra labour and contractor's equipment, based upon the extension. The contractor, however, shall ensure that he is not granted excessive extensions of time on items which carry no money entitlement, as these may prejudice his subsequent claims for reimbursement. An extension of time does not always lead naturally to a claim for entitlement. The claim may be for an item which is not on the critical path, or the time may be regained at extra cost and expense.

The A/E is often obliged to specify the types and forms of record in which he requires the contractor to keep. The contractor must maintain adequate and true records to establish the extra cost and expense which he has incurred and/or is to incur. Quite many claims fail because of inadequate or incorrect records. Records must be accurate and consistent, confirmed by the resident engineer, and often supported by photographs.

8.7.5 Key points in preparation of a claim

Whether or not a claim can be dealt with by the A/E depends, to a great extent, upon the quality of the records, correct, accurate and dated clearly. It is also sensible to notify the A/E, timely as specified in the contract, of the situations causing his claim whenever they arise.

Every claim should be produced and presented carefully, showing the details as necessary.

The claim could conveniently be organized in the following logical sequence:[15]
(1) the particulars of the works-details of the site (as contained in the preliminaries and general items) and those of the contract (as contained in the form of agreement and Appendix to Tender);
(2) the particulars of the claim-a summary of the bases or heads of claim, stating all facts and details, together with full particulars of the specific contract clauses by which the claim can be justified;
(3) evaluation of the claim-a summary of the contractor's financial loss and/or expense; and
(4) appendices-a section that collates all the back-up information described in (2) and (3).

The claim is often based upon a logical argument of loss rather than the proof of it. The cause of this may stem primarily from the variability in the quality and preciseness of the contractor's procedures. As a result of negotiation and compromise the less well-prepared claims are frequently amicably settled or withdrawn, but it is far better for the contractor to start from a sound and realistically based claim. It is also beneficial for the contractor to reach a reasonable settlement with the employer's advisers, by means of a well-presented and carefully considered claim, than to resort to costly arbitration.

In summary, when preparing a claim the contractor shall first determine the extent of his obligations under the contract and then obtain details of the matters that hindered or prevented him from executing the work with expedition and economy.

It is not wise for the contractor to submit inflated claims being prepared for part of his entitlement. Claims shall be realistic and practical attitude adopted by all the contractor's team members, be his discipline.

Words, Phrases and Expressions 8.7

[1] preparation of claims 准备索赔
[2] initiation of claims 提出索赔
[3] invoke v.t. 祈求帮助或保护,诉诸于,求助于(call upon(God, the power of the law, etc.) for help or protection)
[4] Being on-site during course of works 于工程进展过程中到现场
[5] warrant vt. 证明……为正当,说明……合理(give justification for)
[6] multiplicity of A/E's instructions 建筑师/工程师指示多而重复
[7] cause digging 挖掘原因
[8] working drawings (供一线施工人员使用的)施工图

Notes 8.7

疑难词语解释

[1] In each of these circumstances immediate investigation is warranted to identify the cause and to determine whether it provides a realistic basis for a claim and, if so, to give notice of the claim.

【译文】 每当遇到这种情况,有理由立即调查,弄清原因并判断此种情况是否可作为提出索赔的根据,若是,向(建筑师/工程师)发出索赔通知。

【解释】 to identify the cause and to determine... 是 warranted 的逻辑主语,immediate investigation 是逻辑宾语。若将这句话改成主动句,则有 To identify the cause and to determine... warrants immediate investigation. 译成汉文就是:每当遇到这种情况,弄清原因并判断此种情况是否是提出索赔的根据是立即调查的理由。另外,provides 在这句话中的意思是:构成,形成,成为,是。

[2] If the contract is running late the contractor tends to claim for an extension of time for completion, which shall be scrutinized by the A/E.

【译文】 工程若有拖延,承包商往往要求延长竣工时间,建筑师/工程师应当仔细研究这一要求,不放过任何细节。

【解释】 the contract 指合同标的,即工程。

[3] If the A/E does not accept the claim even in principle it nevertheless still is a manifestation of the contractor's the intention to seek a formal decision of the A/E.

【译文】 如果建筑师/工程师原则上也不接受索赔,这一索赔仍然是承包商嘱意前者正式决定的宣言。

【解释】 the contract 指合同标的,即工程。

[4] Claims are submitted principally to cover the extra cost and/or expense resulting from disruption of the work or prolongation of the contract.

【译文】 (承包商)提出的索赔基本上用于支付因工程中断或合同延期而带来的额外费用与开销。

【解释】 cover 意思是:(钱)用于……。

[5] It seems straightforward at first sight but can have more important implications.

【译文】 表面上一看就清楚了,但其可能有更重要的影响。

【解释】 implications 不宜译成"含义"。

[6] Invoking relevant clause or sub-clause in the conditions of contract to justify a claim is essential for it to become successful. The contractor can use the past judgements and/or appropriate literature as guide in selecting appropriate clause or sub-clause.

he should invoke the provisions of sub-clause 52.3 of the *Red Book of Fourth Edition*.

【译文】 索赔要想成功,援引合同条件中有关条款,说明索赔的根据和理由是至关重要的。承包商在斟酌哪些合同条款适合于该项索赔时,可参照过去的判断和恰当的文献。

8.8 Reducing of claims[15]

The only certain way of avoiding payment for contractors' claims is to eliminate the issue of variations and the disruption to regular progress that accompanies them, and to operate the contract so efficiently that the contractor will not incur additional loss and/or expense, but this ideal situation is rarely if ever achievable.

The submission of a claim by the contractor does not automatically create a right to reimbursement. The contractor must first demonstrate that a real loss or expense has been caused directly by one of the criteria recognized in the contract. On the other hand, the employer should not avoid making proper recompense to the contractor should the A/E issue instructions varying the work or otherwise disrupting the contractor's progress.

The number of claims can be reduced by the preparation of accurate and complete contract documents. The design team should take particular care to:

(1) eliminate discrepancies;
(2) resolve details of all nominations; and
(3) coordinate the work of any other contractors who are directly appointed by the employer.

The A/E should have full regard to the following elements when assessing the contractor's entitlement, namely: materials, labour disruption, attraction money and bonus payments; preliminaries, general items and contractor's equipment; inflation; head office overheads and profit; and interest charges.

Chapter 9   Contract Closure

Parties to a contract are always at liberty, by mutual consent, to create obligations which are binding and for which some sort of legal enforcement will be available if performance is not forthcoming as promised. They are also at liberty to agree to rescind, alter or terminate the obligations in various ways.

The elements necessary for contract formation apply also to the agreement on its rescission, alteration and termination.

9.1   Discharge of a contract

Once each party's contractual obligations have been fulfilled, the contract is considered to be discharged; this is known as discharge by performance. However, there are ways in which a contract may be discharged without the parties fulfilling their obligations; these are by agreement, frustration, and breach of contract.

Discharge of a contract by agreement, frustration or breach of contract belongs to premature or early termination of performance of a contract.

9.1.1   Discharge by performance

A contract may be discharged by performance, each party completely fulfilling their obligations under the contract so that nothing remains to be done. However, where one party has done all that is required and the other has not, the contract is not discharged.

Assume that Mr. Bush, a builder, undertakes by a contract to build a house for Mr. Smith for $368,000. Mr. Bush built in full and complete compliance with plans and specifications. The house is completed when it is found that the pipe in the walls is not the brand specified. Assume that the defect is not the product of a willful breach or bad faith.

This case shows that Mr. Bush has not completely performed his obligations under the contract.

As required in an ordinary building contract, Mr. Bush must replace the pipe with the brand as specified in the contract, or he shall not get paid.

One more example, assume, a food supplier enters into a contract with a grocer for a quantity of tinned fruit which contained thirty tins to a case. On delivery, half the cases only contained twenty-four tins, although the overall total was correct. The grocer refused to accept delivery, and the supplier sued for specific performance (a court order forcing acceptance). The court held that grocer was justified in rejecting the delivery because the supplier had not performed his exact obligation.

In the case of a building contract, if a supplier fails to provide the contractor with the exact goods that have been ordered, the contractor may refuse to accept them and also refuse to pay the supplier.

This is a very harsh rule which depends very much on interpretation. Certain exceptions have therefore evolved; these are:

☐ Part performance[20]

If one party accepts the part performance of the other, the law brings in a promise to pay on a quantum meruit basis, meaning to be paid as much as is deserved for the work completed. Once part performance is accepted, both parties' liabilities are discharged and the original agreement cannot be enforced. However, part performance must be willingly accepted as was established in the following case:

Assume a builder in the U.K. undertook by a contract to erect stables and houses on the farmer's land for $ 980,000. The builder did part of the work valued at $ 650,000 and then abandoned the contract. The farmer completed the work himself. The builder claimed on a quantum meruit basis for the amount of work done, on the basis that the farmer had accepted part performance by completing the work. The court would have held that the builder could not recover the $ 650,000 because the farmer had not willingly accepted the part performance. The only reason the farmer had completed the buildings himself, was that in their partly completed state they would have been a nuisance on his land.

Therefore, if the contractor abandons the contract without the employer's consent, the contractor is not entitled to payment for completed work.

☐ Prevention of performance

If a party performs part of his obligation and is prevented from completing it by the other party, then he can sue on a quantum meruit basis for the amount of work completed.

Assume again, a landowner intended to erect a roofed marketplace on his lot to lease stands to vegetable vendors. A builder had agreed to do so and was to be paid on a cost-plus fee basis. The builder spent a good sum of money on the works before the landowner changed his mind and sold the lot to a large developer. The court held that the builder could recover his expenses on a quantum meruit basis, that is, he should be paid for the cost of the partially completed marketplace.

Therefore, the contractor is entitled to claim on a quantum meruit basis if the employer prevents performance by, for example, not giving the contractor possession of the site on the agreed date.

☐ Substantial performance[20]

If a party substantially performs the contract, then he can sue for the contract price and a sum will be deducted to cover the small defects in performance.

An illustration is like this: Mr. Zhang agreed to paint and decorate a flat and provide certain items of furniture. He did all this but a wardrobe door and a shelf did not fit properly. The owner of the flat refused to pay any money at all because the contract had not been properly performed. Mr. Zhang sued the owner for payment. The court held that the plaintiff had substantially performed the contract; therefore he could recover the contract price less a sum to cover the defects.

With regard to building contracts, this is one of the reasons why retention money is withheld by the employer. The employer can use this money to compensate for any defective work that the contractor has failed to rectify.

□ Time of performance

Time is very important in a contract, and therefore delivery at a later date need not be accepted. This is particularly so if the parties stipulate a specific date in a contract.

In 1950 there was a case in which a car vendor sued a client for honoring an order, the client ordered with the car vendor a Rolls Royce to be delivered by a specified date. The car was delivered several months after the date for delivery specified and he refused to accept it. The car vendor sued for specific performance, asking the court to order the client to accept the car and therefore honour the contract. However, the court held that the client had stressed the importance of time and was therefore justified in refusing to accept the Rolls Royce.[20]

9.1.2 Discharge by agreement[20]

A contract is made by agreement and it is also possible to end it by subsequent agreement, provided there is consideration. Where the contract is a promise for a promise and there has been no performance, the mutual agreement of the parties provides consideration and is called bilateral discharge. Where the contract has been partly or entirely performed by one party, then the other party must provide some consideration; this method of discharge is called unilateral discharge.

Sometimes a contract makes provision for its own discharge. For example, contracts of employment provide their own discharge by requiring a minimum notice to be given by the employer or employee, usually one week or one month. Longer periods can of course be provided for expressly in particular contracts.

With regard to the form of discharge, a contract which is made in writing may be discharged by oral or written agreement.

9.1.3 Discharge by frustration[20]

This applies where an agreement is possible to perform when made, but due to unforeseeable circumstances, the fundamental purpose of the contract becomes impossible to perform. The court will decide whether the circumstances responsible for the frustration were foreseeable or unforeseeable at the time the contract was entered into. Where such circumstances are determined by the court to be unforeseeable, both parties may be discharged from their obligation of further performance. Such circumstances may be: total or partial destruction of some object necessary to the performance of the contract; change in the law which renders any attempted performance illegal; where an event fundamental to the contract does not occur.

In the case of Davis Contractors v. Fareham UDC (1956) in the U.K., the plaintiff agreed to build seventy-eight houses for 92,000 pounds, all of which were to be completed within eight months. There was a shortage of materials and labour and the plaintiff took twenty-two months and incurred an extra expenditure of 17,000 pounds.

However, the U.D.C. paid only the price according to the contract and refused to pay the extra costs that Davis had incurred. The plaintiff claimed that the original contract was frustrated by the shortages and that he was therefore entitled to be paid on a quantum meruit basis for the extra 17,000 pounds. The court held that under the circumstances the shortages were foreseeable, and the fact that performance of the contract had become more onerous for the contractor, did not result in the contract being frustrated.[20]

The term '*frustration*' as used in the *Red Book of Third Edition* is virtually unknown outside the common law system. *The Red Book of Fourth Edition* replaces this term by the words '*release from performance*'. As far as the *Red Book* is concerned, both '*frustration*' and '*release from performance*' have disappeared.

☐ Contracts of employment

The death or illness of a person in employment under a contract to provide a personal service (such as an entertainer may provide), may frustrate the contract if consequently the contract is impossible to perform.

A contract of employment may also be frustrated by imprisonment of the employee or employer; contractual obligations undertaken at the time the contract was entered into become impossible to perform.

The basic rules governing frustrated contracts are:

☐ All money paid before discharge must be returned.

☐ All money which was due to be paid need not be paid.

☐ Recovery on a quantum meruit basis may be allowed for part performance before frustration.

It should be noted that the doctrine of frustration will not apply if the contract contains an absolute undertaking to be performed in any event.

9.1.4 Breach of contract[20]

A breach of contract may take one of two forms: either anticipatory breach — an express declaration of intent not to perform, before time for performance arrives; or executionary breach-failure to fulfill obligations during the execution of the contract. In both cases, the injured party may sue immediately the breach becomes apparent. Even in the case of anticipatory breach, it is not necessary to wait until time for performance has passed in order to take legal action. A claim for breach of contract may be made for the whole contract, or of a part only.

A breach of contract entitles the injured party to an action for damages, amongst other remedies, and possibly the right to treat the contract as discharged. Whether the contract can be considered discharged depends on how serious the breach is; whether it is a breach of condition or whether it is a breach of warranty. For a breach of condition, the injured party may treat the contract as being discharged and claim damages. Warranty is a less important term, a breach of which only gives rise to a claim for damages; the contract may not be discharged.

Words, Phrases and Expressions 9.1

[1] rescission *n*. 撤销(Rescission is: 'the cutting down or terminating of a contract by the parties or one of them. It may be done by agreement, or by one party who is entitled to do so by reason of the repudiation or material default of another or by reason of the contract having been induced by fraud or misrepresentation by the other party. It is effected by taking proceedings to have the contract judicially set aside, or by giving notice to the other party of intention to treat the contract as at an end'. 撤销是"合同双方或其中一方删改或终止合同。撤销可能是经双方商定，也可以是一方因为对方拒绝承担义务或实质违约，或者是因为在对方欺诈或歪曲意图的情况下签定合同而有权这样做。撤销可以通过诉讼程序宣布合同在法律上失效而实现，也可以将终止合同的打算通知给对方"。)
[2] alteration *n*. 改变
[3] termination *n*. 终止
[4] discharge of a contract 解除合同，合同解除
[5] discharge by performance （合同）履行（后）解除
[6] discharge by agreement 协议解除
[7] discharge by frustration （中途）受挫解除
[8] breach of contract 违约解除
[9] premature or early termination of performance of a contract 提前终止合同（的履行）
[10] builder *n*. 建筑公司，土建公司
[11] undertakes by a contract to build a house 根据合同建造房屋
[12] specific performance (a court order forcing acceptance)（英国衡平院强迫违反合同者的）强制履行令
[13] This is a very harsh rule which depends very much on interpretation. 这条规则十分苛刻，很大程度上视如何解释而定。

[14]　Certain exceptions have therefore evolved 因此，就积累了一些例外
[15]　part performance 部分履行，局部履行
[16]　quantum meruit 合理的数额
[17]　on a quantum meruit basis 按合理的数额
[18]　prevention of performance 妨碍履行合同
[19]　substantial performance 实质（上）履行（了）合同
[20]　time of performance 履行时间
[21]　honoring an order 兑现订单
[22]　honour the contract 兑现合同
[23]　contracts of employment 雇用合同
[24]　anticipatory breach 预先违约，事前违反
[25]　executionary breach 执行中违约，事后违反
[26]　whether it is a breach of condition or whether it is a breach of warranty. 是违反基本条款，还是违反保障条款。

Notes 9.1

疑难词语解释

[1]　The Latin phrase '*quantum meruit*' means 'what it is worth' and the term is applied in those cases where no specific price has been agreed, or in the very rare cases where the specified price has ceased to be applicable. The convenient English phrase is a 'reasonable sum'.

A 'reasonable sum' is a fair price for doing the work as at the date the work is done. A starting point is of course the actual cost of the work, but clearly the duty to pay a reasonable price does not oblige the building owner to pay for inefficiency.

There are three contractual situations where payment is to be of a reasonable sum: (1) where there is an express agreement to pay a reasonable sum for the work; (2) where there is an agreement to do work and, whilst it is obviously intended that payment should be made, no sum is stipulated by way of payment; (3) where the contract pricing mechanism has been vitiated by the events which have occurred; this is very rare.

A 'reasonable sum' settlement will be imposed under the Law Reform (Frustrated Contracts) Act 1943 where the payment of such sums as the court considers just may be awarded for work done under the contract before it was frustrated and thereby ended. In such a case the payments and the claim for such a sum will normally be ascertained by reference to what is reasonable, having regard to the value to the employer of the work done.

In the rare cases where extra work is ordered and done outside the ambit of the contract, it will give rise to a right on the contractor's part to be paid a reasonable sum. In very special circumstances work which is done in advance of the conclusion of an agreement may fall to be paid for on a reasonable sum basis and may not be subject to valuation in accordance with these subsequently agreed terms.

【译文】 拉丁术语 *quantum meruit* 意味"实际所值",在事先未商定具体价钱或事先虽然规定但现在不再适用的极罕见情况下使用。若图方便,英语"合理数额"可与之对应。

合理数额就是工程在完成之日的公平价钱。确定合理数额当然要以工程实际费用为出发点,但是,房主显然不能因为要支付合理数额就不考虑(承建者)工作效率低。

合同有三种情况要求支付合理数额:(1)合同双方明确规定应为工程支付合理数额;(2)双方商定了要完成的工程且都认为要为之付款,但未规定支付的数额;(3)实际事件使合同定价机制失去了效力,当然,这种情况不多见。

对于在受挫而终止前已完成的合同规定的工作,法院若认为合理并可能裁决给予支付的此类数额,应根据1943年(英国)(受挫合同)法律改革法案,以"合理数额"方式解决。在这种情况下,支付的数额与索赔额一般在考虑已完成工作对于业主的价值之后,根据所有合理的因素确定。

按指示完成合同之外额外工作这种罕见情况,会使承包商方面获得取得合理数额的权利。在极特殊情况下,在达成协议前即已完成的工作可能要按照合理数额支付且无须按照随后商定的条款估价。

[2] Assume a builder in the U.K. undertook by a contract to erect stables and houses on the farmer's land for $980,000. The builder did part of the work valued at $650,000 and then abandoned the contract. The farmer completed the work himself. The builder claimed on a quantum meruit basis for the amount of work done, on the basis that the farmer had accepted part performance by completing the work. The court would have held that the builder could not recover the $650,000 because the farmer had not willingly accepted the part performance. The only reason the farmer had completed the buildings himself, was that in their partly completed state they would have been a nuisance on his land.

Therefore, if the contractor abandons the contract without the employer's consent, the contractor is not entitled to payment for completed work.

【译文】 假定英国某房屋建造者根据合同在该农户土地上建造马厩与房屋,索价98万美元。该建造者在完成价值65万美元的工作后就放弃了这一合同。农户随后自己将其竣工。该建造者就按照合理价钱原则索取了已完工程之所值,理由是农户自己将其完成等于承认合同的部分履行。法院本来会因为农户是不情愿地认可部分履行合同而认为该建造者不能收取这65万美元。农户自己完成这些建筑物的唯一理由就是只干了一部分的这些建筑物若不完成就只能空占土地而无用。

所以,承包商若未经业主同意就放弃合同,就无权要求向其支付已完工作的款项。

【解释】 请注意这段话中的虚拟语气 would have held ... had not ... accepted.

[3] A contract is made by agreement and it is also possible to end it by subsequent agreement, provided there is consideration. Where the contract is a promise for a promise and there has been no performance, the mutual agreement of the parties provides consideration and is called bilateral discharge. Where the contract has been partly or entirely performed by one party, then the other party must provide some consideration; this method of discharge is called unilateral discharge.

【译文】 合同经协商而订立，只要有约因，随后同样可经协商而结束。当合同属于相互间许诺且尚未履行时，合同双方之间的协议就构成了约因，因而叫做双边解除。当一方当事人部分或全部履行了合同时，对方就必须给予某种报酬，这种解除合同的方法叫做单方解除。

【解释】 注意，consideration 在这段话中出现两次，意思一样。但是，译成汉语时，字面可以不一样。前者译成"约因"，后者译成"报酬"。另外，provide 的意思是"形成"或"构成"。

[4] Sometimes a contract makes provision for its own discharge. For example, contracts of employment provide their own discharge by requiring a minimum notice to be given by the employer or employee, usually one week or one month. Longer periods can of course be provided for expressly in particular contracts.

With regard to the form of discharge, a contract which is made in writing may be discharged by oral or written agreement.

【译文】 有时候，合同规定了自己的解除。例如，雇用合同对于雇主或雇员就彼此提前通知规定最短时间就等于这种合同自己解除，一般是一个星期或一个月。当然，具体的合同可以明文规定更长的时间。

至于解除的方式，书面订立的合同可在口头或书面商定之后解除。

9.2 Discharge of contract under FIDIC *Red Book*

9.2.1 Termination by employer

Sub-clause 15.2 of the *Red Book* entitles the employer to terminate the contract if the contractor:

(a) fails to obtain at his cost a performance security for proper performance and then deliver it to the employer within 28 days after receiving the Letter of Acceptance, and send a copy to the engineer or fails to comply with the notice by which the engineer may require him to make good the failure and to remedy it within a specified reasonable time when he fails to carry out any obligation under the contract,

(b) abandons the works or otherwise plainly demonstrates the intention not to continue performance of his obligations under the contract,

(c) without reasonable excuse fails:
(i) to proceed with the works in accordance with clause 8, or
(ii) to make good promptly the defect or nonconformity with the contract which the engineer has found in any plant, materials or workmanship as a result of an examination, inspection, measurement or testing and notified of and ensure that the rejected item complies with the contract,
(iii) to comply with the engineer's instruction requiring him, within a reasonable time, to:
☐ remove from the site and replace any plant or materials which is not in accordance with the contract,
☐ remove and re-execute any other work which is not in accordance with the contract, and
☐ execute any work which is urgently required for the safety of the works, whether because of an accident, unforeseeable event or otherwise.
(d) subcontracts the whole of the works or assigns the contract without the required agreement,
(e) becomes bankrupt or insolvent, goes into liquidation, has a receiving or administration order made against him, compounds with his creditors, or carries on business under a receiver, trustee or manager for the benefit of his creditors, or if any act is done or event occurs which (under applicable laws) has a similar effect to any of these acts or events, or
(f) gives or offers to give (directly or indirectly) to any person any bribe, gift, gratuity, commission or other thing of value, as an inducement or reward:
(i) for doing or forbearing to do any action in relation to the contract, or
(ii) for showing or forbearing to show favour or disfavour to any person in relation to the contract, or if any of the contractor's personnel, agents or subcontractors gives or offers to give (directly or indirectly) to any person any such inducement or reward as is described in this sub-paragraph (f). However, lawful inducements and rewards to contractor's personnel shall not entitle termination.

In any of these events or circumstances, the employer may, upon giving 14 days' notice to the contractor, terminate the contract and expel the contractor from the site. However, in the case of sub-paragraph (e) or (f), the employer may by notice terminate the contract immediately.

The employer's election to terminate the contract shall not prejudice any other rights of the employer, under the contract or otherwise.

The contractor shall then leave the site and deliver any required goods, all the contractor's documents, and other design documents made by or for him, to the engineer. However, the contractor shall use his best efforts to comply immediately with any reasonable instructions included in the notice (i) for the assignment of any subcontract, and (ii) for the protection of life or property or for the safety of the works.

After termination, the employer may complete the works and/or arrange for any other entities to do so. The employer and these entities may then use any goods, contractor's documents and other design documents made by or on behalf of the contractor.

The employer shall then give notice that the contractor's equipment and temporary works will be released to the contractor at or near the site. The contractor shall promptly arrange their removal, at the risk and cost of the contractor. However, if by this time the contractor has failed to make a payment due to the employer, these items may be sold by the employer in order to recover this payment. Any balance of the proceeds shall then be paid to the contractor.

As soon as practicable after a notice of termination has taken effect, the engineer shall proceed to agree or determine the value of the works, goods and contractor's documents, and any other sums due to the contractor for work executed in accordance with the contract.

After a notice of termination has taken effect, the employer may:
(a) claim for any payment and/or any extension of the defects notification period he considers himself to be entitled to under any clause of the conditions of the contract or otherwise in connection with the contract,
(b) withhold further payments to the contractor until the costs of execution, completion and remedying of any defects, damages for delay in completion (if any), and all other costs incurred by the employer, have been established, and/or
(c) recover from the contractor any losses and damages incurred by the employer and any extra costs of completing the works, after allowing for any sum due to the contractor. After recovering any such losses, damages and extra costs, the employer shall pay any balance to the contractor.

The employer shall be entitled to terminate the contract, at any time for his convenience, by giving notice of such termination to the contractor. The termination shall take effect 28 days after the later of the dates on which the contractor receives the notice or the employer returns the performance security. The employer shall not terminate the contract in order to execute the works himself or to arrange for the works to be executed by another contractor or to avoid a termination of the contract by the contractor.

After this termination, the contractor shall cease work and remove contractor's equipment and shall be paid in accordance with the contract.

In addition to clause 15, sub-clause 11.4 of the *Red book* sets out that:

If the contractor fails to remedy any defect or damage within a reasonable time, a date may be fixed by (or on behalf of) the employer, on or by which the defect or damage is to be remedied. The contractor shall be given reasonable notice of this date.

If the contractor fails to remedy the defect or damage by this notified date and this remedial work was to be executed at the contractor's cost, the employer may (at his option):

......

(c) if the defect or damage deprives the employer of substantially the whole benefit of the works or any major part of the works, terminate the contract as a whole, or in respect of such major part which cannot be put to the intended use. Without prejudice to any other rights, under the contract or otherwise, the employer shall then be entitled to recover all sums paid for the works or for such part (as the case may be), plus financing costs and the cost of dismantling the same, clearing the site and returning the contractor's equipment and materials to the contractor.

Sub-clause 15.6 added by the World Bank to the general conditions in the SBDW, in May 2005, states: if the employer determines that the contractor has engaged in corrupt, fraudulent, collusive or coercive practices, in competing for or in executing the contract, then the employer may, after giving 14 days notice to the contractor, terminate the contractor's employment under the contract and expel him from the site, and the provisions of clause 15 shall apply as if such expulsion had been made under sub-clause 15.2.

For the purposes of sub-clause 15.6:

(a) "corrupt practice" means the offering, giving, receiving of soliciting of any thing of value to influence the action of a public official in the procurement process or in the contract execution.

(b) "fraudulent practice" means a misrepresentation of facts in order to influence a procurement process or the execution of the contract to the detriment of the borrower, and includes collusive practice among tenderers (prior to or after tender submission) designed to establish tender prices at artificial non-competitive levels and to deprive the borrower of the benefits of free and open competition.

(c) "collusive practice" means a scheme or arrangement between two or more tenderers, with or without the knowledge of the borrower, designed to establish tender prices at artificial, noncompetitive levels.

(d) "coercive practice" means harming or threatening to harm, directly or indirectly, persons or their property to influence their participation in the procurement process or affect the execution of a contract.

9.2.2　Termination by contractor

Clause 16 of the *Red Book* entitles the contractor suspend work in certain circumstances. For example, if:

(a) the engineer fails to issue to the employer an interim payment certificate which shall state the amount which the engineer fairly determines to be due, with supporting particulars in accordance with sub-clause 14.6;

(b) the employer fails to submit, within 28 days after receiving any request from the contractor, reasonable evidence that financial arrangements have been made and are being maintained which will enable the employer to pay the contract price (as estimated at that time) in accordance with clause 14;

(c) the employer fails to give notice to the contractor with detailed particulars before he makes any material change to his financial arrangements; or

(d) the employer fails to pay to the contractor any amount specified in the contract, the contractor may, after giving not less than 21 days' notice to the employer, suspend work (or reduce the rate of work) unless and until the contractor has received the payment certificate, reasonable evidence or payment, as the case may be and as described in the notice.

The contractor's action shall not prejudice his entitlements to financing charges and to termination as specified in the contract.

If the contractor subsequently receives such payment certificate, evidence or payment (as described in the relevant sub-clause and in the above notice) before giving a notice of termination, the contractor shall resume normal working as soon as is reasonably practicable.

Clause 16 entitles the contractor to terminate the contract, as well, if:

(a) the contractor does not receive the reasonable evidence within 42 days after giving notice in respect of the employer's failure to submit, within 28 days after receiving any request from the contractor, reasonable evidence that financial arrangements have been made and are being maintained which will enable him to pay the contract price (as estimated at that time) as specified in the contract or to give notice to the contractor before he makes any material change to his financial arrangements,

(b) the engineer fails, within 56 days after receiving a statement and supporting documents, to issue the relevant payment certificate,

(c) the contractor does not receive the amount due under an interim payment certificate within 42 days after the expiry of the time stated in the contract within which payment is to be made (except for deductions as described in the contract),

(d) the employer substantially fails to perform his obligations under the contract in such manner as to materially and adversely affect the ability of the contractor to perform the contract,

(e) the employer fails to enter into a contract agreement within 28 days after the contractor receives the Letter of Acceptance, unless they agree otherwise, or fails to bear the costs of stamp duties and similar charges (if any) imposed by law in connection with entry into the Contract Agreement, or assigns the whole or any part of the contract or any benefit or interest in or under the contract,

(f) a prolonged suspension affects the whole of the works as described in the contract, or

(g) the employer becomes bankrupt or insolvent, goes into liquidation, has a receiving or administration order made against him, compounds with his creditors, or carries on business under a receiver, trustee or manager for the benefit of his creditors, or if any act is done or event occurs which (under applicable Laws) has a similar effect to any of these acts or events.

In any of these events or circumstances, the contractor may, upon giving 14 days' notice to the employer, terminate the contract. However, in the case of sub-paragraph (f) or (g), the contractor may by notice terminate the contract immediately.

The contractor's election to terminate the contract shall not prejudice any other rights of the contractor, under the contract or otherwise.

After a notice of termination, the contractor shall promptly:
(a) cease all further work, except for such work as may have been instructed by the engineer for the protection of life or property or for the safety of the works,
(b) hand over contractor's documents, plant, materials and other work, for which the contractor has received payment, and
(c) remove all other goods from the site, except as necessary for safety, and leave the site.
After a notice of termination, the employer shall promptly:
(a) return the performance security to the contractor,
(b) pay the contractor as described in the contract, and
(c) pay to the contractor the amount of any loss or damage sustained by the contractor as a result of this termination.

9.2.3 Force Majeure

9.2.3.1 Definition of Force Majeure

Under clause 19 of the *Red Book*, Force Majeure is defined as an exceptional event or circumstance:
(a) which is beyond a party's control,
(b) which such party could not reasonably have provided against before entering into the contract,
(c) which, having arisen, such party could not reasonably have avoided or overcome, and
(d) which is not substantially attributable to the other party.
Force Majeure may include, but is not limited to, exceptional events or circumstances of the kind listed below, so long as conditions (a) to (d) above are satisfied:
(i) war, hostilities (whether war be declared or not), invasion, act of foreign enemies,
(ii) rebellion, terrorism, sabotage by persons other than the contractor's personnel, revolution, insurrection, military or usurped power, or civil war,
(iii) riot, commotion, disorder, strike or lockout by persons other than the contractor's personnel,
(iv) munitions of war, explosive materials, ionizing radiation or contamination by radio-activity, except as may be attributable to the contractor's use of such munitions,

explosives, radiation or radio-activity, and

(v) natural catastrophes such as earthquake, hurricane, typhoon or volcanic activity.

9.2.3.2 Notice of Force Majeure

Sub-clause 19.2 of the *Red Book* states: if a party is or will be prevented by Force Majeure from performing its substantial obligations under the contract, then it shall give notice to the other party of the event or circumstances constituting the Force Majeure and shall specify the obligations, the performance of which is or will be prevented. The notice shall be given within 14 days after the party became aware, or should have become aware, of the relevant event or circumstance constituting Force Majeure. The party shall, having given notice, be excused performance of its obligations for so long as such Force Majeure prevents it from performing them.

9.2.3.3 Optional termination, payment and release

If the execution of substantially all the works in progress is prevented for a continuous period of 84 days by reason of Force Majeure of which notice has been given as required by the contract, or for multiple periods which total more than 140 days due to the same notified Force Majeure, then sub-clause 19.2 of the *Red Book* entitles either party to give to the other party a notice of termination of the contract. In this event, the termination shall take effect 7 days after the notice is given, and the contractor shall proceed as if the contract had been terminated by the employer, the contractor himself or an option.

Upon such termination, the engineer shall determine the value of the work done and issue a payment certificate which shall include:

(a) the amounts payable for any work carried out for which a price is stated in the contract;

(b) the cost of plant and materials ordered for the works which have been delivered to the contractor, or of which the contractor is liable to accept delivery: this plant and materials shall become the property of (and be at the risk of) the employer when paid for by the employer, and the contractor shall place the same at the employer's disposal;

(c) other costs or liabilities which in the circumstances were reasonably and necessarily incurred by the contractor in the expectation of completing the works;

(d) the cost of removal of temporary works and contractor's equipment from the site and the return of these items to the contractor's works in his country (or to any other destination at no greater cost); and

(e) the cost of repatriation of the contractor's staff and labour employed wholly in connection with the works at the date of termination.

If any event or circumstance outside the control of the parties (including, but not limited to, Force Majeure) arises which makes it impossible or unlawful for either or both parties to fulfill its or their contractual obligations or which, under the law governing the contract, entitles the parties to be released from further performance of the contract, then upon notice by either party to the other party of such event or circumstance:

(a) the parties shall be discharged from further performance, without prejudice to the rights of either party in respect of any previous breach of the contract, and

(b) the sum payable by the employer to the contractor shall be the same as would have been payable if the contract had been terminated under sub-Clause 19.6.

Words, Phrases and Expressions 9.2

[1] termination by employer 由业主终止（合同）
[2] gratuity n. 象征性给付(a gift of money, especially for service, as a tip)
[3] commission n. 佣金，酬金(a sum or percentage paid to an agent for his or her services)
[4] inducement n. 诱因,引诱物(that which induces)
[5] forbear vt. & vi. 克制，忍而不……(refrain, refrain from; not use or mention; be patient)
[6] prejudice vt. 损害(injure or weaken (sb.'s interests, etc))
[7] release vt. 发还，释放，释还(allow to go, set free; unfasten)
[8] Any balance of the proceeds 收入的所有余额(proceeds 与 income 不同, income 是消耗劳动力或资本等要素后的收入，而 proceeds 并非如此)
[9] corrupt practice 腐败行为
[10] fraudulent practice 欺诈行为
[11] collusive practice 阴谋活动
[12] coercive practice 胁迫行为
[13] becomes bankrupt 破产
[14] insolvent 收不抵支，丧失支付能力
[15] goes into liquidation 企业等清算
[16] has receiving or administration order made against him 接到了财产接收或管理令
[17] compounds with his creditors 与债主了结债务
[18] carries on business under a receiver, trustee or manager for the benefit of his creditors 在财产接收者、托管者或管理者监督下为债主继续经营
[19] hand over 移交(与 take over 相对)
[20] Optional termination（遇到不可抗力时）自愿终止合同
[21] repatriation n. 遣返

9.3 Contract closure

The contract closure involves verification that all work and deliverables were acceptable. This process also involves administrative activities, such as updating records to reflect final results and archiving such information for future use. Contract closure addresses each contract applicable to the project or a project phase. In multi-phase projects, the term of a contract may only be applicable to a given phase of the project. In these cases, the contract closure closes the contract(s) applicable to that phase of the project. Unresolved claims may be subject to arbitration and even litigation after contract closure. The contract terms and conditions can prescribe specific procedures for contract closure.

Premature or early termination of a contract is a special case of contract closure, and can result from a mutual agreement of the parties or from the default of one of the parties. The rights and responsibilities of the parties in the event of an early termination are contained in a terminations clause of the contract. Based upon those contract terms and conditions, the employer may have the right to terminate the whole contract or a portion of the project, for cause or convenience, at any time. However, based upon those contract terms and conditions, the employer may have to compensate the contractor for his preparations and for any completed and accepted work related to the terminated part of the works.[16]

9.3.1 Inputs to contract closure[16]

Inputs to contract closure include contract documentation and contract closure procedure.

Contract closure procedure includes all activities and interactions needed to settle and close any contract agreement established for a construction contract, as well as define those related activities supporting the formal administrative closure of the project. This procedure involves both verification of the works and administrative closure (updating of contract records to reflect final works and archiving that information for future use). The contract terms and conditions can also prescribe specifications for contract closure that must be part of this procedure. Early termination of a contract could involve, for example, the inability to complete the works, a budget overrun, or lack of required resources.

Administrative closure procedure details all the activities, interactions, and related roles and responsibilities of the project team members and other stakeholders involved in executing the administrative closure procedure for the project. Performing the administrative closure process also includes integrated activities needed to collect project records, analyze project success or failure, gather lessons learned, and archive project information for future use by the employer's organization.

9.3.2 Tools and techniques for contract closure[16]

9.3.2.1 Records management system (see 8.7.4).
9.3.2.2 Procurement audits

A procurement audit is a structured review by the employer of the procurement process from the procurement planning throughout contract administration as described in chapter 4. The objective of a procurement audit is to identify successes and failures that warrant recognition in the preparation or administration of other procurement contracts on the project, or on other projects within the employer's organization.

9.3.2.3 Final account[15]

(1) Final statement

The engineer shall issue the performance certificate (an sample is shown in Figure 9.1) within 28 days after the latest of the expiry dates of the defects notification periods, or as soon thereafter as the contractor has supplied all the contractor's documents and completed and tested all the works, including remedying any defects.

Performance Certificate

A/E's name and address: Our reference:
                                                                          Date:

Works:
Situated at:

To Main Contractor:

In accordance the Contract, I/We certify that all your obligations to execute and complete the Works and remedy any defects therein have been completed to my/our satisfaction.

I/We declare that a certificate for the balance of the retention monies deducted under previous certificates in respect of the said Works or part thereof is to be issued in accordance with the Conditions of Contract.

Signature: _____ A/E   Date: _____
Original to: Main Contractor
Copies to: A/E's file   Quantity Surveyor
          Clerk of works   Consultant engineers

Figure 9.1 Performance Certificate[33]

A copy of performance certificate shall be issued to the employer under sub-clause 11.9 of the *Red Book*. It states the date on which the contractor completed his obligations under the contract.

Following the issue of the performance certificate, arrangements can proceed for the preparation and settlement of the final account. The procedure is set out in sub-clause 14.11 in the case of the *Red Book*, and the importance of this stage of the contract and its administration is duly emphasized by making the A/E, as opposed to the resident engineer, responsible for this work. Where the interim certificates represent a fair evaluation of all the work contained in the contract, including billed items, variations and extras, the settlement of the final account should not be too difficult. In practice many factors combine to prevent this ideal arrangement from taking place.

Not later than 56 days of the date of the performance certificate, the contractor is required submit to the engineer for consideration six copies of a draft final statement with supporting documents showing in detail, in the form approved by the engineer, the value of all work done in accordance with the contract, together all further sums the contractor considers due to him under the contract or otherwise up to the date of the performance certificate.

If the A/E disagrees with or cannot verify any part of the draft final statement, the contractor shall submit such further information as the A/E may reasonably require and shall make such changes in the draft as may be agreed between them. The contractor shall then prepare and submit to the A/E the final statement as agreed (usually referred to as the Final Statement (as shown in Figure 9.2)). If, following discussions between the A/E and the contractor and any changes to the draft final statement which may be agreed between them, it becomes evident that a dispute exists, the A/E shall deliver to the employer an interim payment certificate for those parts of the draft final statement, if any, which are not in dispute. The dispute may then be settled in accordance with clause 20 if the *Red Book* is used.

Contract No:
Employer's name and address:
A/E's name and address:

SUMMARY

	Omissions $	Additions $
CONTRACT PRICE		2,000,000.00
Deduct contingency sum	50,000.00	
ADJUSTMENTS		
A/E's instructions:	23,456.00	20,145.40
Provisional quantities:	1,090.00	1,210.60
Prime cost sums (including profit and attendance)	421,220.00	410,181.65
Provisional sums	5,000.00	4,681.35
Agreed Contractor's claims	—	2,310.50
Fluctuations	—	55,000.00
	500,775.00	2,493,530.50
		500,775.20
TOTAL FINAL ACCOUNT		$ 1,992,775.30

I/We agree to accept the sum of $ 1,992,775.30 (in words) ONE MILLION, NINE HUNDRED AND NINETY TWO THOUSAND, SEVEN HUNDRED SEVENTY FIVE DOLLARS AND THIRTY CENTS ONLY, as full, final and sufficient payment for work carried out under above the Contract.

Signed: _____ for and on behalf of Main Contractor   Date: _____

Figure 9.2 Final Statement[33]

Most of the information should already be in the A/E's possession in the form of the following documentation:
(1) the contractor's valuation of the measured work as submitted in monthly statements;
(2) the records and other cost data relating to variations which have usually been submitted to and discussed with the resident engineer; and
(3) the full and detailed particulars of all claims which should have been delivered as soon as reasonable to the A/E's site staff, as prescribed by the conditions of contract.

In his final statement, the contractor will take the opportunity to substitute accurate amounts for earlier approximations, correct any discernible errors and, in the case of claims covering activities which continued up to the end of the construction period, it is probable that he can now provide all the necessary supporting financial data.

Since much of the work is often re-measured on completion, the services of a quantity surveyor can prove valuable. In the case of new or varied work, it is necessary to establish rates for items of work that did not appear in the original bills. Sub-clause 12.3 of the *Red Book* prescribes that until such time as an appropriate rate or price is agreed or determined, the Engineer shall determine a provisional rate or price for the purposes of interim payment certificates.

In practice considerable discussion often takes place between the resident engineer and the contractor's representative, in order to establish fair and reasonable prices. The items for which rates are to be fixed normally fall into three categories.
(1) items to be priced pro rata to the contract rates;
(2) items to be priced at rates based on new build-ups which conform to the original method of rate fixing; and
(3) items to be priced at rates which are based on the original billed rates but adjusted to take account of the altered nature of the work or the changed conditions under which it is carried out.

This work of nominated subcontractors is normally re-measured on completion, preferably with a representative of the subcontractor present. It is good practice to agree any new rates before the subcontractor renders his statement to the contractor. The contractor's profit will be computed at the same percentage as inserted in the contract bill, while the sum included for attendance in the original bill should only be adjusted when the quantity, quality or scope of the subcontractor's work is changed. For example, a change in the cost of piling would not justify an alteration to the sum for attendance, unless the number of piles or their lengths are varied, or the method of piling is changed so as to vary the amount or form of attendance required.

346

The final statement normally consists of the following items.
(1) Preliminaries and general items as contained in the original bill with any necessary adjustments arising from changed circumstances or variations.
(2) Measurement of the works following the format contained in the original bill.
(3) Adjustment of prime cost items covering work carried out by nominated subcontractors and goods or materials supplied by nominated suppliers.
(4) Daywork accounts.
(5) Labour and material price fluctuations where applicable.
(6) Variations based on Engineer's instructions.
(7) Claims for loss or expense submitted by the contractor and agreed, possibly in a modified form, by the Engineer.

The settlement of varied work and claims is usually a challenging task at this stage. The values as agreed at an interim stage can be incorporated in the final statement when agreement has previously been reached between the resident engineer and the contractor's representative as to doing so. It is almost inevitable that some rates and price cannot be settled by agreement and must then be determined by the resident engineer. These rates will be considered by the A/E along with the contractor's final statement at the final account stage. The engineer will also consider contractor's claims and supporting particulars and the result of the resident engineer s investigations, before reaching a decision on them.

(2) Discharge

Sub-clause 14.12 under the *Red Book* requires that when submitting the final statement, the contractor shall submit a discharge which confirms that the total of the final statement represents full and final settlement of all moneys due to the contractor under or in connection with the contract. This discharge may state that it becomes effective when the contractor has received the performance security and the outstanding balance of this total, in which event the discharge shall be effective on such date.

(3) Final payment certificate

Sub-clause 14.13 of the *Red Book* requires that within 28 days after receipt of the final statement and all information reasonably required for its verification, the engineer is required to issue a final payment certificate stating:
(a) the amount which he fairly determines is finally due, and
(b) after giving credit to the employer for all amounts previously paid by the employer and for all sums to which the employer is entitled, the balance (if any) due from the employer to the contractor or from the contractor to the employer, as the case may be.

Ideally the final figure should be agreed with the contractor. If the contractor has not applied for a final payment certificate in accordance with sub-clause 1 4.11 and sub-clause 14.12, the engineer shall request the contractor to do so. If the contractor fails to submit an application within a period of 28 days, the engineer shall issue the final payment certificate for such amount as he fairly determines to be due.

Some architects/engineers appear to treat a final account as a technical operation rather than a matter of judgment. In some cases a construction project is subject to such extensive redesign that it bears little resemblance to the original scheme. The A/E may decide to abandon the preparation of a bill of variations, as used extensively on building contracts, and to re-measure the whole of the work, including the varied work, at billed rates. This is tantamount to admission that the entire contract is varied and it is then patently incorrect to use a bill of quantities in the same way as a schedule of rates, and new rates and prices should be negotiated to arrive at a fair valuation of the work.

(4) Cessation of employer's liability

The employer shall not be liable to the contractor for any matter or thing under or in connection with the contract or execution of the works, except to the extent that the contractor shall have included an amount expressly for it:

(a) in the Final Statement, and also

(b) (except for matters or things arising after the issue of the taking-over certificate for the works) in the statement at completion described in sub-clause 14.10.

However, this sub-clause shall not limit the employer's liability under his indemnification obligations, or the employer's liability in any case of fraud, deliberate default or reckless misconduct by the employer.

9.3.3　Outputs of contract closure

9.3.3.1　Closed contracts[16]

The employer, usually through the A/E, his authorized contract administrator, provides the contractor with a formal written notice, the defects notification certificate that the contract has been completed. Requirements for formal contract closure are usually defined in the terms of the contract.

9.3.3.2　Contract documentation[16]

Contract documentation includes, but is not limited to, the contract, along with all supporting schedules, instructed variations and approved claims. Contract documentation also includes any contractor-developed technical documentation and other work performance information, such as the contractor's progress reports, financial documents including invoices and payment records, and the results of contract-related inspections.

Approved claims and instructed variations can include instructions issued by the A/E, or actions taken by the contractor, that the other party considers a constructive change to the contract. Since any of these constructive changes may be disputed by one party and can lead to a claim against the other party, such changes are uniquely identified and documented by correspondence between the parties.

A complete set of indexed contract documentation, including the closed contract, is prepared for inclusion with the final project files.

9.3.3.3 Taking-over of works

The employer, usually through the A/E, his authorized contract administrator, provides the contractor with a formal written notice, the taking-over certificate, that the works have been accepted or rejected. Requirements for formal acceptance of the works, and how to address non-conforming works, are usually defined in the contract.

Clause 10 of the *Red Book* sets the procedures for the employer, through the engineer, to take over the completed the whole of the works, sections or parts.

Sub-clauses 10.1 sets out that:

"...... the works shall be taken over by the employer when (i) the works have been completed in accordance with the contract, including the matters described in sub-clause 8.2 and except as allowed in sub-paragraph (a) below, and (ii) a taking-over certificate for the works has been issued, or is deemed to have been issued in accordance with this sub-clause.

The contractor may apply by notice to the engineer for a taking-over certificate not earlier than 14 days before the works will, in the contractor's opinion, be complete and ready for taking over. If the works are divided into sections, the contractor may similarly apply for a taking-over certificate for each section.

The engineer shall, within 28 days after receiving the contractor's application:
(a) issue the taking-over certificate to the contractor, stating the date on which the works or section were completed in accordance with the contract, except for any minor outstanding work and defects which will not substantially affect the use of the works or section for their intended purpose (either until or whilst this work is completed and these defects are remedied); or
(b) reject the application, giving reasons and specifying the work required to be done by the contractor to enable the taking-over certificate to be issued. The contractor shall then complete this work before issuing a further notice under this sub-clause.

If the engineer fails either to issue the taking-over certificate or to reject the contractor's application within the period of 28 days, and if the works or section (as the case may be) is substantially in accordance with the contract, the taking-over certificate shall be deemed to have been issued on the last day of that period.

Except as otherwise stated in a taking-over certificate, a certificate for a section or part of the works shall not be deemed to certify completion of any ground or other surfaces requiring reinstatement."

Figure 9.3 is a sample taking-over certificate.

TAKING OVER CERTIFICATE

A/E's name and address:_____Job title and no. :

Serial No. :

To Main Contractor:

In accordance the Contract, I/We certify that subject to the finishing with due expedition any outstanding work, and/or rectifying any defects, shrinkage or other faults that appear during the Defects Liability Period, [delete (a) or (b) as necessary]
(a) the Works have been in my/our opinion substantially completed and satisfactorily passed all Tests on Completion as prescribed by the Contract on: _____

and that the said Defects Liability Period will end on: _____

(b) a part of the Works, namely: _____
_____

I/We declare that a certificate for one moiety of the retention monies deducted under previous certificates in respect of the said Works or part thereof is to be issued in accordance with the Conditions of Contract.

Signature: _____ A/E   Date: _____
Original to: Main Contractor
Copies to: A/E's file   Quantity Surveyor
           Clerk of works   Consultant engineers

Figure 9.3 A Sample Taking-over Certificate[33]

Similarly, in accordance with the procedure set out in sub-clause 10.1, sub-clause 10.2 allows the contractor to request and the engineer may, at the sole discretion of the employer, issue a taking-over certificate for any part of the permanent works. The employer shall not use any part of the works (other than as a temporary measure which is either specified in the contract or agreed by both parties) unless and until the engineer has issued a taking-over certificate for this part. However, if the employer does use any part of the works before the taking-over certificate is issued:
(a) the part which is used shall be deemed to have been taken over as from the date on which it is used,
(b) the contractor shall cease to be liable for the care of such part as from this date, when responsibility shall pass to the employer, and
(c) if requested by the contractor, the Engineer shall issue a taking-over certificate for this part.

After the engineer has issued a taking-over certificate for a part of the works, the contractor shall be given the earliest opportunity to take such steps as may be necessary to carry out any outstanding tests on completion. The contractor shall carry out these tests on completion as soon as practicable before the expiry date of the relevant defects notification period.

If the contractor incurs cost as a result of the employer taking over and/or using a part of the works, other than such use as is specified in the contract or agreed by the contractor, the contractor shall (i) give notice to the engineer and (ii) be entitled subject to sub-clause 20.1 to payment of any such cost plus profit, which shall be included in the contract price. After receiving this notice, the engineer shall proceed to agree or determine this cost and profit.

If a taking-over certificate has been issued for a part of the works (other than a section), the delay damages thereafter for completion of the remainder of the works shall be reduced. Similarly, the delay damages for the remainder of the section (if any) in which this part is included shall also be reduced. For any period of delay after the date stated in this taking-over certificate, the proportional reduction in these delay damages shall be calculated as the proportion which the value of the part so certified bears to the value of the works or section (as the case may be) as a whole. The engineer shall proceed to agree or determine these proportions.

Within 84 days after receiving the taking-over certificate for the works, the contractor shall submit to the engineer six copies of a statement (referred to as statement at completion, different from final statement) at completion with supporting documents, showing:

(a) the value of all work done in accordance with the contract up to the date stated in the taking-over certificate for the works,
(b) any further sums which the contractor considers to be due, and
(c) an estimate of any other amounts which the contractor considers will become due to him under the contract. Estimated amounts shall be shown separately in this statement at completion.

The engineer shall then certify the payment due to the contractor in accordance with the contract.

9.3.3.4　Lessons learned documentation

Lessons learned analysis and process improvement recommendations are developed for future contract planning and administration.

9.3.3.5　Recommended corrective actions

A recommended corrective action is anything that needs to be done to bring the contractor in compliance with the terms of the contract.

9.3.3.6　Unfulfilled obligations

The *Red Book* states: Notwithstanding the issue of the Performance Certificate the contractor and the employer shall remain liable for the fulfillment of any obligation (incurred under the provisions of the contract prior to the issue of the performance certificate) which remains unperformed at the time (such performance certificate is issued and), for the purposes of determining the nature and extent of any such obligation, the contract shall be deemed to remain in force (between the parties to the contract).

Words, Phrases and Expressions 9.3

[1] inputs to contract closure 合同结尾依据
[2] tools and techniques for contract closure 合同结尾工具与技术
[3] outputs of contract closure 合同结尾成果
[4] final account 结算（账目）
[5] final statement 最终报表，结算报表
[6] pro rata *adv.* 按比例（分配），成比例 pro rata to the contract rates 按合同单价的比例
[7] build up *vt.* 累加，计算……的总额（develop gradually by increments）
[8] build-up *n.* 累加的结果，累加数额（something produced by building up; the act or process of building up）
[9] billed rates 填入工程量清单的单价
[10] credit *n.* (因所是或所为而获得的) 认可、赞许（approval, honour, that comes to a person because of he is or does）
[11] the sum included for attendance in the original bill 列入原工程量清单用于管理分包商的数额
[12] procurement audits 采购审计
[13] structured review 系统地审查
[14] records management system 记录管理制度
[15] closed contracts 了结的合同
[16] discharge *n.* 结清单（something that discharges or releases; especially a certification of release or payment）

Notes 9.3

疑难词语解释

[1] Final payment certificate stating: (a) the amount which he fairly determines is finally due, and (b) after giving credit to the employer for all amounts previously paid by the Employer and for all sums to which the Employer is entitled, the balance (if any) due from the Employer to the contractor or from the contractor to the Employer, as the case may be.

【译文】 结算证书列明如下事项：(a) 他（工程师）公平确定为最终到期应支付（给承包商）的金额，以及 (b) 在确认业主以前已经支付的所有款项，以及业主有权取得的所有金额之后，业主到期应支付给承包商或承包商到期应支付给业主的可能余额，具体视情况而定。

【解释】 give credit to *sb.* for *sth.* 这一短语在这里的意思是：认可业主以前已经支付了所有款项，业主也有权取得所有金额这一事实。

[2] A Performance certificate is an evidence that the contractor has supplied all the contractor's documents and completed and tested all the works, including remedying any defects. A copy of the performance certificate shall be issued to the employer. Only the performance certificate shall be deemed to constitute acceptance by the employer of the works.

【译文】 完工证书是一份证据，证明承包商已提交所有承包商文件，完成了工程的所有施工和试验，修补了所有缺陷。完工证书的副本应发送给雇主。应当认为，只有完工证书才构成对工程的认可与接收。

【解释】 值得一提的是，在 FIDIC 红皮书 1977 年第三版与 1984 年第四版中，当承包商已经完成了工程的所有施工和试验，修补了所有缺陷之后，工程师签发的证书分别用 maintenance certificate 和 defects liability certificate 表示。请注意，maintenance certificate、defects liability certificate 和 performance certificate 都是在工程基本竣工并在承包商于其后的 12 个月内完成了所有基本竣工时尚未完成的工作并弥补了工程中所有缺陷之后由工程师签发的。这样一来，1977 年第三版的 maintenance certificate 至少有两点使人糊涂：(1) 在工程基本竣工后的 12 个月内，承包商仅应负责弥补工程中应由其负责的缺陷，但 maintenance 这个词却包括修理工程中由他人造成的损坏与磨损，不合理地扩大了承包商应负责的范围；(2) 人们不容易从 maintenance certificate 的字面上弄清这一证书应于何时签发。从第四版的 defects liability certificate 人们也弄不清该证书应于何时签发，还产生了其他一些误解。1999 年出版的红皮书用 performance certificate 代替了 defects liability certificate，实在是一大进步。

[3] defects notification period 缺陷通知期

【解释】 FIDIC 红皮书第三与第四版，在工程或单项工程基本竣工并签发接收证书之后，承包商自费修补其中所有缺陷的一段时间（一般为 12 个月）分别用 period of maintenance 和 defects liability period 表示。1999 年改用 defects notification period。

[4] The contract closure involves verification that all work and deliverables were acceptable.

【译文】 合同结尾需要确认所有的工作与可交付成果均可接受。

[5] The works shall be taken over by the employer when the works have been completed in accordance with the contract. The contractor may apply by notice to the engineer for a Taking-Over Certificate not earlier than 14 days before the works will, in the contractor's opinion, be complete and ready for taking over. If the works are divided into sections, the contractor may similarly apply for a Taking-Over Certificate for each section. The engineer shall, within 28 days after receiving the contractor's application:

(a) issue the Taking-Over Certificate to the contractor, stating the date on which the works or section were completed in accordance with the contract, except for any minor outstanding work and defects which will not substantially affect the use of the works or section for their intended purpose; or

(b) reject the application, giving reasons and specifying the work required to be done by the contractor to enable the Taking-Over Certificate to be issued. The contractor shall then complete this work before issuing a further notice under the contract.

If the engineer fails either to issue the Taking-Over Certificate or to reject the contractor's application within the period of 28 days, and if the works or section (as the case may be) are substantially in accordance with the contract, the Taking-Over Certificate shall be deemed to have been issued on the last day of that period.

【译文】 当工程已按合同规定竣工时，应由雇主接收工程。承包商可在他认为工程将要竣工并做好接收准备时提前14天，向工程师发出通知，申请签发接收证书。如工程分成若干单项工程，承包商可以同样方式申请签发每个单项工程的接收证书。

工程师在收到承包商的申请后28天内，应：

(a) 向承包商颁发接收证书，注明工程或单项工程按照合同要求竣工的日期，任何少量未完工作和缺陷尽管尚未完成或修补，但若对工程或单项工程预期用途没有实质性影响时，则可除外；或

(b) 拒绝申请，说明理由，并指出在能够颁发接收证书前承包商需要做的工作。承包商应在完成此项工作之后，再根据本款重新发出申请通知。

如果工程师在28天期限内既未颁发接收证书，又未拒绝承包商的申请，而工程或单项工程实质上符合了合同要求，则应认为接收证书已在上述规定期限的最后一日颁发。

值得一提的是，在FIDIC红皮书1977年第三版中，当工程基本竣工时由工程师签发的证书叫做Certificate of Completion。从FIDIC红皮书1984年第四版开始，用Taking-Over Certificate代替了Certificate of Completion。

Appendix Clause Headings of the *Red Book*

1. General Provisions 一般规定
1.1 Definitions 定义
1.2 Interpretation 解释
1.3 Communications 沟通
1.4 Law and Language 法律和语言
1.5 Priority of Documents 文件优先次序
1.6 Contract Agreement 合同协议书
1.7 Assignment 转让
1.8 Care and Supply of Documents 文件的照管和提供
1.9 Delayed Drawings or Instructions 延误的图纸或指示
1.10 Employer's Use of Contractor's Documents 雇主使用承包商文件
1.11 Contractor's Use of Employer's Documents 承包商使用雇主文件
1.12 Confidential Details 保密事项
1.13 Compliance with Laws 遵守法律
1.14 Joint and Several Liability 共同的和若干方面的责任
2. The Employer 雇主
2.1 Right of Access to the Site 现场出入权
2.2 Permits, Licenses or Approvals 许可、执照或批准
2.3 Employer's Personnel 雇主人员
2.4 Employer's Financial Arrangements 雇主的资金安排
2.5 Employer's Claims 雇主的索赔
3. The Engineer 工程师
3.1 Engineer's Duties and Authority 工程师的职责和权限
3.2 Delegation by the Engineer 工程师的委托
3.3 Instructions of the Engineer 工程师的指示
3.4 Replacement of the Engineer 工程师的更换
3.5 Determinations 确定
4. The Contractor 承包商
4.1 Contractor's General Obligations 承包商的一般义务
4.2 Performance Security 履约保证
4.3 Contractor's Representative 承包商代表
4.4 Subcontractors 分包商
4.5 Assignment of Benefit of Subcontract 分包合同权益的转让
4.6 Cooperation 合作
4.7 Setting Out 放线
4.8 Safety Procedures 安全程序
4.9 Quality Assurance 质量保证
4.10 Site Data 现场数据
4.11 Sufficiency of the Accepted Contract Amount 中标合同额的赢利可能性
4.12 Unforeseeable Physical Conditions 不可预见物质条件
4.13 Rights of Way and Facilities 道路通行权和设施
4.14 Avoidance of interference 避免干扰
4.15 Access Route 现场出入手段
4.16 Transport of Goods 货物运输
4.17 Contractor's Equipment 施工机具
4.18 Protection of the Environment 环境保护
4.19 Electricity, Water and Gas 电、水和燃气
4.20 Employer's Equipment and Free-issue Material 雇主设备和免费供应的材料
4.21 Progress Report 进展报告
4.22 Security of the Site 现场保安
4.23 Contractor's Operations on Site 承包商的现场作业
4.24 Fossils 化石
5. Nominated Subcontractors 指定分包商
5.1 Definition of "nominated Subcontractor" 指"定分包商"的定义
5.2 Objection to nomination 反对指定
5.3 Payments to nominated Subcontractors 向指定分包商支付款项
5.4 Evidence of Payments 已支付的证据

6. Staff and Labour 员工

6.1 Engagement of Staff and Labour 员工的雇用

6.2 Rates of Wages and Conditions of Labour 工资标准和劳动条件

6.3 Persons in the Service of Employer 为雇主服务的人员

6.4 Labour Laws 劳动法

6.5 Working Hours 工作时间

6.6 Facilities for Staff and Labour 为员工提供设施

6.7 Health and Safety 健康和安全

6.8 Contractor's Superintendence 承包商的监督

6.9 Contractor's Personnel 承包商人员

6.10 Records of Contractor's Personnel and Equipment 承包商人员和施工机具的记录

6.11 Disorderly Conduct 失序行为

7. Plant, Materials and Workmanship 设备、材料和工艺

7.1 Manner of Execution 施工方式

7.2 Samples 样品

7.3 Inspection 检验

7.4 Testing 试验

7.5 Rejection 否决

7.6 Remedial Work 修补工作

7.7 Ownership of Plant and Materials 设备和材料的所有权

7.8 Royalties 土地(矿区)使用费

8. Commencement, Delays and Suspension 开工、延误和暂定

8.1 Commencement of Works 开工

8.2 Time for Completion 竣工时间

8.3 Programme 施工计划

8.4 Extension of Time for Completion 竣工时间的延长

8.5 Delays Caused by Authorities 公共机构造成的延误

8.6 Rate of Progress 工程进展速度

8.7 Delay Damages 误期赔偿费

8.8 Suspension of Work 暂时停工

8.9 Consequences of Suspension 暂停的后果

8.10 Payment for Plant and Materials in Event of Suspension 暂停时设备和材料款的支付

8.11 Prolonged Suspension 继续的暂停

8.12 Resumption of Work 复工

9. Tests on Completion 竣工

9.1 Contractor's Obligations 承包商的义务

9.2 Delayed Tests 延误的试验

9.3 Retesting 重新试验

9.4 Failure to Pass Tests on Completion 未能通过竣工检验

10. Employer's Taking Over 雇主的接收

10.1 Taking Over of the Works and Sections 工程与单项工程的接收

10.2 Taking Over of Parts of the Works 部分工程的接收

10.3 Interference with Tests on Completion 对竣工试验的干扰

10.4 Surfaces Requiring Reinstatement 需要复原的表面

11. Defects Liability 缺陷责任

11.1 Completion of Outstanding Work and Remedying Defects 完成未完工作和修补缺陷

11.2 Cost of Remedying Defects 修补缺陷的费用

11.3 Extension of Defects Notification Period 缺陷通知期限的延长

11.4 Failure to Remedy Defects 未能修补缺陷

11.5 Removal of Defective Work 清除有缺陷的工程

11.6 Further Tests 进一步试验

11.7 Right of Access 出入权

11.8 Contractor to Search 承包商查找

11.9 Performance Certificate 完工证书

11.10 Unfulfilled Obligations 未履行的义务

11.11 Clearance of Site 现场清理

12. Measurement and Evaluation 计量和估价

12.1 Works to be Measured 需计量的工程

12.2 Method of Measurement 计量方法

12.3 Evaluation 估价

12.4 Omissions 工作的取消

13. Variations and Adjustments 变更和调整

13.1 Right to Vary 变更的权力

13.2 Value Engineering 价值工程

13.3 Variation Procedure 变更程序

13.4 Payment in Applicable Currencies 以相应货币支付

13.5 Provisional Sums 暂列金额

13.6 Daywork 计日工作

13.7 Adjustments for Changes in Legislation 因法律改变的调整

13.8 Adjustments for Changes in Cost 因费用改变的调整

14. Contract Price and Payment 合同价和付款

14.1 The Contract Price 合同价

14.2 Advance Payment 预付款

14.3 Application for Interim Payment Certificates 期中付款证书的申请

14.4 Schedule of Payments 付款计划表

14.5 Plant and Materials intended for the Works 拟用于工程的设备和材料

14.6 Issue of Interim Payment Certificates 期中付款证书的颁发

14.7 Payment 付款

14.8 Delayed Payment 延误的付款

14.9 Payment of Retention Money 保留金的支付

14.10 Statement at Completion 竣工报表

14.11 Application for Final Payment Certificate 最终付款证书的申请

14.12 Discharge 结清证明

14.13 Issue of Final Payment Certificate 最终付款证书的颁发

14.14 Cessation of Employer's Liability 雇主责任的终止

14.15 Currencies of Payment 支付的货币

15. Termination by Employer 由雇主终止

15.1 Notice to Correct 通知改正

15.2 Termination by Employer 由雇主终止

15.3 Valuation at Date of Termination 终止日期时的估价

15.4 Payment after Termination 终止后的付款

15.5 Employer's Entitlement to Termination for Convenience 雇主终止合同的权利

16. Suspension and Termination by Contractor 由承包商暂停和终止

16.1 Contractor's Entitlement to Suspend Work 承包商暂停工作的权利

16.2 Termination by Contractor 由承包商终止

16.3 Cessation of Work and Removal of Contractor's Equipment 停止工作和撤离施工机具

16.4 Payment on Termination 终止时的付款

17. Risk and Responsibility 风险与职责

17.1 Indemnities 保障

17.2 Contractor's Care of the Works 承包商对工程的照管

17.3 Employer's Risks 雇主的风险

17.4 Consequences of Employer's Risks 雇主风险的后果

17.5 Intellectual and Industrial Property Rights 知识产权和工业产权

17.6 Limitation of Liability 责任限度

18. Insurance 保险

18.1 General Requirements for Insurances 对保险的一般要求

18.2 Insurance for Works and Contractor's Equipment 工程和施工机具的保险

18.3 Insurance against Injury to Persons and Damage to Property 人身伤害和财产损坏险

18.4 Insurance for Contractor's Personnel 承包商人员的保险

19. Force Majeure 不可抗力

19.1 Definition of Force Majeure 不可抗力的定义

19.2 Notice of Force Majeure 不可抗力的通知

19.3 Duty to Minimise Delay 将延误减至最小义务

19.4 Consequences of Force Majeure 不可抗力的后果

19.5 Force Majeure Affecting Subcontractor 不可抗力影响分包商

19.6 Optional Termination, Payment and Release 自主选择终止、付款和解除

19.7 Release from Performance under the Law 根据合同适用法解除合同

20. Claims, Disputes and Arbitration 索赔、争议和仲裁

20.1 Contractor's Claims 承包商的索赔

20.2 Appointment of the Dispute Adjudication Board 争议裁决委员会的任命

20.3 Failure to Agree the Dispute Adjudication

Board 未能就争议裁决委员会取得一致

20.4 Obtaining Dispute Adjudication Board's Decision 取得争议裁决委员会的决定

20.5 Amicable Settlement 友好解决

20.6 Arbitration 仲裁

20.7 Failure to Comply with Dispute Adjudication Board's Decision 未能遵守争议裁决委员会的决定

20.8 Expiry of Dispute Adjudication Board's Appointment 争议裁决委员会任命期满

Index

A

abrogate v. t. 撤销,废止,废除(合约、法令、习惯等)(1.4.1.1)(说明:(1.4.1.1)的含义是:abrogate 第一次出现在第1章第4节第1段第1小段,下同。)

acceptance 承认(1.4.4.3),承诺(4.1.1.1),自留(5.5.3)

accepted tender amount (由招标人)接受的标价(3.2.2)

accepted contract amount(由招标人)接受的合同价,中标合同价(3.2.2)

accepted practice 公认做法,公认惯例,通行惯例(3.4.2.4)

access 获得,取得,接近(或进入)的方法(或权利、机会等)(3.1)

accounts for 是……的原因(6.8)

Acts of God (= Force majeure) 不可抗力(5.2.2)

actual ascertained quantity 双方实际确认的(工程)数量(7.1)

address 处理(2.5.3)

ad hoc 为此目的安排的,专门,特别(6.2)

admeasurement contracts 计量合同(2.4.3)

advance (payment) (业主在开工时暂时借给承包商的)预付款(3.2.3)

afford 给予,供给(6.2.3)

aftermath responses 事后应对措施(5.5.1.3)

all-in unit rate method 综合单价法(3.5.3)

allocation of risks 分配风险,风险分配(5.4)

alternative operational method 可选用的作业方法(3.5.1.2)

alternative tenders 可供(招标人)选择的标书(3.2.2)

amortisation rate 摊提比例(7.4.4)

anticipatory breach 预先违约,事前违反(9.1.4)

applicable law of a contract 合同适用法(4.1.4)

appreciate 重视(1.5.5);认识(8.4.5.1)

as opposed to 与……不同,与……相反(7.6.1)

assess 估定(税款、罚金等)的数额,估计(8.3.4)

assessment of claims 估计索赔的数额(8.3.4)

attendance (承包商对分包商的)照看与管理(3.5.1.1)

B

bank guarantee 银行保证书(3.2.4)

base date 基准日期(3.2.3)

benchmark n. 基准(1.3.4.1)

billed rates 填入工程量清单的单价(9.3.2.3)

breach of contract 违约解除(1.6.4.2),(9.1.4);违反合同,违约(8.2.2)

break-even analysis 盈亏平衡分析(7.2.4)

brief (业主要求的)简要说明,任务书;就计划、程序等指示某人;提供重要情况(1.5.3.2)

builder 建筑公司,土建公司(9.1.1)

build up 累加,计算……的总额(9.3.2.3)

building and engineering procedure 建筑与工程程序(1.5)

building completion and finishing work 建筑装修工程(1.3.2)

building fabric sundries 建筑物骨架配件(7.1.1)

C

call 要求支付(3.2)

certify 签发证书,以证明……(7.4.1)

cessation 停止(8.2.2)

claims 索款,要求补偿,索补(5.7.2.1)(8.1)

clerks of works 工程管理员,工程检查员(4.6.1.5)

closed contracts 了结的合同(9.3.3.1)

commencement arrangements 开工准备工作(3.4.2.4)

359

commitment on contracts in progress 准备为进行中的合同投入的力量(3.4)
compounds with his creditors 与债主了结债务(9.2.1)
conceptual development 深化概念设计(1.2)
concurrent liability 共存责任(1.6)
condition（合同中的）基本条款(3.2.5.2)
conflict of laws 冲突法,国际私法(4.1.4)
consideration 报酬,对价,约因(2.1.2)
construction n. 构造(1.3.2)
construction and related engineering services 施工及设计服务
consumer sovereignty 买方市场(1.3.1)
contemporary records 当时记录(8.3.1)
context of construction projects 建设项目及其与环境的联系(1.6)
contract administration 合同管理;执行合同;合同执行(1.3.3)
contract formation 合同订立(4.1)
contract liability 合同责任(4.1.3.1)
contracting decisions 发包决策(2.3.3)
contractor's performance 承包商的表现,承包商履行合同的结果(1.4)
contract planning 合同期间规划,履约规划(3.4,2,4)
contract to do something（缔约）承揽（某事）(1.2)
contracts director(负责)合同(的)董事,(负责)项目(的)董事(6.2.2)
contracts manager(负责)合同(的)高层管理人员(6.1.2)
contractual arrangement 合同安排(2.4.3)
contractual undertakings(书于)合同（之中的）承诺(1.3.4)
cost reimbursement contracts 成本加成合同(1.5.7)(2.4.3)
counteroffer 反要约,还价,发还价(4.1.1.2)
cover 包含,包括,包罗(1.1);够用(1.4);足数,(钱)用于,支付费用,负担开支(2.4);保险范围;给……保险(5.7)
cover(ing) letter 首封函(3.6.2)
currency of index 编制指数用的货币(7.5.2)
current workload 目前手头工作量(3.4)

custom-designed equipment 专门设计的设备(3.1)

D

DAB(Dispute Adjudication Board)争议评判(裁决)委员会(3.2)
damages 损失赔偿费(1.6.4.2)
decennial liability 十年责任(5.7)
decommission 停止运行并拆除、处置之(1.2.2)
default 不还债,不负责任(1.2)
defects notification period 缺陷通知期(2.2)
defensive engineering 保守的设计(5.2.2)
delay damages 误期赔偿费(3.2.3)(7.6)
delegated or subordinate legislation 从属立法(1.6.4.2)
descriptive uncertainty 说明不确定性(5.1.1.1)
design input 对设计提出的要求(2.4.1)
designs 式样(1.3.3)
detailed engineering 详细设计(2.1)
develop a team spirit 培养团队精神(6.1.2)
develop systems of work 建立工作制度(6.8)
direct labour 直接人工,自建工程,直接劳务,自营工程 (1.5),(2.3.1)
direct work cost based unit rate method 工料单价法(3.5.3)
discharge. 履行(3.2);结清单(9.3.2.3)
discharge by agreement 协议解除(9.1.2)
discharge by frustration（中途）受挫解除(9.1.3)
discharge by performance(合同)履行(后)解除(9.1.1)
discharge his obligations 履行其义务(3.2.4)
discharge of a contract 解除合同,合同解除(9.1)
document 写入书面文件;将要求、建议或设想用书面文件提出,并说明根据或理由(2.2)
documentation n. 书面凭据(2.3.3)
domestic bidder price preference(给予)本地(国内)投标人(的)价格优惠(3.1.3)
double handling 二次搬运(6.3.2)

E

effect 大意(4.1.1.2);取得,办理保险,购买保险(5.7)
Eichleay method(8.6)埃克雷方法
eligible 符合……要求(条件)的;适于选中的;合格的(3.1.3)
employer's brief and its development 业主设计任务书及其编制(4.8.7.2)
Employer's requirements 业主要求说明书(3.2.5.5)
engineering 设计与制造(1.1)
engineering services 工程设计服务(1.3.3)
enter into 使自己成为一方当事人(4.1)
enter into a bargain 做买卖,参加交易(4.1.1.2)
enter into a contract 缔结合同(4.1)
equipment 施工设备,施工机具(2.4)
equity 衡平法,平衡法(1.6.4.2)
estimate build-up 编制估算(书)(3.4)
ex-contractual claims 非合同索赔,根据合同以外的理由提出的补偿要求(8.2.2)
execute 办理(签字盖章)手续,使……产生法律约束力(3.2.3)
executionary breach 执行中违约,事后违反(9.1.4)
ex gratia payments 道义索赔(8.2.2)
exposures 不利的情况(5.7)
express undertaking 明确承诺(4.2)
extended attendance 对分包商的管理延长了时间,管理分包商的时间延长(8.6.4.3)
extensions [会计用语]转来金额,算出金额(3.6.3)

F

fallback plan 撤退计划,后退计划(5.5)
field services manager 现场设施经理,现场设施负责人(6.2)
final account 结算(账目)(9.3.2.3)
final payment certificate 结算证书,最终付款证书,结算证书(7.4.2)(9.3.2.3)
final statement 最终报表,结算报表(9.3.2.3)
finalizing quotations 最终确定其报价(6.1.2)
financial failure 财务困境,财力不支(5.2.2)
financial risk exposure 财务风险状况(5.5.12)
follow-on step 后继步骤(6.5)
force account 自营工程(2.3.1)
force majeure 不可抗力(5.2)(8.4.3)
form grounds for termination 构成终结合同的理由(4.5.5)
formulate(经深思熟虑之后,正式)提出,正式提出,明确表达(3.4.2.4)
formulate a tender 编制出正式的标书(3.4.2.4)
front loading 前高后低报价,不平衡报价(3.6.3)

G

general description of the arrangements 施工总说明(3.2.1)
general items 一般事项费用,待摊费用(3.2.8)
geotechnical expert 岩土技术专家,岩土工程专家(1.5.4.7)
geotechnical processes 岩土工艺(6.2)
give possession of the site (业主向承包商)交付现场(8.3.1)
governing law 适用(于合同的)法(律)(3.2.3)
governing law of the contract 合同支配法(4.1.4)

H

hands-on 直接插手(2.4)
hedge 设防,两面下注避免损失(4.1.1.2)
honour the contract 兑现合同(9.1.1)

I

incorporate 使……成为一部分(3.1.3)
indemnify 保障……不承担义务、责任或不蒙受损失(3.5.1.1)
indemnity 赔偿保障,补偿保障(5.6.4)
in anticipation of 在……之前(4.2)
in consideration of 为了报答……(3.2.9)
independent estimates 独立估算,标底(3.5.2)
inducement 诱因,引诱物(9.2.1)
industrial economics 产业经济学(1.3.4)
infra dignitatem 有失身份,有失尊严

(8.2.1.1)
　　in-house expertise 本单位专业知识(4.6.1.2)
　　instance 要求,建议(1.1)
　　institution 制度(1.3.1)
　　insurability 受保的可能性(5.6.4)
　　interact 互相作用,互相影响,进而得到某种结果;讨价还价;回合(1.3.1)
　　invitation for negotiation 议标邀请信(书、函)(2.5.2)
　　invitation for tender 投标邀请(书、信、函、广告、电子邮件……)(2.5.2)
　　invite 增加了……可能性(2.4)
　　invoke 祈求帮助或保护,诉诸于,求助于(8.7.2)
　　involve 免不了,需要,不可缺少(1.3);使……参与(2.4)
　　item of work 分项工程(1.2.1)

J

　　judge-made law 法官制定法(4.1)
　　judicial decisions 司法判决(1.6.4.2)
　　jurisdiction 裁决权(1.6)
　　justifying claims by contract 利用合同说明索赔的根据与理由(8.4)

L

　　labour relations 劳资关系(6.2)
　　law of torts 侵权法(1.6.4.2)
　　legislation or statute law 立法或成文法(1.6.4.2)
　　letter addressed to contractors 按地址寄给承包商的信函(3.1.3)
　　letter of tender 投标书,投标函(3.2.2)
　　life cycle 生命期(1.2.2)
　　linked bar chart 连接横道图(4.5.5)
　　liquidated damages for delay 误期赔偿费(7.6.1)
　　lowest evaluated tender 最低评估价标书,最低评标价(3.6.6)

M

　　make-or-buy decision 自制购买决策(2.3)
　　management programme 管理工程进展的计划,进展管理计划(4.5.5)
　　market structures 市场结构(1.3.4)
　　master programme 施工总计划(3.4.2.4)
　　measurement uncertainty 计量不确定性(5.1.1.1)
　　memorandum of a association 联营协议摘要(1.4.1.2)
　　method statement 施工方法说明书(相当我国过去的"施工方案")(3.5.1.2)
　　multiplicity of architect/engineer's instructions 建筑师/工程师指示多而重复(8.7.1.1)

O

　　object(ive) of a contract n. 合同标的(4.1.2)
　　offer 要约(4.1.1.1)
　　oligopolistic industries 寡头垄断行业(1.3.4.3)
　　oligopoly 寡头垄断(1.3.4.3)
　　on a……basis 以……为准则,以……方式(3.5)
　　on an *ad hoc* basis for each project 为每一个项目专门……(3.3.2)
　　on a quantum meruit basis 按合理的数额(9.1)
　　one set of well-referenced documents 彼此配合良好的一套文件(3.2.5.3)
　　optional termination(遇到不可抗力时)自愿终止合同(9.2.3.3)
　　order of magnitude estimate 数量级估算(3.2)
　　outstanding balance of the Retention Money 保留金未支付的剩余部分(7.4.4)
　　overhead 管理费,摊销费(2.4.3)

P

　　package deal contracts 成套交易合同(3.2)
　　part 单位工程(1.2.1)
　　parties to a contract 合同当事人,合同双方(1.5)
　　part performance 部分履行,局部履行(9.1.1)
　　perfect competition 完全竞争(1.3.4.1)
　　perform 执行,履行(2.5)
　　performance bond 履约保函(3.2)
　　performance certificate 完工证书(3.2.10)
　　performance related documentation 履行合同过程产生的文件(4.7.1)

performance reports 履行合同结果报告(4.7.2)

place a subcontract (let a subcontract)发包分合同(4.6)

plant hire firms 设备租赁公司(6.1.1)

plant register 设备登记册(6.6.1)

possession of the site（业主向承包商）移交（施工）现场(6.1.1)

preamble（工程量清单的）前言(3.2.8)

pre-contract arrangements 签约前安排,签约前工作(2.2)

pre-contract planning 签约后规划,开工（前）规划(3.4.1)

prediction programme 预测计划(4.5.5)

preliminaries 开办费,开工准备费,开工动员费(3.2.8)

preliminaries build-up 编制开办费估算（书）(3.4.2.4)

premature or early termination of performance of a contract 提前终止合同（的履行）(9.1)

premium 额外代价,额外费用(5.5.1.2);保险费(5.7)

pre-tender programme 投标施工计划(3.4.2.4)

prime cost sums 指定费用额(3.2.7)

principle of autonomy of parties 当事人自决原则(4.1.4)

principles of allocation of risks 风险分配原则(5.5.2.2)

private international law 国际私法(4.1.4)

procedural law 程序法(1.6.4.2)

procedure for claims 索赔程序(8.3)

procurement documents 采购文件(2.5.2)

procurement management 采购管理(2.2)

procurement methods (procurement systems)采购办法(2.4)

programme security 施工计划的稳妥性(2.4.4.1)

progress meetings 进展会议(4.8.4.5)

progress programme 显示工程实际进展的计划,实际进展计划(4.5.5)

progress reports 进展报告（书）(4.6.1.5)

proper law of the contract 合同适当法(4.1.4)

proposals 建议书（用于咨询或设计服务采购）(2.2)

proprietary rights 专有（技术）权(2.5.3)

pro rata adv. 按比例（分配）,成比例(9.3.2.3)

provide 形成,产生,便于(1.2);给予(2.2);设置(3.2)

provisional quantities 暂定（工作）量(3.2.7)

provisional sums 暂定金额(3.2.3)

purpose 用途(3.1.3)

purpose-built civil engineering projects 专门建造的土木工程项目(1.5)

Q

qualified tenders 附加条件的标书,有保留的标书(3.5)(9.3)

quantity surveyor 工料估算师(4.3.4.2)

quantum meruit 合理的数额(9.1.1)

R

rate(分项工作)单价(2.4)(7.1.1)

rate contract 单价合同(2.4)

reference(关于人品或能力的)证明信（书）、推荐信（书）、介绍信；(关于人品或能力的)证明人、推荐人、介绍人(2.5)

reference 附注；援引注；编号；索引号(3.1.3)

regulatory systems（政府对经济和社会活动）调节的制度(1.3.1)

relief 法律上的补救、补偿(7.6.1)

remeasurement contract 计量合同(1.5)(7.6)(2.4.3)

repudiate 拒绝偿付债务,拒绝履行义务(8.4.5.6)

request for proposal 征求建议书（信、函、广告、电子邮件……)(2.5.2)

request for quotation 询价函（信、广告、电子邮件……)(2.5.2)

rescission 撤消(9.1)

resource scheduling 制订资源使用计划(6.5)

responses to risks 应对风险(5.5)

result uncertainty. 结果不确定性(5.1.1.1)

retention (money)保留金(3.2.3)

right of way(在他人土地上的)通行权(5.2.2)

risk categories 风险分类(5.2.1)

risk response owner 风险应对负责人(5.4)
risk-taking and risk-avoidance 接受风险与回避风险(5.5)
risk transference 转移风险(5.5)
Romano-Germanic systems of law 罗马日耳曼法系(1.6.4.2)
ruling language 裁决语言(3.2.3)
running of the contract 执行合同,履行合同(8.2.2)

S

sanction 权利或许可;批准(1.4.1.1)
satisfy 使确信,使消除疑虑(3.4)
schedule of rates contracts 单价表合同(2.4.3)
seal the bargain 做成买卖,成交(4.1.1.2)
section 单项工程(1.2.1)
section engineer 栋号工程师(6.2)
section manager 单项工程负责人,栋号负责人(6.2)
selective tendering 选择性招标,邀请招标,邀标(3.2.1)
services 公用设施(1.5)
serving of notices 递送通知(6.1)
set out 展示(4.4)
set out the works 工程放线(4.4)
shortlist 决选名单(2.4.4.3)
signify 表示(某人观点、意图、目的等)(1.3.1)
site investigation work 现场勘察工作(3.6.2)
site layout plan 现场布置平面图(6.3)
sources of common law 普通法渊源(1.6.4.2)
specialist contractors 专业承包商(1.4.4)
specialist trades 专业工种(1.3.4)
specific performance (a court order forcing acceptance)(英国衡平院强迫违反合同者的)强制履行令(9.1.1)
specification 技术规格(用于货物采购),技术要求,技术要求说明书,技术条款,设计说明书(用于工程和技术咨询服务采购)(1.4)
speculative risks 投机风险(5.1.1.2)
spread over 将……分摊到……(8.6.4.5)
stakeholders 利害关系者(2.4.4.3)
standing orders 现行规则(3.3.1)
static risks 静态风险(5.1.1.2)
statutory liability 法定责任(4.1.3.3)
statutory notifications 向水、电、消防等部门发出的通知(4.5.2)
statutory undertakings(水电通讯煤气等)公用事业(单位)(4.6.1.3)
strict liability 后果责任,严格赔偿责任,严格责任(4.1.3.4)
structural uncertainty 结构不确定性(5.1.1.1)
subcontract 将……分包出去(1.4)
sublet 分包(1.4)
submission of tenders 递交标书(3.5.6)
substantial performance 实质(上)履行(了)合同(9.1.1)
substantiate *v. t.* 列举事实以支持,证实,证明(8.3.1)
substantive law 实体法(1.6.4.2)
system 办法(3.4)
system charts 系统图(5.2)

T

taking-over certificate 接收证书(3.2.10)
tender documents 招标文件(1.5.6)
tender opening 开标(3.6.1)
tender price 标价(2.4.3)
tender programme 投标时提交的施工计划(3.5.1.1)
tender (adjudication) report 评标报告(3.6.7)
tender security 投标保证,投标保证书,投标保证金(2.5.1)
termination by employer 由业主终止(合同)(9.2.1)
terms and conditions 条款和条件(3.2.5.2)
terms of reference 职权范围,工作职责说明书(1.5.2)
time frame 时间安排(5.3.1)
time obligations 在时间方面的义务(4.5.5)
time-related items 与时间有关的分项(工作)(7.1.2)
to the effect 意思是……(7.6.1)
to the extent that 在这一范围内,在这种情况下,只要(4.6.1.1)
tort liability 侵权行为责任(4.1.3.2)
trade 行业,例如木工、石筑、砌砖等(1.3);工

种,职业,(手工业内)的行业(8.2.1.1)
twelve maxims of equity 衡平法十二准则(1.6.4.2)

U

unit rate contract 单价合同(2.4.3)

V

variation procedure 变更程序(8.4.5.4)

viable 财务上持久的,财务上可行的(1.5.3)

W

warrant 保证,担保(2.5);证明……为正当,说明……合理(8.7.1.1)
warranties 产品质量保证(5.5.1.2)
warranty(合同中的)保证条款(3.2.5.2)
words to that effect 表示上述意思的话,表示那样意思的话(4.1.1.2)

References

[1] World Bank. The Construction Industry: Issues and Strategies in Developing Countries (Discussion paper). Washington D.C.: World Bank, 1984.

[2] China Statistics Bureau. China Statistical Abstract 2006. China Statistic Press, 2006.

[3] 卢有杰编著. 新建筑经济学（第二版）. 北京：中国水利水电出版社，知识产权出版社，2005.

[4] WTO Secretariat. Background Note-CONSTRUCTION AND RELATED ENGINEERING SERVICES. Council for Trade in Services, 8 June 1998.

[5] WTO Secretariat. Background Note-ARCHITECTURAL AND ENGINEERING SERVICES. Council for Trade in Services, 1 July 1998.

[6] Andrew J. Cooke. Economics and Construction. Macmillan, 1996.

[7] Institution of Civil Engineers. Civil Engineering Procedure. London: Thomes Telford, 1986.

[8] www.fidic.org

[9] www.worldbank.org

[10] www.adb.org

[11] FIDIC. Conditions of Contract for Works of Civil Engineering Construction. 4th edition. 1987 amended 1992.

[12] FIDIC. Conditions of Contract for Construction. 1st edition. 1999.

[13] Nael G. Bunni. The Fourth Edition of the Red Book. Second Edition. Blackwell Science, 1997.

[14] International Bank for Reconstruction and Development. Guidelines-selection and employment of consultants by World Bank borrowers. May 2004.

[15] I. H. Seeley. Civil Engineering Contract Administration and Control. Second Edition. Macmillan, 1996.

[16] The PMI. A Guide to the Project Management Body of Knowledge, Third Edition (PMBOK ® Guide) 2004.

[17] Claude D. Rohwer, Gordon D. Schaber. Contract in a Nutshell. 1997.

[18] International Bank for Reconstruction and Development. Guidelines-procurement under IBRD loans and IDA credits. May 2004.

[19] Frank Harris and Ronald McCaffer. Modern Construction Management. Third Edition. Blackwell Scientific Publications Limited, Oxford, 1989.

[20] M. J. Hills. Building Contract Procedure in Hong Kong. Third Edition. 1991.

[21] Franks, James. Building Procurement Systems: a client's guide. Addison Wesley Longman Limited, 1998.

[22] World Bank. Standard Bidding Documents-Procurement of Works and User's Guide. May 2005.

[23] Vincent Powell-Smith. Contract Documentation for Contractors. Second Edition. BSP Professional Books, 1990.

[24] World Bank. Standard Bidding Documents-Procurement for Works. May 2000.

[25] B. Cooke. Contract Planning and Contractual Procedures Second Edition. Macmillan, 1984.

[26] Tong Yizhe and Lu Youjie. Unbalanced bidding on contracts with variation trends in client-provided

quantities. Construction Management and Economics, 1992, 10: 69~80.

[27] Daniel Atkinson. The Use of Programme of work in Construction Contracts. Civil Engineering Surveyor, Feb. 2002:46~47.

[28] Daniel Atkinson. Using Programmes to Manage Projects. Civil Engineering Surveyor, April 2002:28~29.

[29] Robert N. Charette. Software Engineering RISK Analysis and management. McGraw-Hill Book Company, 1989.

[30] Roozbeh Kangari. Risk management perceptions and trends of U.S. construction. Journal of Construction Engineering and Management, Vol.121, No.4, December, 1995.

[31] FIDIC. The FIDIC Contracts Guide. First Edition. 2000.

[32] Brian Cooke and Peter Williams. Construction planning, programming and control. Blackwell Science, 2002.

[33] Derek S. Drew and Stephen Lai. The Quantity Surveying Process-a joint presentation at Tsinghua University. June 1998.

[34] Tang Guangqing. Internationally Accepted Tendering Procedure and Contract Administration-a lecture note. Beijing, 2005.

[35] David Chappell. Contractor's Claims, an architect guide. The Architectural Press, 1990.

[36] Jeremy Winter and Michelle Essen. Extensions of time -Notice as a condition precedent to entitlement. Civil Engineering Surveyor, December/January 2003.

[37] Mark A. Smith. Managing Long-term Contract Disputes presented for the World Bank. Washington D. C., U.S.A., April 14, 1995.

[38] Stephen L. Gruneberg. Construction Economics-an introduction. Macmillan, 1996.

[39] Wang ShouQing and Tiong L K Robert et al. Risk Management Framework for BOT Power Projects in China. Journal of Project Finance, Institutional Investor, Inc., New York, Vol. 4, No. 4, p. 56-67, Winter 1999.

[40] A.S. Hornby, E.V. Gatenby and H. Wakefield. The Advanced Learner's Dictionary of Current English with Chinese Translation(现代高级英汉双解辞典). Oxford University Press, 1982.

[41] Jess Stein and P.Y. Su. The Random House Dictionary. Ballantine Books, New York, 1980.

[42] Merriam-Webster Inc. Merriam-Webster's Collegiate Dictionary, Tenth Edition. Merriam-Webster Inc. 1997.

[43] 《新英汉词典》编写组. 新英汉词典（增补本）. 上海译文出版社, 1985.

[44] 卢有杰, 吴之明. 英汉-汉英对照项目管理词语精选. 清华大学出版社, 2001, 4.